ECO-FRIENDLY NANO-HYBRID MATERIALS FOR ADVANCED ENGINEERING APPLICATIONS

ECO-FRIENDLY NANO-HYBRID MATERIALS FOR ADVANCED ENGINEERING APPLICATIONS

Edited by

S. Ananda Kumar, PhD

AAP | APPLE ACADEMIC PRESS

Apple Academic Press Inc.
3333 Mistwell Crescent
Oakville, ON L6L 0A2
Canada

Apple Academic Press Inc.
9 Spinnaker Way
Waretown, NJ 08758
USA

Exclusive worldwide distribution by CRC Press, a member of Taylor & Francis Group

Printed in the United States of America on acid-free paper

International Standard Book Number-13: 978-1-77188-294-1 (Hardcover)

International Standard Book Number-13: 978-1-77188-295-8 (eBook)

Library and Archives Canada Cataloguing in Publication

Eco-friendly nano-hybrid materials for advanced engineering applications / edited by S. Ananda Kumar, PhD.

Includes bibliographical references and index.
Issued in print and electronic formats.
ISBN 978-1-77188-294-1 (hardcover).--ISBN 978-1-77188-295-8 (pdf)
1. Nanostructured materials--Environmental aspects. 2. Nanostructured materials--Industrial applications. 3. Nanotechnology--Environmental aspects.
I. Kumar, S. Ananda, author, editor

TD196.N36E26 2016 620.1'15 C2016-902515-2 C2016-902516-0

Library of Congress Cataloging-in-Publication Data

Names: Kumar, S. Ananda, 1984- editor.
Title: Eco-friendly nano-hybrid materials for advanced engineering applications / S. Ananda Kumar, PhD, editor.
Description: Toronto : Apple Academic Press, 2016. | Includes bibliographical references and index.
Identifiers: LCCN 2016017364 (print) | LCCN 2016022622 (ebook) | ISBN 9781771882941 (hardcover : alk. paper) | ISBN 9781771882958 ()
Subjects: LCSH: Environmental engineering. | Green technology. | Sustainable engineering. | Nanostructured materials.
Classification: LCC TA170 .E322 2016 (print) | LCC TA170 (ebook) | DDC 620.1/150286--dc23
LC record available at https://lccn.loc.gov/2016017364

ABOUT THE EDITOR

Dr. Ananda Kumar Srinivasan PhD

Srinivasan Ananda Kumar, PhD, is a professor of chemistry at Anna University, Chennai, India. He is a recipient of the Erasmus Mundus External Cooperation Window (European Commission) Visiting Research Fellowship. As a member of staff at Anna University, he has supervised more than 45 postgraduate students (MSc and MTech) and several MPhil and PhD students, reviewing, evaluating, and validating experimental data resulted from their project/PhD work. Dr. Kumar won the Active Researcher Award of Anna University for the year 2014 for his outstanding contribution to the field of nano-hybrid coatings and composites. His research team members won the runners up national award from the Central Ministry, Government of India, in February 2015 for their outstanding contribution to the field of nano-hybrid coatings and composites. The author over 60 scientific papers in international journals and several book chapters, Dr. Kumar has developed many technologies for the prevention and remediation of corrosion, fouling, and flame.

Dr. Kumar's research accomplishments include the development and characterization of phosphorus containing epoxy and siliconized epoxy coatings and nanocomposites as having superior properties. His current research topics comprise a variety of novel organic-inorganic hybrid POSS-epoxy and nano-hybrid coatings and nanocomposite with anticorrosive and anti-bacterial properties.

CONTENTS

LIST OF CONTRIBUTORS

M. Alagar
Department of Chemical Engineering, Alagappa College of Technology, Anna University, Chennai 600025, India. E-mail: mkalagar@yahoo.com

Rajan Choudhary
Materials Chemistry Division, School of Advanced Sciences, VIT University, Vellore, Tamil Nadu 632014, India

Barry Connell
Department of Biomedical Science, Atlantic Veterinary College, University of Prince Edward Island, Charlottetown, PE C1A 4P3, Canada

K. Priya Dasan
Material Chemistry Division, SAS, VIT University, Vellore, Tamil Nadu 632014, India. E-mail: priya-jeenetd@gmail.com

S. Devaraju
Department of Chemical Engineering, Alagappa College of Technology, Anna University, Chennai 600025, India

D. Duraibabu
Department of Chemistry, Anna University, Chennai 600025, India

A. Nirmala Grace
Center for Nanotechnology Research, VIT University, Vellore, Tamil Nadu, India. E-mail: anirmala-gladys@gmail.com

Chang-Sik Ha
Department of Polymer Science and Engineering, Pusan National University, Busan 609 735, Republic of Korea

K. Jayamoorthy
Department of Chemistry, St. Joseph's College of Engineering, Chennai 600119, India

S. Ananda Kumar
Department of Chemistry, Anna University, Chennai 600025, India. E-mail: sri_anand_72@yahoo.com

V. Madhumitha
Department of Chemistry, Anna University, Chennai 600025, India

M. Mandhakini
Department of Chemical Engineering, Alagappa College of Technology, Anna University, Chennai 600025, India

R. Manjumeena
Centre for Advanced Studies in Botany, University of Madras, Chennai 600025, India. E-mail: manju-meena1989@gmail.com

S. Nagappan
Department of Chemistry, College of Engineering, Guindy, Anna University Chennai, Chennai 600 025, India

Tirupattur Srinivasan Natarajan
Department of Physics, Indian Institute of Technology Madras (IITM), Chennai 600 036, India

Kalimuthu Pandi
Department of Chemistry, Anna University, University College of Engineering – Dindigul, Reddiarchatram, Dindigul 624622, India

R. Praveen
The School of Metallurgy and Materials, University of Birmingham, Birmingham, UK

Desikan Rajagopal
Department of Pharmaceutical Chemistry, School of Advanced Sciences, VIT University, Vellore, Tamil Nadu 632014, India. E-mail: rajagopal.desikan@vit.ac.in

G.R. Rajarajeswari
Department of Chemistry, College of Engineering, Guindy, Anna University Chennai, Chennai 600 025, India. E-mail: rajiaravind@gmail.com

R. Ramachandran
Center for Nanotechnology Research, VIT University, Vellore, Tamil Nadu, India

Lakshmi Ravi
Materials Chemistry Division, School of Advanced Sciences, VIT University, Vellore, Tamil Nadu 632014, India

Tarek Saleh
Department of Biomedical Science, Atlantic Veterinary College, University of Prince Edward Island, Charlottetown, PE C1A 4P3, Canada. E-mail: tsaleh@upei.edu

T.M. Sridhar
Department of Analytical Chemistry, University of Madras, Guindy Campus, Chennai 600025, India. Email: tmsridhar23@gmail.com; sridhar@unom.ac.in

Sasikumar Swamiappan
Materials Chemistry Division, School of Advanced Sciences, VIT University, Vellore, Tamil Nadu 632014, India. E-mail: ssasikumar@vit.ac.in

P. Saravanan
Department of Chemistry, St. Joseph's College of Engineering, Chennai 600119, India

Natrayasamy Viswanathan
Department of Chemistry, Anna University, University College of Engineering – Dindigul, Reddiarchatram, Dindigul 624622, India. E-mail: drnviswanathan@gmail.com

LIST OF ABBREVIATIONS

AA	activated alumina
Ach	acetylcholine
ACMN	γ-AlOOH @CS (pseudoboehmite and chitosan shell) magnetic nanoparticles
AFM	atomic force microscope
ALA	alpha lipoic acid
Al_2O_3	aluminum oxide
AlLS	aluminum hydroxide-impregnated lime stone
BAPB	1,4-bis (4-amine-phenoxy)benzene
BBB	blood–brain barrier
BMI	bismaleimide
BRS	baroreceptor reflex sensitivity
C	cubic
CA	carbon black
CA	citric acid
CA	contact angle
CAA	carboxylated alginic acid
CAGR	compound annual growth rate
CaO	calcium oxide
CaO-AA	CaO-modified AA
CaP	calcium phosphate
CB	chitosan beads
CDP	controlled depletion polymer
CE	cyanate esters
CIC	cerium-impregnated chitosan
CLZMOB	chitosan supported lanthanum (III) and zirconium (IV) mixed oxides beads
CNT	carbon nanotube
CNTA	carbon nanotube array
COCA	copper oxide-coated mesoporous alumina
CPs	conducting polymers
CS	chitosan
$CTS/MMT/ZrO_2$	chitosan/montmorillonite/zirconium oxide
CVD	chemical vapor deposition
DC	defluoridation capacity

DDM	diaminodiphenylmethane
DHLA	dihydrolipoic acid
DMA	dynamic mechanical analysis
DMFC	direct methanol fuel cells
DMSO	dimethyl sulfoxide
ECM	extracellular matrix
EDLCs	electrical double layer capacitors
EPC	electrochemical pseudo supercapacitors
EPD	electrophoretic deposition
ESR	equivalent series resistance
EU	European Union
FCNT	functionalized carbon nanotube
Fe_2O_3	iron oxide
FESEM	field emission scanning electron microscope
FHA	fluoridated hydroxyapatite
FRP	fiber-reinforced polymers
FT-IR	Fourier transform infrared spectra
F-ZnO	functionalized ZnO
GAC	granular activated carbon
G-PEDOT	graphene-polyethylenedioxythiophen
GSSG	oxidized glutathione
HAP	hydroxyapatite
HCl	hydrochloric acid
HDPE	high-density polyethylene
HFO	hydrous ferric oxide
HIACMO	hydrated iron (III)–aluminum (III)–chromium (III) ternary mixed oxide
HITMO	Iron(III)–tin(IV) mixed oxide
HMAA	hydroxyapatite-modified activated alumina
H_2O_2	hydrogen peroxide
ILD	interlayer dielectric
I/R	ischemia-reperfusion
IV	intravenous
La	lanthanum
LA	lipoic acid
La-AA	La^{3+}-modified activated alumina
LDH	layered doubled hydroxide
LDPE	low-density polyethylene
LIBs	lithium-ion batteries
LZMO	lanthanum (III)–zirconium (IV) mixed oxide
M	monoclinic

MAAA	magnesia-amended activated alumina
MAP	mean arterial pressure
MB	magnesium incorporated bentonite
MCAO	middle cerebral artery occlusion
M-CAA	metal ions coordinated CAA
M-CSW	metal-loaded cross-linked seaweed
MEPC	marine Environment Protection Committee
MgOC	magnesia/chitosan
MgO	magnesia
MK	micronized kaolinites
MMTNC	montmorillonite nanoclay
MO	metal oxide
MnO_2	manganese oxide
MnO_2-AA	manganese oxide-coated AA
MSC	mesenchymal
MWNTs	multi-walled nanotube
mtDNA	mitochondria DNA
Ni Cd	nickel cadmium
NPs	nanoparticles
NOX	NADPH oxidase
n-HAp	nano-hydroxyapatite
n-HApC	nano-hydroxyapatite/chitosan
n-HApCh	nano-hydroxyapatite/chitin
Ni–Mh	nickel metal hydrate
NMR	nuclear magnetic spectra
NPs	nanoparticles
n-ZnO	nonfunctionalized ZnO
OAPS	octa(Amino Phenyl)Silsesquioxane
Ox	oxalic acid
OCP	octacalcium phosphate
OH	hydroxyl radicals
ONPS	octa(nitrophenyl)silsesquioxane
OPS	octaphenylsilsesquioxane
PANi	polyaniline
PCL	poly(e-caprolactone
PCLCS	poly(e-caprolactone)/starch
PE	polyethylene
PEDOT	poly(3,4-ethylenedioxythiophene)
PEM	proton exchange membrane
PLGA	poly lactic-co-glycolic acid
PLGA-b-PEG	poly-(lactide-co-glycolide)-polyethyleneglycol

PMeT	poly (3-methyl thiophene)
PPy	polypyrrole
PPD	paraphenylenediamine
PSS	poly(sodium 4-styrene sulfonate)
PSZ	partially stabilized zirconia
RFCL	ranbaxy fine chemicals
RGO	reduced graphene oxide
RK	raw kaolinites
ROGSCs	RuO_2/graphene sheet-based composite
ROS	reactive oxygen species
RuO_2	ruthenium oxide
SBF	simulated body fluid
SEM	scanning electron microscopy
SGCS	silica gel/chitosan
SiO_2	silica
SOD	superoxide dismutase
SPC	self-polishing copolymer
SWNTs	single-walled nanotube
T	tetragonal
TBT	tributyltin
TCMTB	thiocyanatomethyl thiobenzothiazole
TCP	tricalcium phosphate
TE	tissue engineering
TGA	thermo gravimetric analysis
TGBAPB	N,N'-Tetraglycidylbis amine-phenoxy benzene
TIA	transitory ischemic attack
TiO_2	titania
TNF	tumor necrosis factor (TNF)
TTCP	tetracalcium phosphate
UHMWPE	Ultra-high molecular weight polyethylene
WHO	World Health Organization
XRD	X-ray diffraction
YSZ	Yttria-stabilized zirconia
ZICNSC	zirconium-impregnated cashew nut shell carbon
ZrIC	zirconium-impregnated cellulose
ZrWP	zirconium(IV) tungstophosphate
ZnO	zinc oxide
ZrO_2	zirconia
3-APTES	3-aminopropyltriethoxysilane

PREFACE

Work on the fabrication and processing of nanomaterials and nanostructures started a long time ago, far earlier than nanotechnology emerged as a new scientific field. Such research has been drastically intensified in the last decade, resulting in an overwhelming literature in many journals across different disciplines. Readers will find that this book has been focused primarily on eco-friendly nanohybrid materials. The aim of this book is to summarize the fundamentals and established techniques of synthesis and processing of eco-friendly nanohybrid materials so as to provide readers with a systematic and coherent picture about synthesis and processing of nanomaterials.

The research on nanotechnology is evolving and expanding very rapidly. Nanotechnology represents an emerging technology that has the potential to have an impact on an incredibly wide number of industries, such as the medical, the environmental, and the pharmaceutical industries. Nanoparticles (NPs) are clusters of atoms in the size range of 1–100 nm. The use of NPs is gaining impetus in the present century as these particles possess defined chemical, optical, and mechanical properties. There is a growing need to develop an environment-friendly process for corrosion control that does not employ any toxic chemicals. Generally, NPs are prepared by a variety of chemical and physical methods, which are not environment-friendly. Nowadays, green chemistry procedures using plant extracts for the synthesis of NPs are commonly employed. TGBAPB epoxy resin was reinforced with 1, 3, and 5 wt% of surface functionalized AgNPs (F-AgNPs), which were synthesized using *Rosa indica wichuraiana* hybrid leaves extract with a view of augmenting the corrosion control property of the epoxy resin and also imparting antimicrobial activity to epoxy coatings on mild steel.Binding interaction studies of 3-aminopropyltriethoxysilane (3-APTES) with ZnO, TiO_2, ZrO_2, Al_2O_3, SiO_2, and Fe_2O_3 nanocrystals have been carried out by UV–visible and fluorescence spectral studies. 3-APTES acts as a host for ZnO and TiO_2 nanocrystals that enhance its fluorescence due to the electron transfer, whereas ZrO_2, Al_2O_3, SiO_2, and Fe_2O_3 do not affect the fluorescence intensity of 3-APTES. 3-APTES gets adsorbed on the surface of all NPs and the surface treatment of NPs by 3-APTES, its mechanism is discussed in detail. Scanning electron microscope (SEM) and Fourier transform infrared spectroscopy (FTIR) spectral studies are carried out to support the biding interaction between 3-APTES with NPs.

Biomaterials plays a major role in the health-care sector and it is highly multidisciplinary in nature as it involves scientists to innovate and prepare the material, engineers to design and manufacture the prosthesis, physicians to implant it in the human body, and to study the response of natural tissues on artificial biomaterials. Global market for biomaterial is huge and it attracts the researchers from various fields hence it remains evergreen among various disciplines of research. Various applications of biomaterial includes joint replacements, bone plates, bone cement, artificial ligaments, dental implants for tooth fixation, blood vessel prostheses, heart valves, skin repair devices, and contact lenses.Biomaterials employed within the body can fall into three categories as inert biomaterial, bioresorbable biomaterial, and active biomaterial. Inert biomaterial like titanium implants remains unchanged in physiological conditions and will not undergo any chemical changes as well as it will not initiate any biological response on its surface. Bioresorbable biomaterials like tricalcium phosphate will dissolve in the physiological conditions of the human body and it will be replaced with the natural biomaterials over a period of time. Active biomaterials like calcium silicate bioceramics are the one that actively take part in physiological processes and at the same time, it will not be replaced by the natural biomaterials. Drinking water is the major source of fluoride intake. The fluoride content in drinking water may be beneficial or detrimental depending on its concentration. Excess fluoride (>1.5 mg/L) level in drinking water leads to fluorosis. The impact of fluoride in drinking water is a major concern in developing countries like India where around 17 states are affected by fluorosis. As there is no remedy for fluorosis, defluoridation of drinking water is one of the options to reduce the fluoride concentration in water. The creation of awareness among the people in rural areas is the best possible way to tackle this problem. Though various treatment techniques, such as precipitation, ion exchange, adsorption, reverse osmosis, electrodialysis, Donnan dialysis, and nanofiltration have been suggested for fluoride removal, adsorption seems to be effective, eco-friendly, and economical one. Various adsorbents have been reported for fluoride removal but each of them possesses own advantages and limitations. Most of the reports are academic oriented but only a few reports are available for technology development. Polymeric materials are widely used for water treatment and they possess various advantages over the other adsorbent materials. This review focuses exclusively on the eco-friendly materials like low-cost inorganic materials, biopolymers, and biopolymeric composites for the defluoridation of water. The fluoride selectivity of the biopolymers/composites will be enhanced by chemical modifications, such as functionalization, metal ion incorporation, etc. A comparison of the adsorption capacity of various low-cost inorganic adsorbents, biopolymers and biopolymeric composites were made.Conclusive evidence from past work has shown that preadministration of apocynin and lipoic acid at subthreshold levels for neuroprotection, enhanced the neuroprotective capacity when injected in combination. Continuing with this strategy, investigation was designed to determine if a codrug consisting of lipoic acid and apocynin functional groups bound by a covalent bond, named UPEI-100, and is capable of similar efficacy using a rodent model of stroke. Male rats were anesthetized with Inactin (100 mg/kg, i.v.) and the middle cerebral artery was occluded for 6 hours (MCAO), or allowed to reperfuse for 5.5 hours following a 30-min occlusion (ischemia/reperfusion; I/R). Preadministration of UPEI-100 dose-dependently decreased infarct volume in the I/R model ($p < 0.05$), but not in the MCAO model of stroke. A time course for this neuroprotective effect showed that UPEI-100 resulted in a decrease in infarct volume following 2 hours of reperfusion compared to vehicle. The time course of this neuroprotective effect was also used to study several mediators along the antioxidant pathway and showed that UPEI-100 increased the level of mitochondrial superoxide dismutase (SOD2) and oxidized glutathione disulfide (GSSG) and decreased a marker of lipid peroxidation due to oxidative stress (HNE-His adduct formation). Taken together, the data suggest that UPEI-100 may utilize similar pathways to those observed for the two parent compounds; however, it may also act through different mechanism of action. With this line of research, it is also proposed to study the sustained release effects of UPEI-100 using poly lactic-co-glycolic acid (PLGA) NPs or squalenoyl nanoassemblies on efficient drug delivery as well as sustained release formulation of developed therapeutics.

Bone is a natural functionally graded material, which exhibits two types of structures. One is a dense and stiff structure and the other a porous, soft load-bearing structure known as cortical and cancellous bone, respectively. This property has inspired the development of functionally graded bioceramic resources as implant materials with the original bio-tissue. The most important considerations for their selection in the human body are their biocompatibility, corrosion resistance, tissue reactions, surface conditions, and osseointegration (a bone bed formed through direct attachment to bone). Surgical grade stainless steel and titanium and its alloys are widely used in orthopedic and dental restorations. The human body is a very hostile environment and metals and alloy implants are unique that they are exposed to this dynamic environment containing living cells, tissues, and biological fluids. Clinical experience has shown that metallic implants are susceptible to localized corrosion in the human body, releasing metal ions into the surrounding tissues. Common failures of metallic implants have led to the application of biocompatible and corrosion resistant coatings, as well as to surface modification of the alloys. A bioactive material is one that elicits a specific biological response at the interface, which results in the formation of a bond between the tissues and the material. Hydroxyapiatite ceramics offer attractive properties, such as lack of toxicity, absence of intervening fibrous tissue, the possibility of forming a direct contact with bone, and the possibility to stimulate bone growth. Deposition of layers of nanobioceramic coatings on 316L SS, titanium and magnesium alloys are a viable solution. Surface engineering, nanobioceramics and functionally graded coatings are the promising techniques to battle corrosion of biomaterials. Nanobioceramics-based orthopedic implants as scaffolds and coatings combined with tissue engineering would lead to the development of new hybrid biomedical devices with the better understanding of the structure–property relationship. Modification of biomaterial surface properties through control of the characteristic length scale is one promising approach to modulate select cell functions.With a fast and tremendous industrial development and escalating human population along with an increase in energy demand, the global energy consumption has been accelerating at a startling rate. The global energy will get exhausted at this present consumption rate. To address this issue caused by energy exhaustion, urgency in the use of renewable energy is needed to suffice the demand. To facilitate the effective usage of renewable energy, it is imperative to develop high-performance, low-cost, and eco-friendly energy conversion and storage systems. Among the various developed protocols, fuel cells and supercapacitors are effective and simple systems that have been adopted for ages toward electrochemical energy conversion and storage. The performance of these systems is dependent on the intrinsic properties of the materials utilized for building the same. Hence, to cater the development of electrochemical energy conversion and storage systems, material technology plays a pivotal and supporting role.

TGBAPB epoxy resin was synthesized using BAPB and epichlorohydrin. The molecular weight was determined by GPC and equivalent weight by means of EEW titration. The amino-functionalized POSS was synthesized via ONPS method and its molecular structure has been confirmed by IR and NMR. The TGBAPB epoxy resin was further reinforced using POSS with varying weight percentages (1–5 wt%) and cured with DDM. Thermomechanical behavior of TGBAPB epoxy matrix and nanocomposites were examined by dynamic mechanical analysis (DMA), TGA, and DSC.

The surface morphology of the epoxy nanocomposites was investigated by X-ray diffraction (XRD), TEM, SEM, and AFM studies.The toughening of brittle epoxy matrix with C8 ether-linked bismaleimide (C8 e-BMI) and the reinforcing effect of carbon black (CB) in enhancing the conducting properties of insulating epoxy matrix are studied. The FTIR and Raman analysis indicate the formation of strong covalent bonds between CB and C8 e-BMI/epoxy matrix. The XRD and field emission scanning electron microscope (FESEM) analysis indicates the event of phase separation in 5 wt% CB-loaded epoxy C8 e-BMI nanocomposites. The impact strength increased up to 5 wt% of CB loading with particle pull and crack deflection to be driving mechanism for enhancing the toughness of the nanocomposite and beyond 5 wt%, the impact strength started to decrease due to aggregation of CB. The DMA also indicates the toughness of the nanocomposites was improved with 5 wt% of CB loading due to the phase segregation between epoxy and C8 e-BMI in the presence of CB. The electrical conductivity was also increased with 5 wt% of CB due to classical conduction by ohmic chain contact.

The pliancy of basic properties of epoxy tailored by appropriate choice of monomers, hardeners, and chemical reactions, impart versatility to these resins. This is one of the prime reasons for the cornucopia of applications, ranging from paints and adhesives to aerospace, electrical and industrial composites. Its characteristic features include excellent adhesion, mechanical properties, and chemical/heat resistance. Several studies on reinforcement of POSS with epoxy resins have shown amelioration in thermomechanical, dielectric, and flame retardancy properties.[7,10] Thus, the scope of this research is to synthesize POSS-reinforced siliconized epoxy and study its thermal and mechanical properties. The composites thus synthesized could be used for high-performance applications like aerospace sector.

Cyanate esters (CEs) are currently in widespread use because of their high thermal stability, excellent mechanical properties, good flame resistance, low outgassing, and good radiation resistance. Applications of CE resins include structural aerospace, electronic, microwave-transparent composites, encapsulant, and as an adhesives. In case of high-speed electronic device, circuits require low-k material to realize the faster signal transmission without crosstalk. To achieve this, it is important to develop the low-k dielectric materials that are been needed for the efficient integrated circuits. It is well established that the low-k silica and other related materials can prevent the signal crossover with low power consumption. In this view, many efforts have been taken to reduce the dielectric constant (<1.8) with the different kind of materials. Especially, nanolevel porous inorganic materials, variety of polymeric materials and their combinations are investigated for low-k dielectrics. Hence, in the present chapter, development of organic–inorganic hybrid polymer nanocomposites based on CEs hybridizing with porous nanoreinforcement like POSS, mesoporous SBA-15 and silica under appropriate conditions to yield low dielectric constant has been discussed. In addition, the thermal and morphological properties of the nanocomposites are discussed to ascertain their high-performance characteristics.

Biofouling has been a matter of huge concern in terms of economy and environment. Though many techniques are known for countering this, the Coatings Technology involving biocides dominates the market. Tributyltin (TBT)-based coatings were the most widely used and popular among the antifouling coatings industry till its toxic nature to marine systems led to a ban by IMO on its application in antifouling coatings. This promoted research for new technologies in antifoul coatings and still the search is at nascent stage. Previous studies have shown that antimicrobial formulations in the form of NPs could be used as effective bactericidal materials. The bactericidal effect of these metal NPs has been attributed to their small sizes. However to understand the potential of nanometals as biocides, a deep understanding of fouling technology and antifouling coatings are very much required. The present chapter looks into the fundamental aspects of biofouling, the antifouling techniques involved, the IMO ban on TBT, and other biocides. The chapter also discusses the potential of nanometals as biocides in place of conventional metal-based coatings in market.

Biocompatible polymeric composite materials are significant in aiding advanced applications for tissue engineering scaffolds and body implants. In the present study, nanoclay-reinforced poly(ε-caprolactone)/starch (PCLCS) hybrid nanofiber was prepared and evaluated for its thermal stability, hydrophobicity, biocompatibility, and storage stability. Solutions of PCL in chloroform and starch in dimethyl sulfoxide (DMSO), both in the range of 10–20% (w/w) were mixed and the nanofibers were obtained by electrospinning at a high voltage power supply. Montmorillonite nanoclay was used as a reinforcement to the fiber mat. SEM studies indicated that uniform fibers with narrow fiber diameter distribution were formed when 15% PCL and starch were used together with 5% nanoclay. Addition of nanoclay enhanced the thermal stability and restored the hydrophobicity of the surface that was lost due to starch. The prepared nanofibers supported the formation of hyroxyapatite films on the surface, when immersed in simulated body fluid and hence were proved to be biocompatible. The reasonable long-time stability of the prepared nanofibers stored at room temperature for over a period of 2 years and above was recorded from the FTIR and XRD spectra post storage.

The electrospun PCLCS hybrid nanofibrous mats possessing uniform fibers with narrow diameter distribution, good thermal stability, superhydrophobicity, biocompatibility, and good long-term storage stability are suitable to be developed into tissue engineering scaffolds and implant materials. This book would serve as a handbook to people just entering the field, and also for experts seeking for information in other subfields. Therefore, this book is well suited as a textbook for upper-level undergraduate, graduate, and professional short courses.

PART I
Nano-Hybrid Materials for Therapeutic and Aesthetic Applications

CHAPTER 1

THERAUPEUTIC EFFECT OF SMALL MOLECULAR ENTITIES

DESIKAN RAJAGOPAL[1], BARRY CONNELL[2], and TAREK SALEH[2]

[1]*Department of Pharmaceutical Chemistry, School of Advanced Sciences, VIT University, Vellore, Tamil Nadu 632014, India. E-mail: rajagopal.desikan@vit.ac.in*

[2]*Department of Biomedical Science, Atlantic Veterinary College, University of Prince Edward Island, Charlottetown, PE, C1A 4P3, Canada. E-mail: tsaleh@upei.edu*

CONTENTS

ABSTRACT

Conclusive evidence from past work has shown that preadministration of apocynin and lipoic acid (LA) at subthreshold levels for neuroprotection, enhanced the neuroprotective capacity when injected in combination. Continuing with this strategy, investigation was designed to determine if a codrug consisting of LA and apocynin functional groups bound by a covalent bond, named UPEI-100 (University Prince Edward Island -100), and is capable of similar efficacy using a rodent model of stroke. Male rats were anesthetized with inactin [100 mg/kg, intravenous (iv)], and the middle cerebral artery was occluded for 6 h [middle cerebral artery occlusion (MCAO)], or allowed to reperfuse for 5.5 h following a 30 min occlusion (ischemia/reperfusion; I/R). Preadministration of UPEI-100 dose dependently decreased infarct volume in the I/R model ($p < 0.05$), but not in the MCAO model of stroke. A time course for this neuroprotective effect showed that UPEI-100 resulted in a decrease in infarct volume following 2 h of reperfusion compared to vehicle. The time course of this neuroprotective effect was also used to study several mediators along the antioxidant pathway and showed that UPEI-100 increased the level of mitochondrial superoxide dismutase (SOD2) and oxidized glutathione (GSSG) and decreased a marker of lipid peroxidation due to oxidative stress (HNE-His adduct formation). Taken together, the data suggest that UPEI-100 may utilize similar pathways to those observed for the two parent compounds; however, it may also act through different mechanism of action. With this line of research, it is also proposed to study the sustained release effects of UPEI-100 using poly lactic-co-glycolic acid (PLGA) nanoparticles (NPs) or squalenoyl nanoassemblies on efficient drug delivery as well as sustained release formulation of developed therapeutics.

1.1 INTRODUCTION

Ischemic stroke due to occlusion of cerebral vasculature, results in hypoxia and hypoglycemia, failure of ATP (adenosine triphosphate)-dependent pumps, disruption of ionic equilibrium and calcium homeostasis, excitotoxicity, and eventual cell death.[1] Reintroduction of blood flow (i.e., with thrombolytic therapy) can arrest and/or reverse these adverse events. However, if recanalization is delayed beyond 4–6 h, further neuronal death, known as reperfusion injury, will occur due to the formation of reactive oxygen species (ROS). Several laboratories have demonstrated that NADPH (nicotinamide adenine dinucleotide phosphate) oxidase (NOX) is a major source of superoxide generation, and that NOX is involved in mediating ischemia-reperfusion (I/R)-induced neuronal death.[2] The pathophysiological importance of NOX in hypoxia has led many researchers to attempt to modify the activity of NOX in both permanent and transient (I/R) animal models of stroke.

Apocynin is regarded as a powerful antioxidant and anti-inflammatory agent, primarily by interfering with the assembly of the cytosolic and membrane-bound subunits to inhibit NOX activation.[3] For this reason, apocynin has been tested and shown to have promise as a neuroprotectant in many rat models of stroke through its ability to attenuate ischemic damage following cerebral I/R.[4] However, the enthusiasm over the potential clinical benefit of apocynin as a neuroprotectant has been dampened due to the narrow therapeutic dose range, in which apocynin is effective.[5] This is likely due to the high dose of apocynin required to measure neuroprotective effects,[5] as well as the fact that apocynin has been shown, under certain circumstances, to stimulate ROS production, and thereby increase cellular oxidative stress.[6]

Our laboratory has provided further support for this hypothesis and demonstrated that when using the same stroke modes as is being used in this publication, coadministration of nonprotective doses of apocynin with nonprotective doses of LA produced significant neuroprotection compared with either drug injected alone.[7]

There has been recent interest in the use of codrugs as a therapeutic approach in various pathologies.[8] In several different animal models of pathology, the development and administration of a codrug, containing LA covalently linked to another therapeutic drug, has been shown to have a greater positive effect compared to the injection of a solution containing a mixture of the two drugs.[8,9] However, to the best of our knowledge, no research has been published examining the use of codrug as a neuroprotectant following I/R injury. Based on our previous observation demonstrating the beneficial effects of coadministration of a solution containing apocynin with LA,[7] our laboratory has developed a codrug that is a covalent conjugate between LA and apocynin (named UPEI-100).

There is a limited window of opportunity following cerebral vascular occlusion for thrombolytic therapy to protect against further I/R-induced cell death (4 h following the onset of clinical signs in humans). This represents a critical time frame within which to study the efficacy of neuroprotectants. Also, most cardiovascular consequences following stroke occur within the first 4–6 h post stroke in humans.[10] Therefore, it is our intention to examine the acute, neuroprotective effects of a codrug, UPEI-100, using a rodent model of I/R recently developed and validated in our laboratory.[11] We examined tissue harvested from the ischemic cortex in animals pretreated with UPEI-100 to determine the effect of UPEI-100 on various cellular antioxidant pathways.

1.2 OXIDATIVE STRESS

ROS generated during cerebral I/R is well-documented as important mechanism through which neuronal apoptosis take place.[12] The close relationship between cerebral ischemia and oxidative stress has generated considerable interest to develop antioxidant agents to fight the harmful consequences of oxidative injuries in ischemia and recirculation injury. NOXis a multisubunit enzyme complex present at the cell plasma membranes.[13] This enzyme is especially well-characterized in immune cells and leukocytes for their involvement of ROS production for host defense purpose. Various protein components of NOX are expressed in neurons and astrocytes as well as in microglia. In the brain, distinct distribution patterns have been described for some of the NOX subunits, for example, gp91phox, p22phox, p40phox, p47phox, and p67phox. A report on hippocampal slices has shown a link between N-methyl-N-aspartic acid, an excitatory neurotransmitters receptor agonist, and production of ROS through NOX.[14] These oxidant radicals contribute to increased neuronal death by oxidizing protein, damaging DNA, and inducing the lipidperoxidation of cellular membranes. As alternatives for reducing cerebral damage, multiple antioxidant therapies have been proposed with variable results.[15] The antioxidants, LA, and various phenolic phytochemicals are present in normal diets. Several lines of evidence in experimental models suggest that the antioxidant effects of LA are achieved when it is used synergistically with other bioactive molecules.[16] We strongly anticipate that LA effects would be enhanced by chemically combining with other bioactive small molecules to extend the biological effects of parent structure. We also believe that such an approach would be greatly beneficial to models of cerebral ischemia.

1.3 PUBLIC HEALTH RELEVANCE

Stroke is one of the leading causes of death and disability in India.[17] The estimated prevalence rate of stroke range is 84–262/100,000 in rural and 334–424/100,000 in urban areas. The incidence rate is 119–145/100,000 based on the population-based studies. About 25% of men and 20% of women can expect to suffer a stroke if they live up to 85 years or beyond. It is one of the leading causes of death or disability around the world. Although stroke is a major cause of death, mortality data underestimate the true burden of stroke. Stroke causes chronic disability and in India, it is number one cause of disability. Since stroke causes disability more often than death, stroke patients often require long hospital stays followed by ongoing care in the community, or nursing-home facility. Stroke is consequently a major drain on health-care funding. The total incidence of stroke is projected to increase considerably over the next two decades. This is because of the rapid increase in the elderly population. It is predicted that stroke will account for 6.2% of the total burden of illness in 2020. Thus, without more effective strategies for the prevention, treatment, and rehabilitation of stroke, the cost of this disease will increase dramatically. Last two decades have seen significant progress in the understanding of mechanism of the pathophysiology of ischemic stroke. Remarkable developments have occurred in stroke trial methodology and in new intervention approaches.[18] Aspirin and other antiplatelet agents are now prescribed in secondary stroke prevention,[19] as well as anticoagulants[20] for the subgroup of patients who experience a stroke from a cardiac source of emboli. Blood pressure lowering after stroke and transitory ischemic attack (TIA) helps to prevent recurrent stroke and cardiac events. Existing evidence also indicates that statins have an important role as adjunct therapy for the prevention of stroke.[19b, 21] The reason that an effective drug has not been developed so far, despite intensive efforts in the pharmaceutical industry, is that most of the drugs do not cross the blood–brain barrier (BBB). Although, several agents are developed for CIF (cerebral infarction), many of them have not crossed beyond laboratory level. Our current approach focuses on new therapies for cerebral I/R, specifically designed to cross the BBB, that are carefully synthesized from natural substances, and it is expected to possess high efficacy and minimal toxicity.

1.4 LIPOIC ACID—A PROMISING THERAPEUTIC MATERIAL

Alpha lipoic acid (ALA) a multifaceted cardiometabolic drug: ALA and its reduced form dihydrolipoic acid are now recognized as compounds that have many biological functions.[22] ALA is a naturally occurring eight-carbon fatty acid that is synthesized by plants and animals, including humans and can be found in very small amounts in foods, such as spinach, broccoli, peas, Brewer's yeast, Brussels sprouts, rice bran, and organ meats. It is chemically named 1,2-dithiolane-2-pentanoic acid, and is also referred to as thioctic acid. It has emerged as a potent antioxidant,[22a, 23] anti-inflammatory molecule,[24] and as a mitochondrial protective agent.[25] ALA is normally found in the mitochondria, an energy producing organelle found inside each cell. The body under optimal conditions may be able to produce enough ALA for its metabolic activities (as a cofactor for enzymes involved in converting fat and sugar to energy), but additional amounts provided as a supplement may allow replenishment. As a dietary supplement, ALA functions as a powerful antioxidant.[26] ALA action is enhanced by other nutrient antioxidants like vitamins C and E1.[22b, 27] ALA has the unique ability to function as both a water, and fat-soluble antioxidant, in direct contrast to conventional antioxidants which are effective in one compartment or the other. ALA supplements are available in capsule form at health food stores and some drugstores. Numerous health benefits have been attributed to externally supplemented ALA. In humans, ALA is rapidly metabolized in the liver with 30% bioavailability. Thus, preparations to extend its half-life are likely to be helpful.[22c] ALA chelates redox-active transition metals, thus inhibiting the formation of toxic hydroxyl radicals[28] and also scavenges ROS, thereby increasing the levels of reduced GSH.[29] Furthermore, ALA can scavenge lipid peroxidation products, such as hydroxynonenal and acrolein.[30] ALA down regulates the expression of redox-sensitive proinflammatory proteins, including tumor necrosis factor (TNF-α) and inducible nitric oxide synthase.[31] ALA also increases acetylcholine (Ach)[32] production by activation of choline acetyltransferase and increases glucose uptake in the cell, thus supplying more acetyl-CoA to enter the Krebs cycle, a pathway critical for energy production by the cell. In experimental diabetes models, ALA enhances glucose utilization and significantly increases insulin-stimulated glucose disposal,[33] increasing metabolic clearance rate for glucose by about 50%. In addition, ALA has powerful weight loss[34] and anti-inflammatory effects in diabetes, and thus by itself could be of benefit in diabetic patients. Recent studies have also suggested mild blood pressure lowering effects.[35] ALA is frequently used for treatment of diabetic polyneuropathy,[36] and in randomized controlled clinical trials has been shown to be of benefit in diabetic neuropathy alleviating symptoms significantly. The American Diabetes Association has suggested that ALA plus vitamin E may be helpful in combating some of the health complications associated with diabetes,[37] including heart disease,[38] vision problem, nerve damage, and kidney diseases.[39] It helps to protect the brain from damage following stroke.[40] On the basis of several published results, there is good precedent for using ALA alone in the treatment of metabolic diseases and its complications.[41] Furthermore, ALA is nontoxic with no serious side effects. Intake of as much as 2000 mg per day has no serious side effects, except of mild gastric discomfort at higher doses. On the positive side, taking ALA can sometimes lead to a mild and relaxing feeling, and lead to a better feel of well-being. The beneficial effects of ALA in cardiometabolic disease can be broadly summarized as effects on insulin signaling/glucose uptake, mitochondrial function[42] and reduction of target organ damage,[43] and is expounded in the following paragraphs.

ALA improves insulin signaling and glucose uptake: An important aspect of the insulin resistance state of prediabetes and overt type 2 diabetes is an impaired capability of insulin to activate glucose transport in skeletal muscle due to defects in.[33c, 44] A rising body of evidence indicates that one likely factor in the multifactorial etiology of skeletal muscle insulin resistance is oxidative stress, an imbalance between the cellular exposure to an oxidant stress and the cellular antioxidant defenses. Exposure of skeletal muscle to oxidant stress leads to impaired insulin signaling and subsequently to reduce glucose transport activity. ALA and its reduced form scavenge hydroxyl radicals, hypochlorous acid, peroxynitrite, and singlet oxygen and can restore thioredoxin,[45] vitamin C and GSH, thereby reducing oxidative stress. The positive metabolic actions of ALA have been established in a variety of experimental models. Evidence from cell culture studies, primarily using 3T3-L1 adipocytes and L6 myocytes, indicates that ALA can protect cells from the deleterious effects of oxidant stress. Because ALA possesses antioxidant properties, it is particularly suited to the improvement of insulin signaling and/or treatment of diabetic complications that occur from an overproduction of reactive oxygen and nitrogen species. Consistent with these predictions, ALA increases glucose uptake through recruitment of the glucose transporter-4 to plasma membranes. It has also been confirmed recently that chronic administration of ALA to diabetes-prone Otsuka Long-Evans Tokushima Fatty rats prevents the age-dependent development of hyperglycemia, hyperinsulinemia, dyslipidemia, and plasma markers of oxidative stress. Another key defect in diabetes and insulin resistance associated with obesity is mitochondrial dysfunction.

Effects of ALA on mitochondrial function: Mitochondrial decay has been postulated to be a considerable part of the metabolic dysfunction in diabetes. Decline in mitochondrial function may lead to cellular energy deficits, chiefly in times of greater energy demand, and compromise vital ATP-dependent cellular operations, including detoxification, repair systems, DNA replication, and osmotic balance. Mitochondrial decay may also lead to enhanced oxidant production, and thus leave the cell more prone to oxidative insult. In particular, the heart may be especially vulnerable to mitochondrial dysfunction due to supply of high-energy phosphates by mitochondria to carry out both contraction and relaxation. To preserve appropriate myocardial function, a constant supply of ATP is required, but few reserves are maintained. Thus, when the energy supply is interrupted (ischemia) or impaired (aging), ATP levels decline rapidly. For both systolic contractions and diastolic relaxation require high levels of ATP because ATP acts as an allosteric effector to disassociate actin from myosin. Thus, any decrement in mitochondrial ATP synthesis affects cardiac stiffness appreciably. Decline in ATP synthesis also compromises Ca^{2+} reuptake into the sarcoplasmic reticulum from the cytosol again disturbing myocardial relaxation. The $Na2^{+}$/Ca^{2+} transporter is also energy dependent, and a decline in myocardial ATP levels would thus slow cardiac relaxation by decreasing the rate of Ca^{2+} removal from the cytosol. Hence, maintenance of mitochondrial function may be imperative to maintain overall myocardial function. Due to steady production of ROS by mitochondria as by-products of electron transport; these not only injure mitochondria but also other cellular biomolecules due to oxidant seepage into cytoplasm. Mitochondria DNA (mtDNA) is predominantly prone to oxidant-induced damage due to its proximity to the source of ROS production and the lack of protecting histones. mtDNA damage, if not repaired, may be converted into mutations. These mutations would affect electron transport, thereby further reducing the capacity to synthesize ATP and increasing ROS production. The oxidized form of ALA is reduced in mitochondria by specific dehydrogenases, and its presence would thus target an antioxidant to the mitochondria, the major location of ROS production. ALA may also boost mitochondrial function because it is a cofactor for pyruvate and alpha-ketoglutarate dehydrogenase. Consistent with this results in experimental models of diabetes have demonstrated an important effect of ALA in improving mitochondrial function and biogenesis. Finally, ALA through the effects outlined previously reduces target organ damage, thereby making a powerful statement of its potential as a therapy in cardiometabolic disease.

ALA improves target organ damage: Endothelial dysfunction predisposes to the initiation and progression of various fatal clinical consequences, including myocardial infarction, stroke, and atherosclerosis. All the recognized cardiovascular risk factors, such as smoking, dyslipidemia, and hypertension, have been shown to cause endothelial dysfunction. Recent studies have clearly shown that ALA is shown to improve acetylcholine-induced endothelium-dependent vasorelaxation in experimental animal models and humans Furthermore, ALA has been shown to have effects, such as the ability to reduce adhesion molecules and chemokines, to lower serum triglycerides and to activate the phosphoinositide 3-kinase/Akt-signalling pathway leading to reduced activation of nuclear factor-kappa B, a key proinflammatory transcription factor. We have demonstrated in humans that the combination of ALA with the angiotensin receptor blocker irbesartan can markedly reduce pro-inflammatory IL-6 and VCAM-1 (vascular cell adhesion molecule-1) levels besides improving endothelial function. We have additionally shown that ALA when administered to humans in conjunction with the ACE (angiotensin converting enzyme) inhibitor, Quinapril, can potentiate effects on endothelial function.

LA is an antioxidant with strong free radical-scavenging abilities and several researchers have shown in different rat models of stroke, that administration of LA produced significant neuroprotection by decreasing infarct size.[46] Further, several laboratories have demonstrated that administration of LA in combination with another compound can produce significant neuroprotective effects exceeding the effect of either drug alone.

1.5 APOCYNIN–LIPOIC ACID CONJUGATES

1.5.1 DESIGN

Drugs targeting more than one mechanism of action could potentially overcome important dilemma in drug designing. The synthesis and development of codrugs using simple yet biologically relevant molecules as building blocks provides the ability to simultaneously target multiple pathways involved in the pathogenesis of neurological diseases, specifically, pathways involved in the initiation of oxidative stress-induced neuronal damage following reperfusion. Chemical combinations of LA with other compounds have previously been demonstrated to provide neuroprotective effects greater than the two compounds administered on their own. Covalent linkage of LA with ibuprofen has been demonstrated to be neuroprotective in rodent models of Alzheimer's disease where administration of the codrug decreased the oxidative damage due to the infusion of Aβ. In addition, a codrug produced by chemically linking LA with L-dopa or dopamine decreased neuronal oxidative damage associated with the administration of L-dopa or dopamine alone. To the best of our knowledge, there have been no other reports where apocynin has been chemically linked to other compounds and no other reports where administration of a codrug was used to attenuate damage due to oxidative stress following ischemia or ischemia/reperfusion.

1.6 RESULTS ON EFFICACIOUS COMBINATION: UPEI-100

1.6.1 THE EFFECT OF PREADMINISTRATION OF UPEI-100 ON INFARCT VOLUME

This experiment was designed to determine the effect of UPEI-100 preadministration on infarct volume following I/R. Preadministration of UPEI-100 resulted in a dose-dependent neuroprotection with a dose of 0.1 and 0.5 mg/kg resulting in a significant decrease in infarct volume compared to the administration of vehicle [$p \leq 0.05$; Fig. 1.1(A)].

To examine the effect of UPEI-100 on ischemia-induced cell death only, injections of UPEI-100 (0.1 or 1.0 mg/kg) or vehicle were made 30 min prior to 6 h of MCAO (sutures left in place for 6 h). Neither concentration of UPEI-100 (the optimal dose of 0.1 mg/kg as determined above or 1.0 mg/kg, a 10-fold increase in the optimal dose) produced significant neuroprotection, when infarct volume was measured 6 h following the start of MCAO [$p \geq 0.05$; Fig. 1.1(B)].

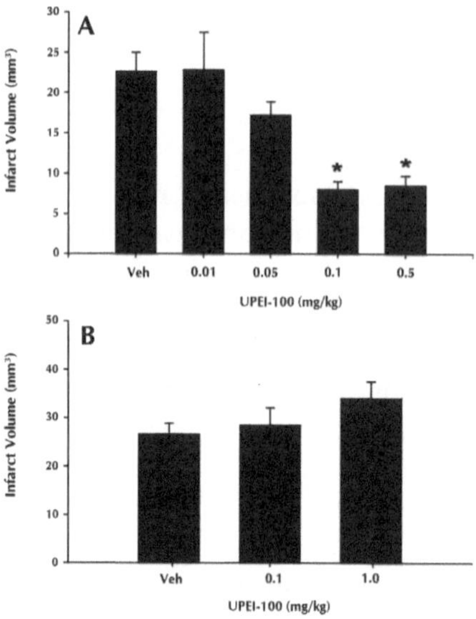

FIGURE 1.1 (A) Effect of pretreatment (30 min prior to occlusion of the middle cerebral artery; MCAO) with UPEI-100 on infarct volume (mm^3) calculated from TTC (tetrazolium chloride) stained 1-mm-thick coronal sections throughout the extent of the infarct following I/R. Each bar represents the mean ± S.E.M (Standard Error of the Mean) and *indicates significance ($p \leq 0.05$) from the vehicle group. (B) Effect of UPEI-100 administered 30 min prior to permanent (6 h) MCAO on infarct volume (mm^3) calculated from TTC stained, 1-mm-thick coronal slices throughout the extent of the infarct. Each bar represents the mean ± S.E.M.

1.6.2 THE EFFECT OF SYSTEMIC UPEI-100 INJECTION ON BLOOD PRESSURE AND HEART RATE

The following experiment was designed to determine the effect of UPEI-I00 on arterial pressure and heart rate for a period of 2 h following administration. Mean baseline arterial pressure and mean heart rate prior to drug administration was 112 ± 11 mmHg and 388 ± 26 bpm. Administration of UPEI-100 (0.1 mg/kg; i.v.) did not significantly alter mean arterial blood pressure or mean heart rate at any time point during the 2 h continuous recording compared to vehicle ($p \geq 0.05$; data not shown).

1.6.3 THE EFFECT OF UPEI-100 PREADMINISTRATION ON CARDIOVASCULAR PARAMETERS FOLLOWING I/R

The following experiment was designed to determine the effect of preadministration of UPEI-100 on mean arterial pressure, mean heart rate, and the mean cardiac baroreceptor reflex sensitivity (BRS) before, during, and following 30 min of MCAO. Mean arterial pressure, heart rate, and BRS prior to UPEI-100 administration were 119 ± 15 mmHg, 402 ± 22 bpm, and 0.55 ± 0.05 bpm/mmHg, respectively and prior to vehicle administration were 110 ± 9 mmHg, 398 ± 19 bpm, and 0.52 ± 0.05 bpm/mmHg, respectively. These values did not change following UPEI-100 or vehicle administration prior to MCAO [$p \geq 0.05$ for all comparisons; Fig. 1.2(A–C)]. During 30 min of MCAO and during 5.5 h of reperfusion, there were no significant differences in the mean arterial pressure or mean heart rate compared to pre-MCAO values [$p \geq 0.05$ for both UPEI-100 and vehicle; Fig. 1.2(A and B)]. However, mean BRS values in both the UPEI-100 and vehicle groups were equal and significantly decreased within 5 min of the beginning of MCAO [0.23 ± 0.05 bpm/mmHg and 0.28 ± 0.08 bpm/mmHg, respectively; $p \leq 0.05$ for both groups compared to pre-MCAO values; Fig. 1.2(C)] and remained significantly depressed throughout the 30 min of MCAO [$p \leq 0.05$ for both groups compared to pre-MCAO values; Fig. 1.2(C)]. The mean BRS for both groups remained significantly decreased compared to pre-MCAO values throughout the 5.5 h of reperfusion [$p \leq 0.05$ at all time points measured for both groups compared to pre-MCAO values; Fig. 1.2(C)].

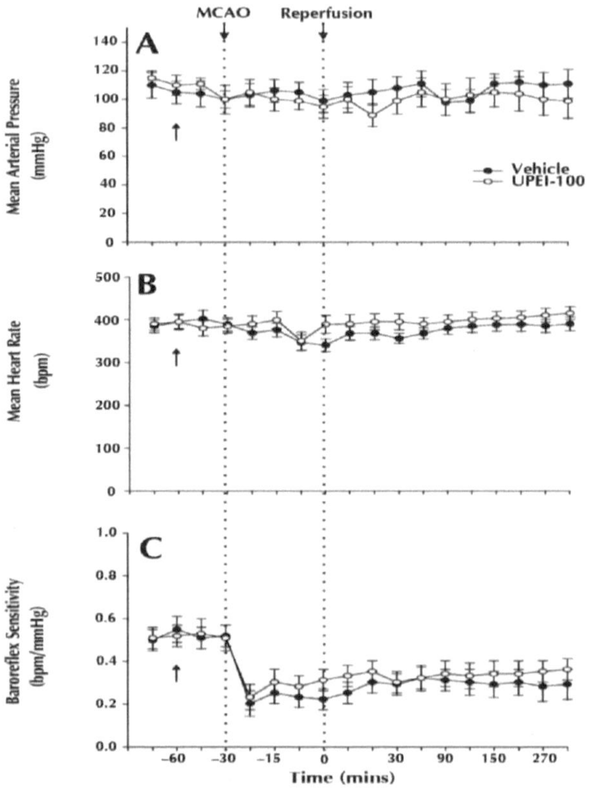

FIGURE 1.2 Cardiovascular responses to UPEI-100 (0.1 mg/kg) or vehicle (25% EtOH (Ethanol) pretreatment (i.v.) 30 min prior to I/R. Graphs represent average changes in (A) mean arterial pressure (MAP); mmHg; (B) heart rate (HR; bpm) and (C) baroreflex sensitivity (BRS; bpm/mmHg) following phenylephrine injection (0.0025 μg/ml; i.v.). The up arrow indicates time of UPEI-100 or vehicle injection. The first dashed lines represents the times at which the MCA was occluded and the second line indicates when blood flow was returned (reperfusion). Each data point represents the mean ± S.E.M. and * indicates significance ($p \leq 0.05$) from the pre-MCAO values.

1.6.4 THE EFFECT OF UPEI-100 ON INFARCT VOLUME WHEN ADMINISTERED EITHER DURING MCAO OR FOLLOWING THE START OF REPERFUSION

UPEI-100 or vehicle was injected (iv) at various time points during MCAO or following the start of reperfusion. There were no significant differences in the mean infarct volumes when vehicle was injected during MCAO or at any time point during reperfusion ($p \geq 0.05$); therefore, the vehicle data for all time points were pooled ($n = 29$) in Figure 1.3. All statistical comparisons were made between the infarct volumes measured following vehicle and UPEI-100 administration at each time point.

Administration of UPEI-100 (0.1 mg/kg; iv) 15 min into a 30 min period of MCAO (15 min prior to the start of reperfusion) produced significant neuroprotection compared to vehicle when infarct volume was measured following 5.5 h of reperfusion ($p \leq 0.05$; Fig. 1.3).

We determined the effect of UPEI-100 on reperfusion injury only, by measuring the infarct volume following drug administration immediately prior to suture removal, or 30, 60, 90, 120, or 180 min following the start of reperfusion. Administration of UPEI-100 (0.1 mg/kg) at time 0 (start of reperfusion), and 30, 60, and 90 min following the start of reperfusion resulted in significant decreases in infarct volume compared to vehicle administration ($p \leq 0.05$ at each time point; Fig. 1.3). Administration of UPEI-100 (0.1 mg/kg) 120 and 180 min following suture removal did not result in a significant difference in infarct volume compared to the administration of vehicle at those time points ($p \geq 0.05$; Fig. 1.3).

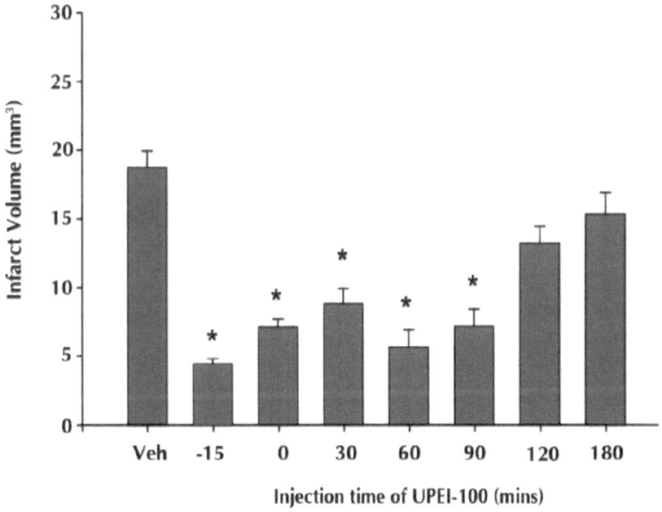

FIGURE 1.3 Effect of UPEI-100 (0.1 mg/kg) or vehicle (0.0125% EtOH) administered 15 min prior to the beginning of reperfusion (−15), immediately prior to suture removal, and the start of reperfusion (0), or 30, 60, 90, 120, and 180 min following reperfusion on infarct volume (mm^3). Each bar represents the mean ± S.E.M. and * indicates significance ($p \leq 0.05$) from the vehicle control group at the same time point. There were no significant differences in the mean infarct volumes when vehicle was injected during MCAO or at any point during reperfusion ($p \geq 0.05$); therefore, the vehicle data for all time points were pooled and is represented by a single bar.

1.6.5 THE EFFECT OF INTRACORTICAL INJECTIONS OF UPEI-100 ON INFARCT VOLUME FOLLOWING I/R

This experiment was designed to determine the effect of direct cortical pre-injection with UPEI-100 on infarct volume following I/R. UPEI-100 pre-injection resulted in a dose-dependent neuroprotection with a dose of 1.0 μM resulting in a significant decrease in infarct volume compared to the intracortical preinjection of vehicle ($p \leq 0.05$; Fig. 1.4).

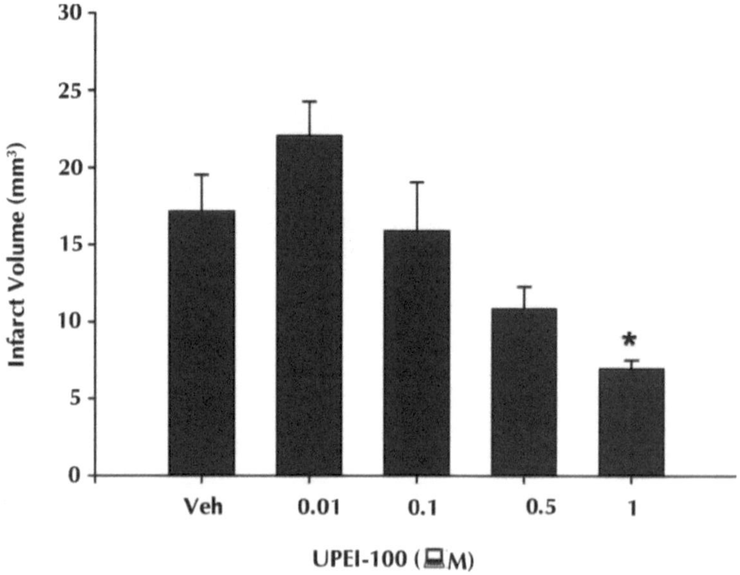

FIGURE 1.4 Effect of direct cortical injections of UPEI-100 or vehicle (0.0125% EtOH) 10 min pre-MCAO on infarct volume (mm^3) calculated from TTC stained 1-mm-thick coronal sections throughout the extent of the infarct following I/R. Each bar represents the mean ± S.E.M. and * indicates significance ($p \leq 0.05$) from the vehicle control group.

1.6.6 TIME COURSE FOR THE EFFECT OF UPEI-100 ON INFARCT VOLUME

In animals pretreated with UPEI-100 (0.1 mg/kg) or vehicle and undergoing MCAO for 30 min, we observed a rapid increase in infarct volume from 30 min into the reperfusion period until 2 h of reperfusion (Fig. 1.5). After 2 h of reperfusion, the mean infarct volume continued to increase significantly in the vehicle group. In contrast to the vehicle pretreated group, the mean infarct volume in the UPEI-100 group did not increase further during the remaining reperfusion period (Fig. 1.5). In addition, the mean infarct volume in the UPEI-100 group was significantly smaller than the vehicle pretreated group when measured at 4 and 6 h ($p \leq 0.05$ at each time point; Fig. 1.5). By the end of the 5.5 h of reperfusion, the mean infarct volume of the UPEI-100 group was approximately 53% smaller than the vehicle group.

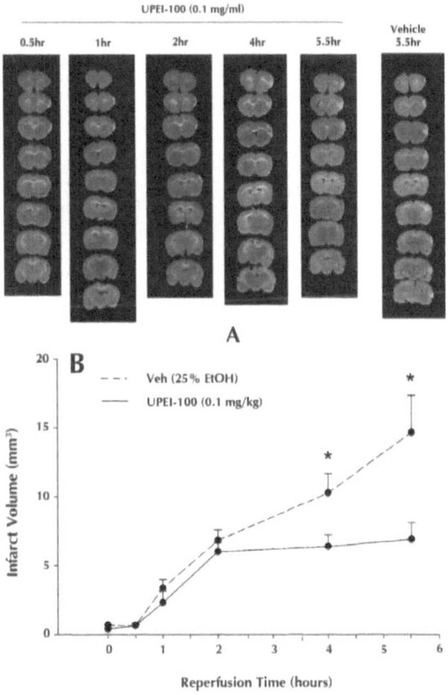

FIGURE 1.5 (A) Representative digital photomicrographs of TTC-stained 1-mm-thick coronal sections following 30 min of MCAO illustrating the extent of the infarct size within the prefrontal cortex at progressive time points during 5.5 h of reperfusion in vehicle (25% EtOH) and UPEI-100 (0.1 mg/kg) treated animals. (B) Graphic representation of the change in infarct volume (mm^3) at progressive time points of reperfusion following 30 min of ischemia. Each data point represents the mean ± S.E.M. and * indicates significance ($p \leq 0.05$) between vehicle and UPEI-100 treated rats at the same time point.

1.6.7 EFFECT OF UPEI-100 ON SOD ACTIVITY

In animals pretreated with UPEI-100 or vehicle and undergoing MCAO for 30 min, we did not observe any significant differences in cytoplasmic SOD1 activity levels between these two groups at any time point during 5.5 h of reperfusion [$p \geq 0.05$ for each time point measured; Fig. 1.6(A)].

In contrast, mitochondrial SOD2 activity levels at the 4 h time point were significantly higher in the UPEI-100 group compared to vehicle [$p \leq 0.05$;

Fig. 1.6(B)]. There were no other significant differences measured at any time point; however, the SOD2 activity levels were consistently higher in the UPEI-100 group compared to those in the vehicle pretreated group at each time point [Fig. 1.6(B)].

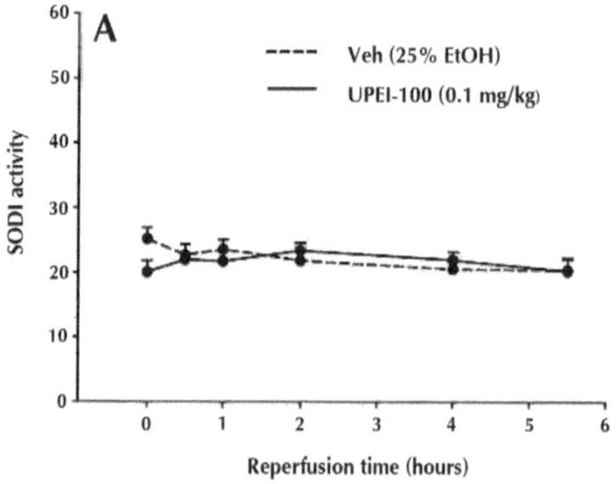

FIGURE 1.6(A) Effect of preadministration of UPEI-100 (0.1 mg/kg) or vehicle (25% EtOH) on (A) cytoplasmic superoxide dismutase (Cu/ZnSOD; SOD1) activity within the infarct area following I/R at progressive time points following the beginning of reperfusion. SOD1 activity measured as % inhibition of chromogenic reduction. Each data point represents the mean SOD1 activity ± S.E.M. and * indicates significance ($p \leq 0.05$) between vehicle and UPEI-100 treated rats at the same time point.

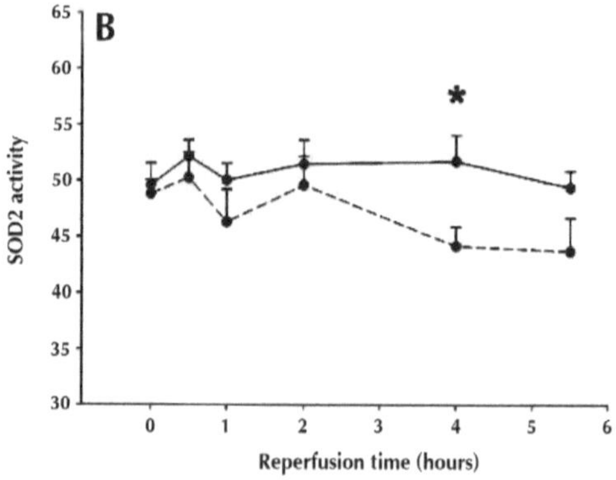

FIGURE 1.6(B) Effect of preadministration of UPEI-100 (0.1 mg/kg) or vehicle (25% EtOH) on mitochondrial MnSOD (SOD2) activity within the infarct area following I/R at progressive time points following the beginning of reperfusion. SOD2 activity measured as % inhibition of chromogenic reduction. Each data point represents the mean SOD2 activity ± S.E.M. and * indicates significance ($p \leq 0.05$) between vehicle and UPEI-100 treated rats at the same time point.

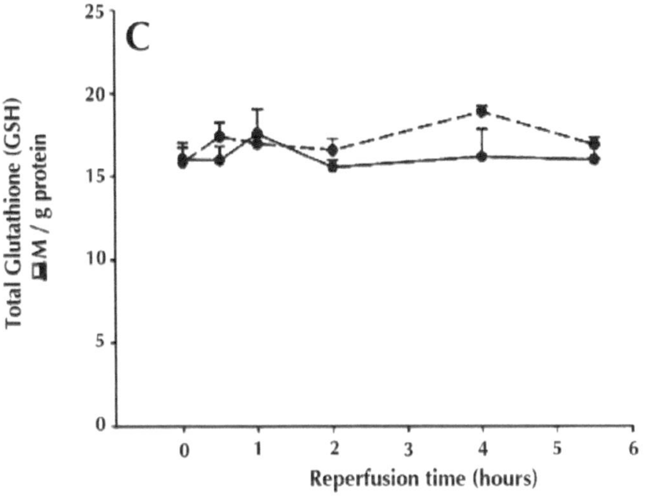

FIGURE 1.6(C) Effect of preadministration of UPEI-100 (0.1 mg/kg) or vehicle (25% EtOH) on (A) total glutathione (GSH + GSSG) levels within the infarct area following I/R at progressive time points following the beginning of reperfusion. GSH and GSSG levels were measured in μM/g protein. Each data point represents the mean total GSH or GSSG activity level ± S.E.M. and * indicates significance ($p \leq 0.05$) between vehicle and UPEI-100 treated rats at the same time point.

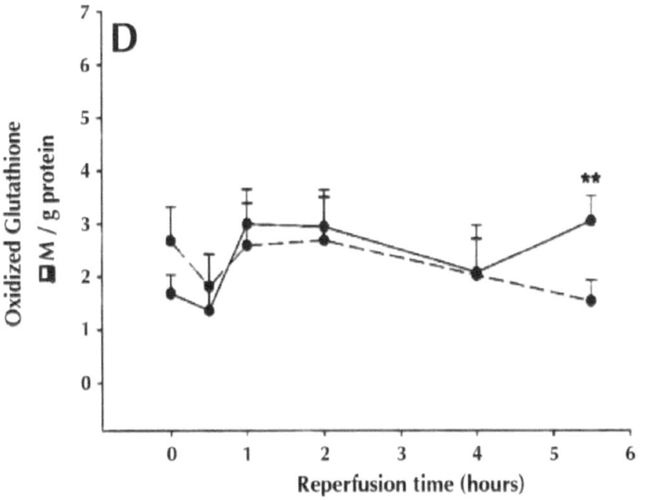

FIGURE 1.6(D) Effect of preadministration of UPEI-100 (0.1 mg/kg) or vehicle (25% EtOH) GSSG levels within the infarct area following I/R at progressive time points following the beginning of reperfusion. Each data point represents the mean total GSH or GSSG activity level ± S.E.M. and * indicates significance ($p \leq 0.05$) between vehicle and UPEI-100 treated rats at the same time point.

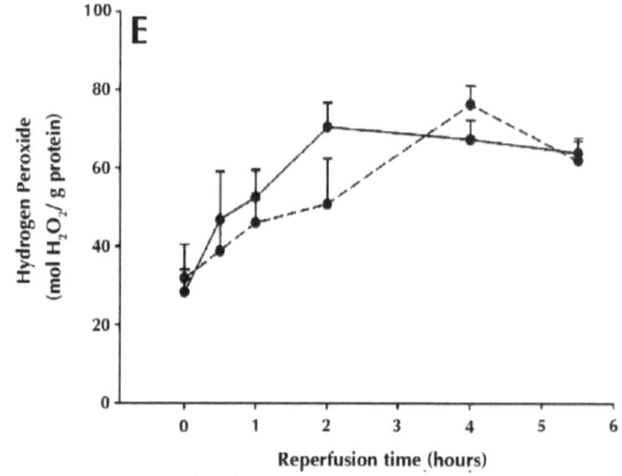

FIGURE 1.6(E) Effect of preadministration of UPEI-100 (0.1 mg/kg) or vehicle (25% EtOH) on hydrogen peroxide (H_2O_2) levels within the infarct area following I/R at progressive time points following the beginning of reperfusion. Data points represent mean H_2O_2 levels measured as μmol H_2O_2 per gram of protein.

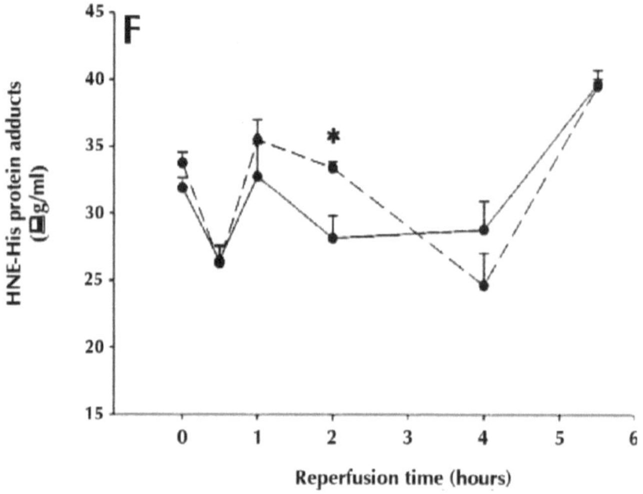

FIGURE 1.6(F) Effect of preadministration of UPEI-100 (0.1 mg/kg) or vehicle (25% EtOH) on total HNE-His (4-hydroxynonenal-histidiine) protein adduct levels (μg/mL) within the infarct area following I/R at progressive time points following the beginning of reperfusion. Data points represent the mean HNE-His adduct level (μg/mL) ± S.E.M. and * indicates significance ($p \leq 0.05$) between vehicle and UPEI-100 treated rats at the same time point.

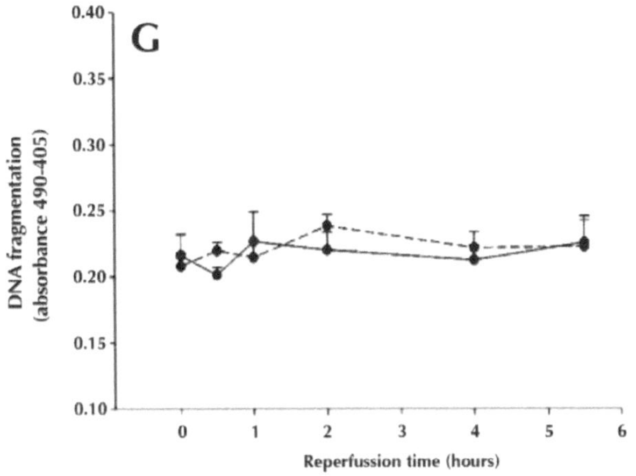

FIGURE 1.6(G) Graphic representation of the effect of preadministration of UPEI-100 (0.1mg/kg) or vehicle (25% EtOH) on the amount of DNA fragmentation (apoptosis) within the infarct area following I/R at progressive time points following the beginning of reperfusion. Data points represent the mean amount of DNA fragmentation measured as the mean absorption at 405 nm ± S.E.M. of cytoplasmic nucleosomes.

1.6.8 EFFECT OF UPEI-100 ON GLUTATHIONE GSH LEVELS

We did not measure any significant differences in the total amount of GSH (includes both oxidized and reduced forms) in brain samples taken from either UPEI-100 or vehicle treated rats at any time point [$p \geq 0.05$ for each comparison; Fig. 1.6(C)]. In contrast, the amount of the oxidized form of GSH (GSSG) was significantly greater in the UPEI-100 treated rats than the amount of GSSG measured in vehicle treated rats after 5.5 h of reperfusion [$p \leq 0.05$; Fig. 1.6(D)]. There was no difference in the level of GSSG between the two groups at any other time point [$p \geq 0.05$ at each time point; Fig. 1.6(D)].

1.6.9 EFFECT OF UPEI-100 ON HYDROGEN PEROXIDE LEVELS

In UPEI-100 pretreated rats, the level of hydrogen peroxide was not significantly greater than the level of hydrogen peroxide in vehicle treated rats at any time point over the 5.5 h of reperfusion [$p > 0.05$; Fig. 1.6(E)]. However, in both UPEI-100 and vehicle pretreated groups, the level of hydrogen peroxide increased significantly over time [$p < 0.05$; Fig. 1.6(E)].

1.6.10 EFFECT OF UPEI-100 ON PROTEIN (HNE-HIS) ADDUCT LEVELS

The following experiment was designed to determine whether UPEI-100 mediated neuroprotection resulted in a change in HNE-His adduct levels. HNE-His adduct levels following UPEI-100 administration 30 min prior to occlusion were significantly attenuated compared to vehicle administered following 2 h of reperfusion [$p \leq 0.05$; Fig. 1.6(F)]. HNE-His adduct levels were not significantly different in the two groups at any other time point [$p \geq 0.05$ for each time point measured; Fig. 1.6(F)].

1.6.11 EFFECT OF UPEI-100 ON DNA FRAGMENTATION (APOPTOTIC CELL DEATH)

The following experiment was designed to determine whether the UPEI-100 mediated neuroprotection observed was mediated by an alteration in the extent of apoptotic cell death. DNA fragmentation was quantified as an indicator of apoptotic cell death. No significant changes in the extent of DNA fragmentation in animals pretreated with either UPEI-100 or vehicle throughout the experimental time course were observed [$p \geq 0.05$ for each time point measured; Fig. 1.6(G)].

1.7 DISCUSSION—PERSPECTIVE ANALYSIS

Oxidative stress associated with excessive production of ROS is a fundamental mechanism of brain damage in reperfusion injury following ischemic stroke. The multiplicity of mechanisms involved in I/R-induced neuronal damage following an occlusive stroke remains an obstacle in providing treatment in clinical settings. Drugs targeting more than one mechanism of action could potentially overcome this dilemma. The synthesis and development of codrugs using simple yet biologically relevant molecules as building blocks provides the ability to simultaneously target multiple pathways involved in the pathogenesis of neurological diseases, specifically, pathways involved in the initiation of oxidative stress-induced neuronal damage following reperfusion. Chemical combinations of LA with other compounds have previously been demonstrated to provide neuroprotective effects greater than the two compounds administered on their own. Covalent linkage of LA with ibuprofen has been demonstrated to be neuroprotective in rodent models of Alzheimer's disease where administration of the codrug decreased the oxidative damage due to the infusion of Aβ. In addition, a codrug produced by chemically linking LA with L-dopa or dopamine decreased neuronal oxidative damage associated with the administration of L-dopa or dopamine alone. To the best of our knowledge, there have been no other reports where apocynin has been chemically linked to other compounds and no other reports where administration of a codrug was used to attenuate damage due to oxidative stress following ischemia or ischemia/reperfusion.

In this study, we determined that UPEI-100, a chemical combination of two naturally occurring antioxidants, LA and apocynin, produced dose-dependent neuroprotection against neuronal cell death as observed in a previously validated, novel model of I/R injury. The results demonstrated that UPEI-100 produced dose-dependent, short-term neuroprotection (within 5.5 h of reperfusion) in a model of focal ischemia, which is restricted to the cerebral cortex. Further, the dose of UPEI-100 required to produce significant neuroprotection (0.1 mg/kg) was many fold less compared with the doses required for either apocynin or LA on their own. Also, this optimal dose of UPEI-100 produced significant neuroprotection when administered 15 min prior to the start of reperfusion, just prior to the induction of reperfusion, and 30, 60, and 90 min following the onset of reperfusion. The reason for administering UPEI-100 during the occlusion was to mimic the clinical situation where a patient would present during a stroke. This result does not indicate if UPEI-100 or an intermediate metabolite produced as a result of hepatic biotransformation was responsible for mediating the neuroprotective effect. We were able to reproduce the dose-dependent neuroprotection with direct cortical injections into the ischemic area, which suggests that the parent compound UPEI-100, and not a metabolite(s), was responsible for producing the neuroprotection observed.

There have been other reports which demonstrated an attenuation of infarct volume when apocynin was administered during an occlusion of the MCA (middle cerebral artery), but these benefits were lost when administration of apocynin was delayed following the onset of reperfusion. However, one laboratory demonstrated that administration of apocynin in gerbils 5 min following 5 min of global ischemia decreased neuronal degeneration and delayed neuronal death and microglial activation when assessed 4 days later. It is not known if these apocynin-induced beneficial effects translated to a decrease of infarct size as it was not measured in that study. We have also demonstrated that administration of the most effective dose of LA (5 mg/kg) did not result in significant neuroprotection when administered just prior to the beginning of reperfusion, and only produced neuroprotection when administered prior to both occlusion and reperfusion. It therefore appears that UPEI-100, a chemical combination of apocynin and LA, is far superior then either compound alone in its ability to provide neuroprotection when administered during reperfusion. Interestingly, UPEI-100 did not produce neuroprotection when administered prior to a 6 h permanent occlusive stroke (no reperfusion). These results support the suggestion that UPEI-100 produced neuroprotection against reperfusion injury alone, perhaps via decreasing ROS production, as ROS-induced growth of ischemic volume due to reperfusion injury has been demonstrated by many labs, and/or possibly via free radical scavenging.

Clinically, elevated sympathetic tone (sympathoexcitation) and abnormal electrocardiograms have been observed within 1–2 h following thrombolytic or hemorrhagic stroke involving the MCA. Such autonomic dysfunction increases the risk of sudden cardiac death and can be mimicked in rat models of MCAO (8). Arrhythmogenesis and sudden cardiac death, which can occur following MCAO in humans, is associated with depressed baroreflex sensitivity (BRS; 5). Our results have demonstrated that during the 30 min occlusion, BRS was significantly depressed and administration of UPEI-100 did not alter the level of BRS depression despite the UPEI-100-induced neuroprotection. The increase in neuronal survival represented by an attenuated ischemic volume was not associated with recovery of autonomic function. Our laboratory has demonstrated a similar dissociation between changes in infarct volume and autonomic function following drug intervention prior to MCAO. We suggest that for an agent injected systemically, it may be required to have multiple sites of action within the CNS (central nervous system) as well as in the periphery (such as the myocardium) to decrease or reverse the sympathoexcitation and subsequent autonomic function observed following stroke. Many studies have demonstrated significant neurochemical and electrophysiological alterations in extracortical autonomic and cardiovascular regulatory nuclei shortly following MCAO. This suggests that any drug, which demonstrates functional cardiovascular protection may be required to act extracortically to prevent an abnormal sympathetic outflow and to restore or prevent changes in autonomic function following MCAO.

Neuronal hypoxia and the ensuing mitochondrial response are involved in both the initiation of both necrotic and apoptotic pathways leading to cell death. Severe cerebral ischemia causes neuronal mitochondria to be unable to produce adenosine triphosphate leading to necrotic cell death. The mitochondrial antioxidant enzyme, manganese SOD2, is the primary cellular defense enzyme involved protecting cells from oxidative stress and has been shown to reduce oxidative stress following cerebral I/R as SOD2-deficient mice have enhanced infarct size following cerebral ischemia and overexpression of SOD2 provided neuroprotection following cerebral ischemia. We observed enhanced SOD2 activity throughout the 5.5 h of reperfusion (reaching significance following 4 h of reperfusion) when UPEI-100 was administered 30 min prior to I/R. We have previously demonstrated using the same model of I/R that LA administered 30 min prior to I/R also increased SOD2 activity, while apocynin was ineffective in altering SOD2 activity. UPEI-100 did not alter cytosolic copper-zinc SOD (CuZnSOD or SOD1) activity, which is consistent with previous results from our laboratory where neither of the two parent compounds, apocynin or LA, altered SOD1 activity. There appears to be support in the literature for a selective effect of LA on mitochondrial function as LA is a protein bound cofactor for mitochondrial α-ketoacid dehydrogenase and serves a critical role in mitochondrial energy metabolism. In addition, exogenous LA is reduced to dihydro-lipoic acid (DHLA) within the mitochondria and both LA and DHLA have been shown to have powerful antioxidant activity and ROS scavenger abilities. Therefore, we conclude that that at least part of UPEI-100-induced neuroprotection may be due to the actions of the LA functional group of UPEI-100 on SOD2 activity.

GSH is another key intracellular antioxidant and protects cells by scavenging free radicals. GSH is involved with the breakdown of peroxides, regulating the nitric acid cycle, DNA synthesis and repair, and maintenance of protein disulfide bonds. In addition to its role in the prevention of oxidative stress, GSH also helps to maintain exogenous antioxidants, such as vitamins C and E. Within cells, GSH exists as reduced (GSH) and oxidized states (GSSG). In healthy cells, more than 90% of the total GSH pool is in the reduced form while less than 10% exists in the oxidized or disulfide form. An increased level of GSSG is generally indicative of enhanced oxidative stress. Interestingly, we have reported above that UPEI-100 administration resulted in a significant increase in the amount of GSSG at the 5.5 h of reperfusion time interval. It is possible that the UPEI-100-induced enhancement in the activity of SOD2 during the 5.5 h of reperfusion resulted in higher levels of hydrogen peroxide production. However, the measured levels of hydrogen peroxide were not different between vehicle and UPEI-100 groups indicating that the neurons were able to adequately deal with this excess hydrogen peroxide. Hydrogen peroxide may be fully reduced to water, but may also form hydroxyl radicals in the presence of ferrous or cuprous ions. A slow increase in hydroxyl radical levels over the 5.5 h of reperfusion following UPEI-100 administration could have attenuated the activity of GSH reductase, the enzyme responsible for reducing GSSG into GSH, resulting in enhanced levels of GSSG as measured following 5.5 h of reperfusion.

Lipid peroxidation in models of I/R occurs very quickly and byproducts of lipid peroxidation can form adducts with proteins and DNA and thus may play an important role in the underlying mechanism for oxidative stress-induced neuronal apoptosis. Our data suggest that UPEI-100 pretreatment resulted in a decrease in I/R-induced lipid and protein peroxidation (HNE-HIS adducts) at 2 h of reperfusion, suggesting that UPEI-100 was effective in preventing oxidative stress at this time point. Interestingly, the 2 h time point represents the time when the infarct volume of the UPEI-100 and vehicle treated groups diverge (Fig. 1.5). Since the level of apoptotic death, measured as the amount of DNA fragmentation, remained the same at all time points during reperfusion, this would suggest that the growth in infarct volume in the vehicle-treated group may have been primarily due to necrotic cell death. We speculate that the increase in neuronal survival measured when rats were pretreated with UPEI-100 was due to the ability of UPEI-100 to attenuate reperfusion-induced oxidative stress and subsequent necrotic cell death.

We therefore conclude that during the initial period following stroke and reperfusion, UPEI-100-induced neuroprotection was primarily due to increased SOD2 activity, thereby causing an increase in the neurons ability to immediately deal with reperfusion-induced mitochondrial superoxide production. This, along with reduced peroxidation of lipids and proteins as a result of oxidative stress, combined to produce our observed neuroprotective capacity of UPEI-100. We should mention that although we conducted time-dependent measures of molecular mediators of cellular stress pathways, these changes do not necessarily suggest a cause-and-effect relationship between UPEI-100 and these proteins. This is particularly true since the temporal changes in these proteins do not correspond to all time a point in which neuroprotection was observed. Therefore, studies are currently underway utilizing siRNA (small interfering RNA) technology to establish this relationship.

1.8 BIOCOMPATIBLE ENGINEERED NANOSTRUCTURES FOR EFFICIENT DRUG CARRIER AND DELIVERY: A PERSPECTIVE

The use of engineered tunable nanostructured in stroke therapy is now gaining considerable interest. Nanoparticles (NPs) are an attractive option for the drug delivery into the brain. The advantages of using NPs also include: (1) protect therapeutic agents from degradation in physiological condition; (2) minimizing exposure to non-target tissue when tissue-selective NPs being used; (3) providing a protective shell for the encapsulated contents, which prevents drugs from non-specific binding and enzymatic digestion; (4) being able to afford a high intraparticles drug concentration; (5) increase the translocation efficiency across the blood-brain-barrier (BBB). Polymeric NPs are nanosized carriers (1–1000 nm), made of natural or synthetic polymers including f-CNT specifically meant for dual route of administration for both chronic disease and acute condition, in which the drug can be loaded in the solid state or in solution, or adsorbed or chemically linked to the surface. Most popular materials are poly lactides (PLA), poly glycolides (PGA), poly(lactide-co-glycolides) (PLGA), polyanhydrides, polycyanoacrylates, and polycaprolactone (PCL). NPs for brain drug delivery will be prepared by membrane emulsification to tune particles size and particles size distribution while the specificity against the central nervous system will be obtained by NPs surface modification. Membrane emulsification technique is a unique technique which is especially useful in preparing uniform-sized particles (with average size from tens of nano-meters to one hundred of micro-meters). Lidietta Giorno *et al* have done extensive work on the preparation and application of uniform-sized particles which can be applied in bio-engineering application. Figure 1.7 represents schematic representation of emulsification technique to prepare uniform-sized nano particles.

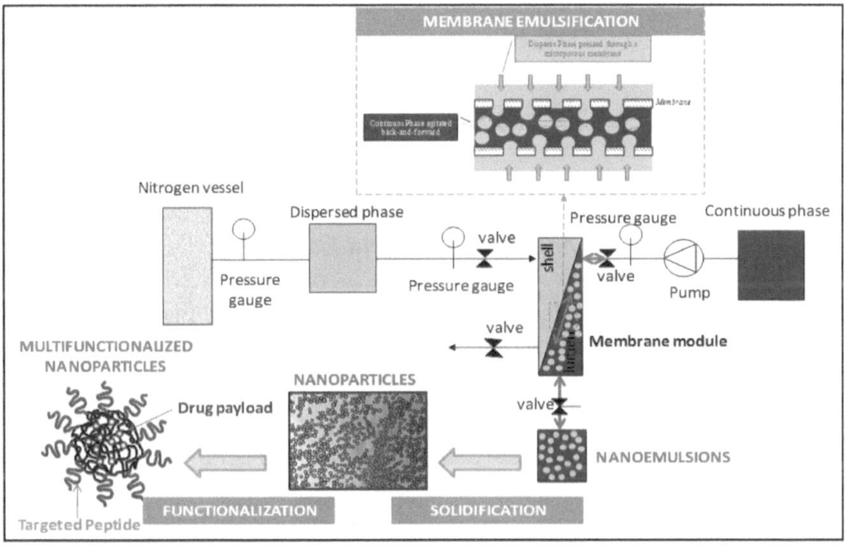

FIGURE 1.7 Schematic representation of membrane emulsification technique with various components associated with the process.

The advantages of the membrane emulsification technique mainly includes: (1) The size of prepared droplets and formed particles is easy to control and the size distribution is narrow; (2) break-up and coalescence between droplets rarely occur due to the mono-disparity of droplets; (3) The preparation process is mild because no high shear force needs to be used. Biocompatible polymers will be selected considering the solubility of the synthesized active compounds (such as ALA-SPN conjugates). Effective conjugation strategies will be studied to combine, in a highly controlled way, specific biomolecules (such as apolipoprotein A) to the surface of NPs. The presence of specific receptors (such as low-density lipoprotein receptor) at the BBB able to recognize the molecule signal (lipoprotein) will be exploited to obtain more efficient systems of brain drug delivery. Adsorptive-mediated transcytosis will also be evaluated as an alternative strategy to enhance the brain drug delivery. NPs surface will be functionalized with positively charged biomolecules allowing electrostatic interaction with the luminal surface of BBB due to the presence of negative charges on endothelial cells. Supramolecular assemblies of functionalized lipid squalenoyl NP or PLA in targeted drug delivery is gaining wide attention due to ease of loading and efficient delivery. Their versatile physicochemical features enable them as a carrier of pharmaceutically relevant entities and allow them for novel nanoscale candidates for DDS. Once the cargos are carried into multiple organs they are able to perform their biological function. Owing to their high carrying capacity, biocompatibility, it is highly advantageous to use biocompatible polymers in neurological disorders and to evaluate its pharmacokinetic and pharmacodynamic parameters for efficient drug delivery.

1.8.1 LIPID SQUALENOYL NANOPARTICLE

The application of nanotechnology in biomedical research is rapidly growing and this has led to the development of novel prospective engineered nanomaterials for drug formulation, targeted drug delivery, and gene therapy. Novel materials widely used for this purpose are nanoparticles of metals, quantum dots, fullerenes, carbon nanotubes, and dendrimers. There is a great deal of work, which needs to be explored on the development of new pathways in the therapeutic intervention for severe neurological trauma, such as stroke and spinal cord injuries. Several available therapies with efficient neuropharmacological activity are ineffective during systemic circulation because of fast metabolization and low bioavailability with rapid excretion from circulating fluids. Recent reports on the formation of nanoassemblies via conjugation of neuroprotective agent, adenosine, with lipid squalene have been shown to slowdown rapid metabolization and extend circulation time and rendered neuroprotection in experimental animal models in stroke and spinal cord injuries. Conjugation of small molecules, like UPEI-100, possessing chemical linkers is expected to form nano assemblies with lipid squalene. The purpose of nano-conjugation is to prevent an easily hydrolysable ester bond in UPEI-100, so that the molecule will not be cleaved apart upon entering bloodstream. In order to retain the pharmacological activity and keeping UPEI-100 intact, it is imperative to provide stability to the molecule. The conjugation of UPEI-100 with lipid squalene, as in the case of adenosine squalenoyl nanoparticle, is expected to impart stability and enhance the neuropharmacological activity in stroke model. Further modifications on UPEI-100 or similar type of derivatives on nano-conjugation with lipid squalene are expected impart stability and sustained pharmacological activity. This nano-conjugation approach will open up new avenues for efficient target drug delivery in neuroprotection study.

1.8.2 POLY-(LACTIDE-CO-GLYCOLIDE)-POLYETHYLENEGLYCOL NANOPARTICLES

Encapsulation of pharmacologically active ingredient in suitable nanoparticles for efficient drug transport to the site of the action without undergoing degradation in circulation can be of great valuable process in drug development for neurodegenerative disease. Nano-encapsulation using poly-(lactide-co-glycolide)-polyethyleneglycol (PLGA-b-PEG) has been shown to effectively moderate the process. Nano encapsulation of thyroid hormone (T3) appears to show in controlling ischemic damage in MCAO model of ischemic brain stroke. With this line of research, we plan to create nano-assemblies using PLGA-b-PEG and study the potential application in neuroprotection.

KEYWORDS

- **ischemic stroke**
- **hypoxia**
- **hypoglycemia**
- **oxidative stress**
- **apocyanin**
- **lipoic acid conjugate**
- **reactive oxygen species**

BIBLIOGRAPHY

1. Lipton, P., Ischemic Cell Death in Brain Neurons. *Physiol. Rev.* **1999,** *79* (4), 1431–568.
2. (a) Bedard, K.; Krause, K.H. The NOX Family of ROS-Generating NADPH Oxidases: Physiology and Pathophysiology. *Physiol. Rev.* **2007,** *87* (1), 245–313; (b) Chan, P.H. Reactive Oxygen Radicals in Signaling and Damage in the Ischemic Brain. *J. Cereb. Blood Flow Metab.* **2001,** *21* (1), 2–14.
3. Sun, A.Y.; Wang, Q.; Simonyi, A.; Sun, G.Y. Botanical Phenolics and Brain Health. *Neuromolecular Med.* **2008,** *10* (4), 259–74.
4. (a) Connell, B.J.; Saleh, M.C.; Khan, B.V.; Saleh, T.M., Apocynin May Limit Total Cell Death Following Cerebral Ischemia and Reperfusion by Enhancing Apoptosis. *Food Chem. Toxicol.* **2011,** *49* (12), 3063–3069; (b) Kelly, K.A.; Li, X.; Tan, Z.; VanGilder, R.L.; Rosen, C.L.; Huber, J.D., NOX2 Inhibition with Apocynin Worsens Stroke Outcome in Aged Rats. *Brain Res.* **2009,** *1292*, 165–72; (c) Tang, L.L.; Ye, K.; Yang, X.F.; Zheng, J.S. Apocynin Attenuates Cerebral Infarction After Transient Focal Ischaemia in Rats. *J. Int. Med. Res.* **2007,** *35* (4), 517–22; (d) Wang, Q.; Tompkins, K.D.; Simonyi, A.; Korthuis, R.J.; Sun, A.Y.; Sun, G.Y. Apocynin Protects Against Global Cerebral Ischemia-Reperfusion-Induced Oxidative Stress and Injury in the Gerbil Hippocampus. *Brain Res.* **2006,** *1090* (1), 182–189.
5. Tang, X.N.; Cairns, B.; Cairns, N.; Yenari, M.A. Apocynin Improves Outcome in Experimental Stroke with a Narrow Dose Range. *Neuroscience* **2008,** *154* (2), 556–562.
6. Castor, L.R.; Locatelli, K.A.; Ximenes, V.F. Pro-Oxidant Activity of Apocynin Radical. *Free Radic. Biol. Med.* **2010,** *48* (12), 1636–1643.
7. Connell, B.J.; Saleh, T.M., Co-administration of Apocynin with Lipoic Acid Enhances Neuroprotection in a Rat Model of Ischemia/Reperfusion. *Neurosci. Lett.* **2012,** *507* (1), 43–46.
8. Das, N.; Dhanawat, M.; Dash, B.; Nagarwal, R.C.; Shrivastava, S.K. Codrug: An Efficient Approach for Drug Optimization. *Eur. J. Pharm. Sci.* **2010,** *41* (5), 571–588.
9. Di Stefano, A.; Sozio, P.; Cerasa, L.S.; Iannitelli, A.; Cataldi, A.; Zara, S.; Giorgioni, G.; Nasuti, C. Ibuprofen and Lipoic Acid Diamide as Co-Drug with Neuroprotective Activity: Pharmacological Properties and Effects in Beta-Amyloid (1-40) Infused

Alzheimer's Disease Rat Model. *Int. J. Immunopathol. Pharmacol.* **2010,** *23* (2), 589–599.
10. Myers, M.G.; Norris, J.W.; Hachinski, V.C.; Weingert, M.E.; Sole, M.J. Cardiac Sequelae of Acute Stroke. *Stroke* **1982,** *13* (6), 838–842.
11. Connell, B.J.; Saleh, T.M., A Novel Rodent Model of Reperfusion Injury Following Occlusion of the Middle Cerebral Artery. *J Neurosci. Methods* **2010,** *190* (1), 28–33.
12. Zeldich, E.; Chen, C.D.; Colvin, T.A.; Bove-Fenderson, E.A.; Liang, J.; Tucker Zhou, T.B.; Harris, D.A.; Abraham, C.R. The Neuroprotective Effect of Klotho is Mediated via Regulation of Members of the Redox System. *J. Biol. Chem.* **2014,** *289* (35), 24700–24715.
13. (a) Chen, H.; Kim, G.S.; Okami, N.; Narasimhan, P.; Chan, P.H., NADPH Oxidase is Involved in Post-Ischemic Brain Inflammation. *Neurobiol. Dis.* **2011,** *42* (3), 341–348; (b) Zhang, Q.G.; Laird, M.D.; Han, D.; Nguyen, K.; Scott, E.; Dong, Y.; Dhandapani, K.M.; Brann, D.W. Critical Role of NADPH Oxidase in Neuronal Oxidative Damage and Microglia Activation Following Traumatic Brain Injury. *PloS One* **2012,** *7* (4), e34504; (c) McCann, S. K.; Roulston, C. L. NADPH Oxidase as a Therapeutic Target for Neuroprotection against Ischaemic Stroke: Future Perspectives. *Brain Sci.* **2013,** *3* (2), 561–598.
14. (a) Kishida, K.T.; Pao, M.; Holland, S.M.; Klann, E. NADPH Oxidase Is Required for NMDA Receptor-Dependent Activation of ERK in Hippocampal Area CA1. *J. Neurochem.* **2005,** *94* (2), 299–306; (b) Park, K.W.; Baik, H.H.; Jin, B.K. Interleukin-4-Induced Oxidative Stress via Microglial NADPH Oxidase Contributes to the Death of Hippocampal Neurons in vivo. *Curr. Aging Sci.* **2008,** *1* (3), 192–201; (c) Pestana, R.R.; Kinjo, E.R.; Hernandes, M.S.; Britto, L.R. Reactive Oxygen Species Generated by NADPH Oxidase Are Involved in Neurodegeneration in the Pilocarpine Model of Temporal Lobe Epilepsy. *Neuroscience letters* **2010,** *484* (3), 187–191.
15. Saleh, M.C.; Connell, B.J.; Rajagopal, D.; Khan, B.V.; Abd-El-Aziz, A.S.; Kucukkaya, I.; Saleh, T.M., Co-administration of Resveratrol and Lipoic Acid, Or Their Synthetic Combination, Enhances Neuroprotection in a Rat Model of Ischemia/Reperfusion. *PloS One* **2014,** *9* (1), e87865.
16. (a) Dokuyucu, R.; Karateke, A.; Gokce, H.; Kurt, R.K.; Ozcan, O.; Ozturk, S.; Tas, Z.A.; Karateke, F.; Duru, M., Antioxidant Effect of Erdosteine and Lipoic Acid in Ovarian Ischemia-Reperfusion Injury. *Eur. J. Obstet. Gynecol. Reprod. Biol.* **2014,** *183*, 23–7; (b) Thomas, S.; Vieira, C.S.; Hass, M.A.; Lopes, L.B. Stability, Cutaneous Delivery, and Antioxidant Potential of a Lipoic Acid and Alpha-Tocopherol Codrug Incorporated in Microemulsions. *J. Pharm. Sci.* **2014,** *103* (8), 2530–2538.
17. (a) Gourie-Devi, M. Epidemiology of Neurological Disorders in India: Review of Background, Prevalence and Incidence of Epilepsy, Stroke, Parkinson's Disease and Tremors. *Neurol. India* **2014,** *62* (6), 588–598; (b) Sethi, N. K. Meeting the Challenges of Stroke in India. *Neurology* **2014,** *82* (1), 96; (c) Sachan, D. Large Cohort Study is Launched to Find Causes of Stroke and Dementia in India. *BMJ* **2014,** *348*, g1325.
18. (a) Boden-Albala, B.; Edwards, D.F.; St Clair, S.; Wing, J.J.; Fernandez, S.; Gibbons, M.C.; Hsia, A.W.; Morgenstern, L.B.; Kidwell, C.S., Methodology for a Community-Based Stroke Preparedness Intervention: the Acute Stroke Program of Interventions Addressing Racial and Ethnic Disparities Study. *Stroke* **2014,** *45* (7), 2047–2052; (b) A Systems Approach to Immediate Evaluation and Management of Hyperacute Stroke. Experience at Eight Centers and Implications for Community Practice and Patient Care.

The National Institute of Neurological Disorders and Stroke (NINDS) rt-PA Stroke Study Group. *Stroke* **1997,** *28* (8), 1530–1540.

19. (a) Lin, G.; Ren, D.; Guo, S.; Geng, Y. Effectiveness of Cilostazol in Transient Ischemic Attack Refractory to Aspirin: A Report of Two Cases. *Exp. Ther. Med.* **2014,** *7* (3), 739–741; (b) Lavallee, P.; Amarenco, P., Stroke Subtypes and Interventional Studies for Transient Ischemic Attack. *Front. Neurol. Neurosci.* **2014,** *33*, 135–146.
20. (a) Culebras, A.; Messe, S.R.; Chaturvedi, S.; Kase, C.S.; Gronseth, G. Summary of Evidence- Based Guideline Update: Prevention of Stroke in Nonvalvular Atrial Fibrillation: Report of the Guideline Development Subcommittee of the American Academy of Neurology. *Neurology* **2014,** *82* (8), 716–724; (b) Reiffel, J.A. Novel Oral Anticoagulants. *Am. J. Med.* **2014,** *127* (4), e16–e17; (c) Kirkman, M.A.; Citerio, G.; Smith, M. The Intensive Care Management of Acute Ischemic Stroke: An Overview. *Intensive Care Med.* **2014,** *40* (5), 640–653.
21. (a) Arnan, M.K.; Burke, G.L.; Bushnell, C. Secondary Prevention of Stroke in the Elderly: Focus on Drug Therapy. *Drugs Aging* **2014,** *31* (10), 721–730; (b) Amarenco, P.; Callahan, A.; Campese, V.M.; Goldstein, L.B.; Hennerici, M.G.; Messig, M.; Sillesen, H.; Welch, K.M.; Wilson, D.J.; Zivin, J.A. Effect of High-Dose Atorvastatin on Renal Function in Subjects with Stroke or Transient Ischemic Attack in the SPARCL Trial. *Stroke* **2014,** *45* (10), 2974–2982.
22. (a) Park, S.; Karunakaran, U.; Jeoung, N.H.; Jeon, J.H.; Lee, I.K. Physiological Effect and Therapeutic Application of Alpha Lipoic Acid. *Curr. Med. Chem.* **2014,** *21*(32), 3636–3645; (b) Gomes, M.B.; Negrato, C.A. Alpha-Lipoic Acid as a Pleiotropic Compound with Potential Therapeutic Use n Diabetes and Other Chronic Diseases. *Diabetol. Metab. Syndr.* **2014,** *6* (1), 80; (c) Brufani, M.; Figliola, R. (R)-Alpha-Lipoic Acid Oral Liquid Formulation: Pharmacokinetic Parameters and Therapeutic Efficacy. *Acta Biomed.* **2014,** *85* (2), 108–115.
23. Ying, Z.; Kherada, N.; Farrar, B.; Kampfrath, T.; Chung, Y.; Simonetti, O.; Deiuliis, J.; Desikan, R.; Khan, B.; Villamena, F.; Sun, Q.; Parthasarathy, S.; Rajagopalan, S. Lipoic Acid Effects on Established Atherosclerosis. *Life Sci.* **2010,** *86* (3–4), 95–102.
24. (a) Fasano, E.; Serini, S.; Mondella, N.; Trombino, S.; Celleno, L.; Lanza, P.; Cittadini, A.; Calviello, G. Antioxidant and Anti-Inflammatory Effects of Selected Natural Compounds Contained in a Dietary Supplement on Two Human Immortalized Keratinocyte Lines. *BioMed Res. Int.* **2014,** *2014*, 327452; (b) Li, G.; Fu, J.; Zhao, Y.; Ji, K.; Luan, T.; Zang, B. Alpha-Lipoic Acid Exerts Anti-Inflammatory Effects on Lipopolysaccharide-Stimulated Rat Mesangial Cells via Inhibition of Nuclear Factor Kappa B (NF-kappaB) Signaling Pathway. *Inflammation* **2014**; (c) Piechota-Polanczyk, A.; Fichna, J., Review Article: the Role of Oxidative Stress in Pathogenesis and Treatment of Inflammatory Bowel Diseases. *Naunyn Schmiedebergs Arch. Pharmacol.* **2014,** *387* (7), 605–620; (d) Stankovic, M.N.; Mladenovic, D.; Ninkovic, M.; Ethuricic, I.; Sobajic, S.; Jorgacevic, B.; de Luka, S.; Vukicevic, R.J.; Radosavljevic, T.S. The Effects of Alpha-Lipoic Acid on Liver Oxidative Stress and Free fatty Acid Composition in Methionine-Choline Deficient Diet-Induced NAFLD. *J. Med. Food* **2014,** *17* (2), 254–261.
25. (a) Sung, M.J.; Kim, W.; Ahn, S.Y.; Cho, C.H.; Koh, G.Y.; Moon, S.O.; Kim, D.H.; Lee, S.; Kang, K.P.; Jang, K.Y.; Park, S.K. Protective Effect of Alpha-Lipoic Acid in Lipopolysaccharide-Induced Endothelial Fractalkine Expression. *Circ. Res.* **2005,** *97* (9), 880–890; (b) Li, X.; Liu, Z.; Luo, C.; Jia, H.; Sun, L.; Hou, B.; Shen, W.; Packer, L.; Cotman, C. W.; Liu, J. Lipoamide Protects Retinal Pigment Epithelial Cells from

Oxidative Stress and Mitochondrial Dysfunction. *Free Radic. Biol. Med.* **2008,** *44* (7), 1465–1474.

26. (a) Rochette, L.; Ghibu, S.; Richard, C.; Zeller, M.; Cottin, Y.; Vergely, C. Direct and Indirect Antioxidant Properties of Alpha-Lipoic Acid and Therapeutic Potential. *Mol. Nutr. Food Res.* **2013,** *57* (1), 114–125; (b) Yi, X.; Maeda, N. Alpha-Lipoic Acid Prevents the Increase in Atherosclerosis Induced by Diabetes in Apolipoprotein E-Deficient Mice Fed High-Fat/Low-Cholesterol Diet. *Diabetes* **2006,** *55* (8), 2238–2244.
27. Al-Rasheed, N.M.; Abdel Baky, N.A.; Faddah, L.M.; Fatani, A.J.; Hasan, I.H.; Mohamad, R.A. Prophylactic Role of Alpha-Lipoic Acid and Vitamin E against Zinc Oxide Nanoparticles Induced Metabolic and Immune Disorders in Rat's Liver. *Eur. Rev. Med. Pharmacol. Sci.* **2014,** *18* (12), 1813–1828.
28. (a) Ghibu, S.; Lauzier, B.; Delemasure, S.; Amoureux, S.; Sicard, P.; Vergely, C.; Muresan, A.; Mogosan, C.; Rochette, L. Antioxidant Properties of Alpha-Lipoic Acid: Effects on red Blood Membrane Permeability and Adaptation of Isolated Rat Heart to Reversible Ischemia. *Mol. Cell. Biochem.* **2009,** *320* (1–2), 141–148; (b) Serviddio, G.; Bellanti, F.; Vendemiale, G. Free Radical Biology for Medicine: Learning from Nonalcoholic Fatty Liver Disease. *Free Radic. Biol. Med.* **2013,** *65*, 952–68.
29. (a) Jia, L.; Zhang, Z.; Zhai, L.; Bai, Y. Protective Effect of Lipoic Acid against Acrolein-Induced Cytotoxicity in IMR-90 Human Fibroblasts. *J. Nutr. Sci. Vitaminol.* **2009,** *55* (2), 126–130; (b) Kleinkauf-Rocha, J.; Bobermin, L.D.; Machado Pde, M.; Goncalves, C.A.; Gottfried, C.; Quincozes-Santos, A. Lipoic Acid Increases Glutamate Uptake, Glutamine Synthetase Activity and Glutathione Content in C6 Astrocyte Cell Line. *Int. J Dev. Neurosci.* **2013,** *31* (3), 165–170; (c) Moraes, T.B.; Dalazen, G.R.; Jacques, C.E.; de Freitas, R.S.; Rosa, A. P.; Dutra-Filho, C. S. Glutathione Metabolism Enzymes in Brain and Liver of Hyperphenylalaninemic Rats and the Effect of Lipoic Acid Treatment. *Metab. Brain Dis.* **2014,** *29* (3), 609–615.
30. Jia, L.; Liu, Z.; Sun, L.; Miller, S.S.; Ames, B.N.; Cotman, C. W.; Liu, J. Acrolein, a Toxicant in Cigarette Smoke, Causes Oxidative Damage and Mitochondrial Dysfunction in RPE Cells: Protection by (R)-Alpha-Lipoic Acid. *Invest. Ophthalmol. Vis. Sci.* **2007,** *48* (1), 339–348.
31. (a) Byun, C.H.; Koh, J.M.; Kim, D.K.; Park, S.I.; Lee, K.U.; Kim, G.S., Alpha-Lipoic Acid inhibits TNF-Alpha-Induced Apoptosis in Human Bone Marrow Stromal Cells. *J. Bone Miner. Res.* **2005,** *20* (7), 1125–1135; (b) Safa, J.; Ardalan, M.R.; Rezazadehsaatlou, M.; Mesgari, M.; Mahdavi, R.; Jadid, M.P. Effects of Alpha Lipoic Acid Supplementation on Serum Levels of IL-8 and TNF-Alpha in Patient with ESRD Undergoing Hemodialysis. *Int. Urol. Nephrol.* **2014,** *46* (8), 1633–1638; (c) Lee, C.K.; Lee, E.Y.; Kim, Y.G.; Mun, S.H.; Moon, H.B.; Yoo, B. Alpha-Lipoic Acid Inhibits TNF-Alpha Induced NF-Kappa B Activation through Blocking of MEKK1-MKK4-IKK Signaling Cascades. *Int. Immunopharmacol.* **2008,** *8* (2), 362–370; (d) Kohler, H.B.; Knop, J.; Martin, M.; de Bruin, A.; Huchzermeyer, B.; Lehmann, H.; Kietzmann, M.; Meier, B.; Nolte, I. Involvement of Reactive Oxygen Species in TNF-Alpha Mediated Activation of the Transcription Factor NF-kappaB in Canine Dermal Fibroblasts. *Vet. Immunol. Immunopathol.* **1999,** *71* (2), 125–42.
32. (a) Haugaard, N.; Levin, R.M., Activation of Choline Acetyl Transferase by Dihydrolipoic Acid. *Mol. Cell. Biochem.* **2002,** *229* (1-2), 103–106; (b) Haugaard, N.; Levin, R.M., Regulation of the Activity of Choline Acetyl Transferase by Lipoic Acid. *Mol. Cell. Biochem.* **2000,** *213* (1–2), 61–63.

33. (a) Packer, L.; Kraemer, K.; Rimbach, G. Molecular Aspects of Lipoic Acid in the Prevention of Diabetes Complications. *Nutrition* **2001,** *17* (10), 888–895; (b) Jacob, S.; Henriksen, E. J.; Schiemann, A. L.; Simon, I.; Clancy, D. E.; Tritschler, H. J.; Jung, W.I.; Augustin, H.J.; Dietze, G.J. Enhancement of Glucose Disposal in Patients with Type 2 Diabetes by Alpha-Lipoic Acid. *Arzneimittelforschung* **1995,** *45* (8), 872–874; (c) Bitar, M.S.; Al-Saleh, E.; Al-Mulla, F., Oxidative Stress—Mediated Alterations in Glucose Dynamics in a Genetic Animal Model of Type II Diabetes. *Life Sci.* **2005,** *77* (20), 2552–2573.
34. (a) Huerta, A.E.; Navas-Carretero, S.; Prieto-Hontoria, P.L.; Martinez, J.A.; Moreno-Aliaga, M. J. Effects of Alpha-Lipoic Acid and Eicosapentaenoic Acid in Overweight and Obese Women during Weight Loss. *Obesity* **2014,** *23*, 313–321; (b) Kim, H.; Park, M.; Lee, S.K.; Jeong, J.; Namkoong, K.; Cho, H.S.; Park, J.Y.; Lee, B.I.; Kim, E. Phosphorylation of Hypothalamic AMPK on Serine(485/491) Related to Sustained Weight Loss by Alpha-Lipoic Acid in Mice Treated with Olanzapine. *Psychopharmacology* **2014,** *231* (20), 4059–4069; (c) Ratliff, J.C.; Palmese, L.B.; Reutenauer, E.L.; Tek, C. An Open-Label Pilot Trial of Alpha-Lipoic Acid for Weight Loss in Patients with Schizophrenia without Diabetes. *Clin. Schizophr. Relat. Psychoses* **2015,** *8* (4), 196–200.
35. (a) Vasdev, S.; Gill, V.D.; Parai, S.; Gadag, V. Effect of Moderately High Dietary Salt and Lipoic Acid on Blood Pressure in Wistar-Kyoto Rats. *Exp. Clin. Cardiol.* **2007,** *12* (2), 77–81; (b) Kocak, G.; Aktan, F.; Canbolat, O.; Ozogul, C.; Elbeg, S.; Yildizoglu-Ari, N.; Karasu, C. Alpha-Lipoic Acid Treatment Ameliorates Metabolic Parameters, Blood Pressure, Vascular Reactivity and Morphology of Vessels Already Damaged By Streptozotocin-Diabetes. *Diabetes Nutr. Metab.* **2000,** *13* (6), 308–318; (c) Vasdev, S.; Ford, C.A.; Parai, S.; Longerich, L.; Gadag, V. Dietary Alpha-Lipoic Acid Supplementation Lowers Blood Pressure in Spontaneously Hypertensive Rats. *J. Hypertens.* **2000,** *18* (5), 567–573.
36. (a) Ziegler, D.; Reljanovic, M.; Mehnert, H.; Gries, F.A. Alpha-Lipoic Acid in the Treatment of Diabetic Polyneuropathy in Germany: Current Evidence from Clinical Trials. *Exp. Clin. Endocrinol. Diabetes* **1999,** *107* (7), 421–430; (b) Biewenga, G.; Haenen, G.R.; Bast, A. The Role of Lipoic Acid in the Treatment of Diabetic Polyneuropathy. *Drug Metab. Rev.* **1997,** *29* (4), 1025–1054.
37. (a) Udupa, A.; Nahar, P.; Shah, S.; Kshirsagar, M.; Ghongane, B. A Comparative Study of Effects of Omega-3 Fatty Acids, Alpha Lipoic Acid and Vitamin E in Type 2 Diabetes Mellitus. *Ann. Med. Health Sci. Res.* **2013,** *3* (3), 442–446; (b) Sadi, G.; Yilmaz, O.; Guray, T. Effect of Vitamin C and Lipoic Acid on Streptozotocin-Induced Diabetes Gene Expression: mRNA and Protein Expressions of Cu-Zn SOD and Catalase. *Mol. Cell. Biochem.* **2008,** *309* (1–2), 109–116.
38. (a) Vodoevich, V.P. [Effect of Lipoic Acid, Biotin and Pyridoxine on Blood Content of Saturated and Unsaturated Fatty Acids in Ischemic Heart Disease and Hypertension]. *Vopr. Pitan.* **1983,** 14–16; (b) Grigor'ian, V.A.; Klantsa, N.A.; Prishliak, V.D.; Klantsa, P.N. [Effect of Prodectin and Lipoic Acid on the Lipid Metabolism Indicators in Chronic Ischemic Heart Disease]. *Vrach. Delo* **1981,** (10) 13–15.
39. (a) Ramos, L.F.; Kane, J.; McMonagle, E.; Le, P.; Wu, P.; Shintani, A.; Ikizler, T.A.; Himmelfarb, J. Effects of Combination Tocopherols and Alpha Lipoic Acid Therapy on Oxidative Stress and Inflammatory Biomarkers in Chronic Kidney Disease. *J. Ren. Nutr.* **2011,** *21* (3), 211–218; (b) Teichert, J.; Tuemmers, T.; Achenbach, H.; Preiss, C.; Hermann, R.; Ruus, P.; Preiss, R. Pharmacokinetics of Alpha-Lipoic Acid in Subjects

with Severe Kidney Damage and End-Stage Renal Disease. *J. Clin. Pharmacol.* **2005,** *45* (3), 313–328.

40. (a) Zaitone, S.A.; Abo-Elmatty, D.M.; Shaalan, A.A., Acetyl-L-Carnitine and Alpha-Lipoic Acid Affect Rotenone-Induced Damage in Nigral Dopaminergic Neurons of Rat Brain, Implication for Parkinson's Disease Therapy. *Pharmacol. Biochem. Behav.* **2012,** *100* (3), 347–360; (b) Suchy, J.; Chan, A.; Shea, T.B. Dietary Supplementation with a Combination of Alpha-Lipoic Acid, Acetyl-L-Carnitine, Glycerophosphocoline, Docosahexaenoic Acid, and Phosphatidylserine Reduces Oxidative Damage to Murine Brain and Improves Cognitive Performance. *Nutr. Res.* **2009,** *29* (1), 70–74; (c) Bagh, M.B.; Maiti, A.K.; Roy, A.; Chakrabarti, S. Dietary Supplementation with N-Acetylcysteine, Alpha-Tocopherol and Alpha-Lipoic Acid Prevents Age Related Decline in Na(+),K (+)-ATPase Activity and Associated Peroxidative Damage in Rat Brain Synaptosomes. *Biogerontology* **2008,** *9* (6), 421–428; (d) Muthuswamy, A.D.; Vedagiri, K.; Ganesan, M.; Chinnakannu, P. Oxidative Stress-Mediated Macromolecular Damage and Dwindle in Antioxidant Status in Aged Rat Brain Regions: Role of L-Carnitine and DL-Alpha-Lipoic Acid. *Clin. Chim. Acta* **2006,** *368* (1–2), 84–92; (e) Samuel, S.; Kathirvel, R.; Jayavelu, T.; Chinnakkannu, P., Protein Oxidative Damage in Arsenic Induced Rat Brain: Influence of DL-Alpha-Lipoic Acid. *Toxicol. Lett.* **2005,** *155* (1), 27–34.
41. Sancheti, H.; Kanamori, K.; Patil, I.; Diaz Brinton, R.; Ross, B.D.; Cadenas, E. Reversal of Metabolic Deficits by Lipoic Acid in a Triple Transgenic Mouse Model of Alzheimer's Disease: A 13C NMR Study. *J. Cereb. Blood Flow Metab.* **2014,** *34* (2), 288–296.
42. Padmalayam, I.; Hasham, S.; Saxena, U.; Pillarisetti, S. Lipoic Acid Synthase (LASY): A Novel Role in Inflammation, Mitochondrial Function, and Insulin Resistance. *Diabetes* **2009,** *58* (3), 600–608.
43. (a) Kates, S.A.; Lader, A.S.; Casale, R.; Beeuwkes, R., 3rd, Pre-Clinical and Clinical Safety Studies of CMX-2043: A Cytoprotective Lipoic Acid Analogue for Ischaemia-Reperfusion Injury. *Basic Clin. Pharmacol. Toxicol.* **2014**; (b) Carocci, A.; Rovito, N.; Sinicropi, M.S.; Genchi, G. Mercury Toxicity and Neurodegenerative Effects. *Rev. Environ. Contam. Toxicol.* **2014,** *229*, 1–18.
44. (a) Henriksen, E.J. Exercise Training and the Antioxidant Alpha-Lipoic Acid in the Treatment of Insulin Resistance and Type 2 Diabetes. *Free Radic. Biol. Med.* **2006,** *40* (1), 3–12; (b) Evans, J.L.; Maddux, B.A.; Goldfine, I.D. The Molecular Basis for Oxidative Stress-Induced Insulin Resistance. *Antioxid. Redox Signal.* **2005,** *7* (7–8), 1040–1052; (c) Saengsirisuwan, V.; Perez, F.R.; Sloniger, J.A.; Maier, T.; Henriksen, E.J. Interactions of Exercise Training and Alpha-Lipoic Acid on Insulin Signaling in Skeletal Muscle of Obese Zucker Rats. *Am. J. Physiol. Endocrinol. Metab.* **2004,** *287* (3), E529–536.
45. (a) Snider, G.W.; Dustin, C.M.; Ruggles, E.L.; Hondal, R J. A Mechanistic Investigation of the C-Terminal Redox Motif of Thioredoxin Reductase from Plasmodium Falciparum. *Biochemistry* **2014,** *53* (3), 601–609; (b) Lothrop, A.P.; Snider, G.W.; Ruggles, E.L.; Hondal, R.J. Why is Mammalian Thioredoxin Reductase 1 So Dependent upon the Use of Selenium? *Biochemistry* **2014,** *53* (3), 554–565.
46. (a) Connell, B.J.; Saleh, M.; Khan, B.V.; Saleh, T.M. Lipoic Acid Protects Against Reperfusion Injury in the Early Stages of Cerebral Ischemia. *Brain Res.* **2011,** *1375*, 128–136; (b) Panigrahi, M.; Sadguna, Y.; Shivakumar, B.R.; Kolluri, S.V.; Roy, S.; Packer, L.; Ravindranath, V. Alpha-Lipoic Acid Protects against Reperfusion Injury Following Cerebral Ischemia in Rats. *Brain Res.* **1996,** *717* (1–2), 184–188.

CHAPTER 2

PREPARATION, CHARACTERIZATION, BIOACTIVITY, AND LONG-TERM STABILITY OF ELECTROSPUN, CLAY-REINFORCED POLY (ε- CAPROLACTONE)/STARCH HYBRID NANOFIBERS

SARAVANAN NAGAPPAN[1,3], G. R. RAJARAJESWARI[1*], TIRUPATTUR SRINIVASAN NATARAJAN[2], and CHANG-SIK HA[3]

[1]*Department of Chemistry, College of Engineering, Guindy, Anna University Chennai, Chennai 600 025, India*

[2]*Department of Physics, Indian Institute of Technology Madras (IITM), Chennai 600 036, India*

[3]*Department of Polymer Science and Engineering, Pusan National University, Busan 609 735, Republic of Korea*

**Corresponding author. E-mail: rajiaravind@gmail.com*

CONTENTS

ABSTRACT

Biocompatible polymeric composite materials are significant in aiding advanced applications for tissue engineering scaffolds and body implants. In the present study, nanoclay-reinforced poly(ε-caprolactone)/starch (PCLCS) hybrid nanofiber was prepared and evaluated for its thermal stability, hydrophobicity, biocompatibility, and storage stability. Solutions of PCL in chloroform and starch in dimethyl sulfoxide (DMSO), both in the range of 10–20% (w/w) were mixed and the nanofibers were obtained by electrospinning at a high voltage power supply. Montmorillonite nanoclay was used as a reinforcement to the fiber mat. Scanning electron microscope (SEM) studies indicated that uniform fibers with narrow fiber diameter distribution were formed when 15% of PCL and starch were used together with 5% nanoclay. Addition of nanoclay enhanced the thermal stability and restored the hydrophobicity of the surface that was lost due to starch. The prepared nanofibers supported the formation of hyroxy apatite films on the surface, when immersed in simulated body fluid and hence, were proved to be biocompatible. The reasonable long-time stability of the prepared nanofibers stored at room temperature for over a period of 2 years and above was recorded from the FTIR and XRD spectra post storage. The electrospun PCLCS hybrid nanofibrous mats possessing uniform fibers with narrow diameter distribution, good thermal stability, superhydrophobicity, biocompatibility, and good long-term storage stability are suitable to be developed into tissue engineering scaffolds and implant materials.

2.1 INTRODUCTION

Research on biodegradable and biocompatible polymers and their composites has gained significant attention in the past couple of decades, due to their potential enduse in biomedical applications like, surgical sutures, drug delivery devices and tissue supports, food packaging, sensors, coating, textile applications, and the fabrication of superhydrophobic surfaces.[1–6] The main criteria required for polymers to be used in biomedical applications are the possession of three-dimensional and highly porous structures, good biocompatibility, and non-toxicity, combined with good mechanical characteristics, such as high modulus, tensile strength, and good flexibility. Among the commercially available biodegradable polymers, poly(ε-caprolactone) (PCL) has received considerable attention for biomedical applications due to its biodegradability, biocompatibility, and high flexibility, though it is slightly expensive than other biopolymers.[7] The main limitations of PCL are its low melting temperature (~65°C) and slow consumption of microorganisms.[8] These can be overcome by blending this biodegradable polymer with other polymers, such as poly(L-lactic acid), starch, cellulose, chitin/chitosan, gelatin, and sodium alginate to create new materials with desired properties.[9–17]

In the present work, starch was chosen to be blended with PCL on account of its wide natural abundance, renewability, and biodegradability. Starch has been widely investigated in bone and tissue engineering. Starch-based biodegradable bone cements could provide immediate structural support and degrade from the site of application. This property is particularly useful when starch is combined with bioactive particles, since it will allow new bone growth to be induced in both the interface of cement–bone and the volume left by polymer degradation.[18]

Mixing of hydrophobic PCL with starch is expected to lead to poor adhesion due to the poor compatibility between the two phases. Such an operation therefore, requires a compatibilizer to enhance the miscibility of the phases in the blend and to improve the mechanical properties of the composite. Addition of montmorillonite nanoclay (MMTNC) has been established to be effective in significantly improving the mechanical properties, thermal deformation temperature, and gas barrier characteristics of several biopolymer blends.[19–23] Reinforcement with clay is expected to increase the viscosity of the PCL/starch blend and induce better compatibility between PCL and starch in the blend.[24] The crystallization behavior of PCL/organoclay nanocomposites has illustrated that well-dispersed organoclay platelets act as nucleating agents and dramatically increase the crystallization rate of PCL.[19,25]

The present work aimed to prepare nanofibers of clay-reinforced PCL/starch blends through electrospinning. The biocompatibility of nanofibers was studied by immersion in stimulated body fluid (SBF).[26] The long-term storage stability of the fibers was assessed over 2 years and changes in the functional groups and structural properties of the stored nanofiber was evaluated by FTIR and XRD studies.

2.2 EXPERIMENTAL

2.2.1 MATERIALS

Poly(ε-Caprolactone) [PCL] (Mn=80,000) and montmorillonite nanoclay (MMTNC) were obtained from Sigma-Aldrich. Starch, sodium sulphate (Na_2SO_4), sodium bicarbonate ($NaHCO_3$), potassium hydrogen phosphate ($K_2HPO_4.3H_2O$), hydrochloric acid (HCl), and tris-hydroxy methyl amino methane were purchased from Sisco Research Laboratory. Chloroform ($CHCl_3$) and dimethyl sulfoxide (DMSO) were bought from Ranbaxy Fine Chemicals (RFCL) Ltd. Sodium chloride (NaCl) and magnesium chloride ($MgCl_2.6H_2O$) were purchased from Qualigens Fine Chemical Ltd. Potassium chloride (KCl) and calcium chloride ($CaCl_2$) were purchased from Merck Chemical Ltd. All the chemicals were used as received.

2.2.2 PREPARATION OF NANOCLAY-REINFORCED PCL/STARCH (PCLCS) COMPOSITE SOLUTION

The formation of nanofibers depends mainly on the concentration of the solution used for electrospinning. The surface may produce uneven nanofibers with micro beads on the surface at lower concentrations. This could be overcome by increasing the concentration of the solution.[27,28] In the present study, the concentrations of polymer solutions were optimized in order to obtain continuous nanofibers. Solutions of 10–20% (w/w) concentration of PCL in chloroform and starch in DMSO were separately prepared and then mixed. A constant weight percentage of MMTNC (5wt%) was added to PCL solution before it was mixed with the starch solution. All the solutions were stirred for 24 h with magnetic stirrer to obtain homogeneous solution. The obtained solution was called as PCLCS. The solutions (10–20% (w/w)) were also prepared in the absence of MMTNC and called as PCLS. The relative concentration of PCL and starch in the ratio 80:20 was found to be optimum to produce continuous nanofibers and hence, all the other studies were carried out with 80:20 ratio of PCL/starch. The schematic of the fiber formation by electrospinning is depicted in Figure 2.1.

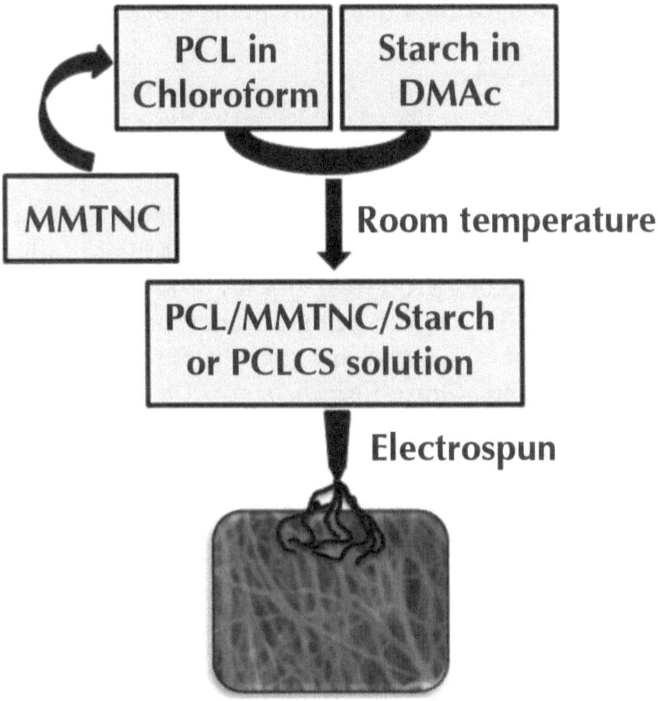

FIGURE 2.1 A schematic diagram of the fabrication of PCLCS nanofibers by electrospun processing.

2.2.3 *ELECTROSPINNING METHOD*

The electrospinning set-up that was used in the study was the same as reported by Sundaray et al. and Bhattarai et al.[29,30] It consisted of a disposable plastic syringe (2 mL) placed above a rotating drum wrapped with aluminum foil. The positive terminal of a high voltage 15 kV DC power supply was connected to the needle and the negative terminal was connected to the aluminum foil. The distance between the tip of the needle and aluminum foil collector was 18 cm. The flow rate of the solution in the syringe was computer controlled to be around 0.5 mL/h during the electrospinning process. The solution was taken into the syringe without air bubbles and electrospinning was carried out. Due to the electrostatic field, the ejected solution was drawn toward the drum. During the time of travel, the solvent quickly evaporated leaving solidified continuous fibers that accumulated on the collector target (Fig. 2.1). Orientation of the fibers can be controlled through rotation and translation of the collector drum. These nonwoven mats can have fiber diameters in the range of micro meters to nanometers.[31]

2.2.4 CHARACTERIZATION

The surface morphology of the electrospun nanofibers of PCLCS blends and SBF doped PCLCS fibers were observed using high-resolution scanning electron microscope (HRSEM—Hitachi) at 15 kV under the magnification of 1000–15,000. The fibrous mats were sputtered with gold using a current of 30 mA for 60 s with a fine gold coater prior to the analysis of the fibrous surface. The average fiber diameter of the electrospun nanofibers were measured up to 100 individual fibers from the HRSEM pictures by using UTHSCSA image tool software. The Fourier transform infrared spectroscopic (FTIR) measurement was performed using a Perkins-Elmer spectrum RX 100 in the range of 400–4000 cm^{-1}. The crystallinity and phase structure of the nanofibers were analyzed from the X-ray diffraction (XRD) patterns. The XRD patterns were carried out with (X'pert pro PANalytical Instrument) using Cu radiation (λ=1.5418 Å) at the scanning angle from 5° 2θ to 40° 2θ. The thermal degradation studies were conducted with a thermo gravimetric analyzer (TGA-Water SDT Q 600) at a heating rate of 10°C/min from 37 to 800°C under constant nitrogen flow of 20 cm^3 min^{-1}. The apatite formation at the nanofibrous mat was studied by HRSEM using the samples immersed in simulated body fluids. The surface properties of the nanofibrous mat was studied by contact angle (CA) measurement (Krüss drop shape analysis system) using a water droplet of 3 μL at 600 rpm water flow rate.[32–34]

2.3 RESULTS AND DISCUSSION

2.3.1 HIGH-RESOLUTION SCANNING ELECTRON MICROSCOPY (HR-SEM)

The HRSEM images of PCLCS nanofibers with 10, 15, and 20% (w/w) are shown in Figure 2.2a–c. The fibers are oriented well in all the three different concentrations, revealing the formation of highly cross-linked continuous fibers with the diameter ranging from 50 to 450 nm. At 10% w/w concentration, over flow of polymer solution resulted in some broad overlapping and the resultant fiber mat consisted of nanofibers with higher fiber diameters compared to 15% PCLCS fibers and large beads. At a higher concentration of 20% w/w, the flow behavior of the polymer solution was minimized due to the higher viscosity of the polymer solution, but the fiber diameter increased compared to those e-spun with lower polymer concentrations. The fibers were highly cross-linked with less broadening for 15% PCLCS. The histogram peaks also affirmed this trend, wherein, 15% PCLCS fibers were shown to have the best of fiber diameter distribution, with the maximum percentage of the fibers being in the range 50–100 nm. Hence, 15% w/w PCLCS was taken as the optimized concentration in the present study.

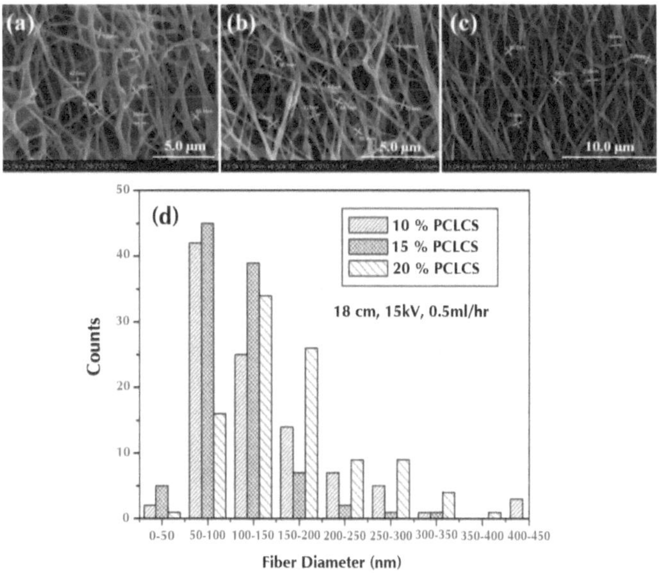

FIGURE 2.2 SEM images of (a) 10% PCLCS, (b) 15% PCLCS, and (c) 20% PCLCS nanofibers. (d) Histogram peaks for the SEM images of 10–20% PCLCS nanofibers.

2.3.2 FOURIER TRANSFORM INFRARED (FTIR) SPECTROSCOPY

The FTIR spectra of PCL (10% (w/w)), PCLS (10–20% (w/w)), MMTNC (5 wt%), and PCLCS (10–20% (w/w)) nanofibers are shown in Figure 2.3. PCL spectrum showed the main peaks at 2942, 2863, 1723, 1416, 1290, and 1239 cm^{-1} due to the presence of asymmetric and symmetric stretching and bending vibrations of CH_2, C=O, C—O, C—C, and C—O—C bonds (Fig. 2.3a).[33,35] The spectrum of PCL/starch (10%(w/w)) showed an intense peak in the region of 3200–3700 cm^{-1} due to the O—H stretching vibration of starch (Fig. 2.3b).[36,37] The expanded spectrum for PCLS showed the C=O stretching vibration as a strong broad band at 1725–1736 cm^{-1} similar to that reported by Fang et al. and Wang et al.[38,39] The FTIR spectrum of PCL /starch obtained was almost similar in characteristics to the native PCL and starch compounds.[40] The FTIR spectrum of nanoclay is shown in Figure 2.3e. The band at 3632 cm^{-1} corresponded to O—H stretching of the Si—OH group and the peaks at 1018, 525, and 465 cm^{-1}, respectively were assigned for Si—O, Si—O—Al, and Si—O—Si bending. The FTIR spectrum of 10% (w/w) PCLCS is shown in Figure 2.3f. All the regions of the band correlated to those reported by Wang et al. and Kizil et al.[39,40] FTIR spectra of 15 and 20% PCLS and PCLCS (Fig. 2.3c, d, g, and h) were also similar, with minor variations in the peak positions.

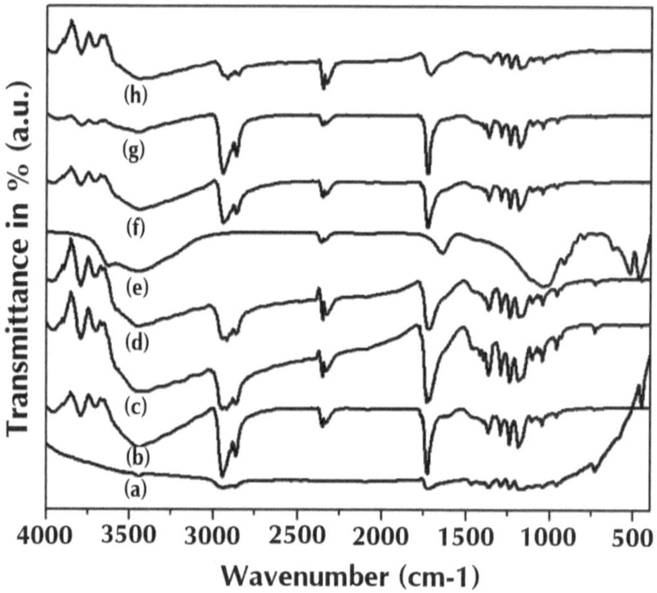

FIGURE 2.3 FTIR spectra of (a) 10% PCL, (b–d) 10–20% PCLS, (e) MMTNC, and (f–h) 10–20% PCLCS nanofibers.

2.3.3 X-RAY DIFFRACTION (XRD) STUDIES

XRD patterns of PCL (10% (w/w)), PCL/starch (10% (w/w)) with and without MMTNC reinforcement, and plain nanoclay are shown in Figure 2.4a–d. The pristine PCL nanofibers were semicrystallinne as indicated by two sharp peaks at 2θ=21.52° (110) and 23.84° (200) (Fig. 2.4b).[8,41] When starch was incorporated into PCL, the crystallinity of PCL/starch was slightly increased due to the crystalline nature of the starch (Fig. 2.4b). The sharp intense peaks of PCL nanofibers at 21.52° 2θ (110) and 23.84° 2θ (200) were also slightly shifted toward higher degree by mixing starch with PCL. The XRD pattern of MMNTC (Fig. 2.4c) showed the diffraction peaks at 2θ of 6.29° and 28.70°. Mixing of MMTNC into PCL and starch seemed to produce small changes in the crystallinity of the nanofibrous mats, as can be seen from Figure 2.4d–f.

The nanofiber formation was uniform and controlled only when the concentrations of PCL and starch were 15% w/w in the PCLCS fibers (Fig. 2.4e). For both 10 and 20% PCLCS, XRD patterns showed peaks of decreased intensity (Fig. 2.4d and f). The results suggested that starch and MMTNC interacted with the PCL strongly when the concentrations of the polymers were above 10% w/w. The fact that the crystalline peaks of plain nanoclay and starch were significantly suppressed in the PCLCS nanofibers lead us to propose that the prepared nanofibrous mat possessed a semicrystalline surface formed from the effective blending of all the three components of the fibers.

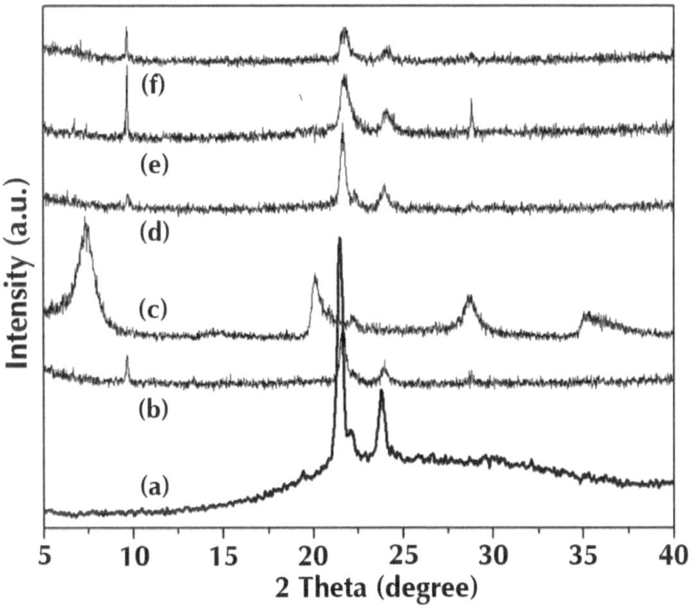

FIGURE 2.4 XRD patterns of (a) 10% PCL, (b) 10% PCLS, (c) MMT nanoclay, and (d–f) 10, 15, and 20% of PCLCS nanofibers.

2.3.4 THERMAL STABILITY (TGA)

PCL nanofibers have been shown to possess excellent thermal stability till around ~410°C (Fig. 2.5a).[42] The thermal degradation of 10 and 15%(w/w) of PCLS and PCLCS nanofibers obtained are shown in (Fig. 2.5b–e). The addition of 10% (w/w) starch/DMAc solution to the 10% (w/w) PCL/chloroform solution resulted in PCLS nanofibers with partial enhancement in the thermal stability (Fig. 2.5b–c). On the other hand, the initial weight loss for PCLS nanofibers was reduced slightly due to the poor compatibility of PCL and the starch in chloroform/DMAc solvent. A maximum weight loss of 93.67–96.85% from 390 to 435°C showed that all the organic groups present in PCLS were lost. The remaining compounds degraded at higher temperatures and the total weight loss of PCLS was found to be 99% at 800°C. These results showed that most of the organic groups vanished below 500°C and formed least amount of residue, only up to 0.6068%. Addition of MMTNC (5 wt% each to PCL/chloroform and starch/DMAc solutions) enhanced the thermal stability (450–480°C) of the PCL–starch nanofibers (Fig. 2.5d–e).[20] The increase in thermal stability of PCLS and PCLCS nanofibers may have been due to the increase of hydrophobic properties of the nanofibers as well as the introduction of small amount (about 5wt%) of MMTNC. Introduction of nanoclay improved the thermal stability of the polymer composite and delayed the polymer mass loss to a minimum (Fig. 2.6). The inset images of Figure 2.6 clearly supported the above observations. The observed results corroborated well with those reported by Zanetti et al.[43]

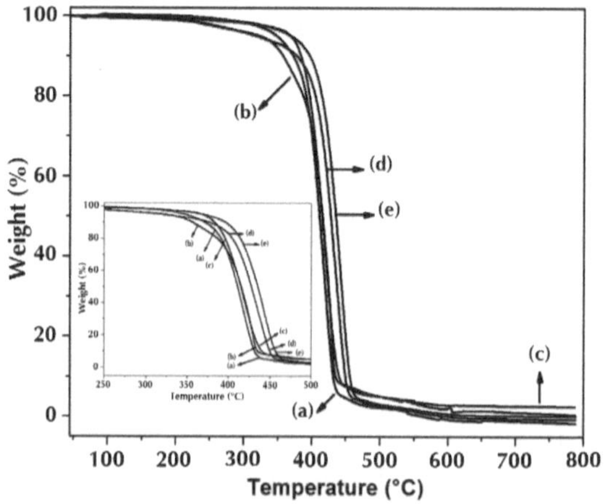

FIGURE 2.5 TGA of (a) 10% PCL (w/w), (b and c) 10 and 15% PCLS, and (d and e) 10 and 15% PCLCS nanofibers.

2.3.5 CONTACT ANGLE (CA) MEASUREMENTS

The surface wettability of the PCL and PCLCS nanofibers are shown in Figure 2.6a–d. PCL (10% (w/w)) nanofiber nonwoven mat showed almost superhydrophobic property with the CA 148.60° ± 3.0° (Fig. 2.6a) due to the inherent hydrophobicity of the polymer. The porosity and hydrophobic properties of the nonwoven mat enhanced the contact angle, with the air trapped in the pores aiding the phenomenon. Addition starch/DMAc solution to PCL (10% (w/w)) reduced the hydrophobicity (CA =130.40° ± 3.0°) due the presence of hydrophilic starch. However, introduction of nanoclay into the fibrous mat compensated for this reduction and the hydrophobicity of the mats increased to 138.20° ± 4.0° for 10% and 145.70° ± 5.6° for 15% (w/w) PCLCS (Fig. 2.6c and d). In 15% w/w PCLCS, the addition of clay resulted in the restoration of superhydrophobicity, closer to that of pristine PCL mats. These results indicated that nanoclay played a very significant role in not only enhancing the thermal and mechanical properties of the PCL–starch blended fibers, but also in imparting better hydrophobicity to the mats.

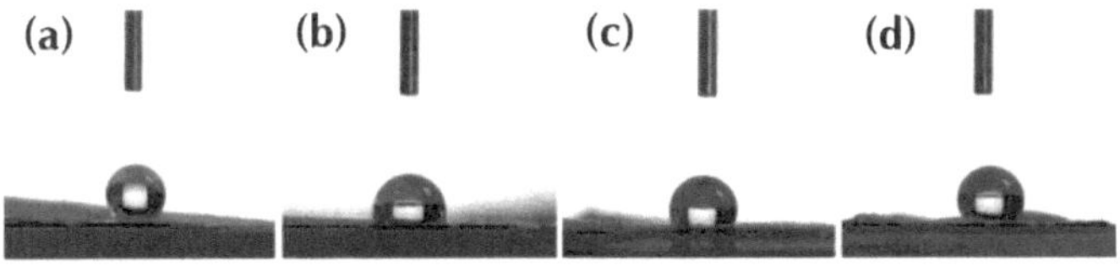

FIGURE 2.6 Contact angle images of (a) 10% PCL, (b) 10% PCLS, (c–d) 10 and 15 % w/w PCLCS nanofibrous mats.

2.3.6 BIOACTIVITY TEST

The biocompatibility of the electrospun fibers was evaluated in an *in vitro* study by immersing the prepared fiber mats in simulated body fluid (SBF) and analyzing the hydroxy apatite (HA) formation on them. A biomimetic method for forming a bone-like apatite layer developed by Kokubo et al.[26] which is shown to be applicable for any material, not only at the human body temperature (37°C) but also at room temperature was adopted for this purpose. The formation of apatite layer on materials immersed in SBF generally indicated its ability to form interfacial bonds with tissues when used in implants that are in contact with physiological fluid.[44] The deposition of apatite can facilitate osteoinductivity and osteoconductivity of polymeric scaffolds and hence, is a prerequisite for a biomaterial to bond to living bone. There are several reports about the potential of combining the bioactive materials and polymers to form bioactive composite materials for bone repair biomaterials.[45]

SBF is a salt mixture with an ionic concentration nearly equal to that of human body fluids. SBF solution was proposed initially by Kokubo et al. and applied in an *in vitro* study by Oyane et al.[46] The solution was prepared by dissolving NaCl, $NaHCO_3$, $MgCl_2$, Na_2SO_4, KCl, $K_2HPO_4.3H_2O$ in distilled water and buffered with tris-hydroxy methyl amino methane and hydrochloric acid (HCl) to bring the pH down to 7.4 at room temperature (Table 2.1). A comparison of ionic concentrations of the SBF and human blood plasma is illustrated in Table 2.2. The amounts of salts added for SBF preparation is shown in Table 2.1.

TABLE 2.1 Preparation of SBF solution

Order	Reagent	Amount
		SBF (1000 mL)
01	Ultrapure water	750 mL
02	NaCl	7.996 g
03	$NaHCO_3$	0.350 g
04	KCl	0.224 g
05	$K_2PO_4.3H_2O$	0.228 g
06	$MgCl_2.6H_2O$	0.305 g
07	1 kmol/m^3 HCL	40 cm^3
08	$CaCl_2$	0.278 g
09	Na_2SO_4	0.071 g
10	$(CH_2OH)_3CNH_2$	6.057 g
11	1 kmol/m^3 HCL	Appropriate amount for adjusting pH

The preweighed composite materials were kept immersed in SBF at 37°C for 1 week and the formation of HA on their surface was monitored. The apatite-coated fibers were removed from SBF, gently washed with deionized water, and dried at room temperature. The morphology and apatite formation on the PCLCS nanofibers were assessed by SEM (Fig. 2.7a–c). The histogram of PCLCS-SBF fibers is shown in Figure 2.7d. After soaking for 1 week, the PCLCS nanofibers were found to have some scattered and discrete deposits on the surface that could be attributed to apatite layer growth on the PCLCS nano fibrous surface. The formation apatite layer on

TABLE 2.2 Ion concentration of SBF and human blood plasma (from Kokubo et al.)

Ion	Concentration ($mmol/dm^3$)	
	(SBF)	Human blood plasma
Na^+	142.0	142.0
K^+	5.0	5.0
Mg^{2+}	1.5	1.5
Ca^{2+}	2.5	2.5
Cl^-	147.8	103.0
HCO_3^-	4.2	27.0
HPO_4^{2-}	1.0	1.0
SO_4^{2-}	0.5	0.5

the PCLCS nanofibers suggested that the espun PCLCS fibers could be useful as scaffolds in bone and tissue engineering applications. An increase in the fiber diameter in the range of 100–800 nm was observed and that could be attributed to the absorption of fluid on the surface of the nanofibers (Fig. 2.7d). The deposition of hydroxyapatite from SBF solutions on the PCL nanofibers prepared in 2,2,2trifluoroethanol (TFE) has also been reported by Yang et al.[47]

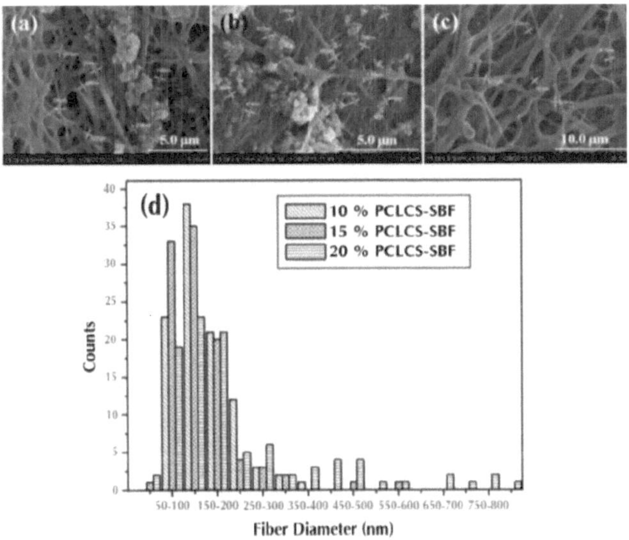

FIGURE 2.7 SEM images of (a) 10% PCLCS, (b) 15% PCLCS, and (c) 20% PCLCS nanofibrous mats soaked in SBF solution for 7 days. (d) Histogram peaks for SEM images of the nanofibrous mats soaked in SBF solution for 7 days.

2.3.7 LONG-TIME STABILITY

While the prepared nanoclay-reinforced PCL–starch fibers were proved to be biocompatible, their suitability for functional tissue engineering scaffolding could be predicted only from their long-term storage stability characteristics. It was quite evident from Figures 2.8 and 2.9 that the PCLCS fibers were very stable for over 2–3 years during storing at room temperature.[8] The FTIR spectra of PCLCS (10–20% (w/w)) nanofibers stored at room temperature for about 2 years showed almost similar peaks. The asymmetric and symmetric C—H stretching and bending vibration peaks, C=O and other peaks were similar to those of fresh PCLCS nanofibers (Figs. 2.3 and 2.8a–c). The PCLCS nanofibers stored for over 3 years also showed the appearance of above peaks (Fig. 2.8d), indicating that the chemical composition was almost unaltered and no major degradation occurred during storage. The XRD pattern of the 10% (w/w) PCLCS nanofibers stored at room temperature for 2 years showed the maintenance of semicrystalline property of PCLCS nanofibers (Figs. 2.4 and 2.9a). However, 15 and 20% w/w PCLCS nanofibers lost some of their crystallinity during storage for about 2 years

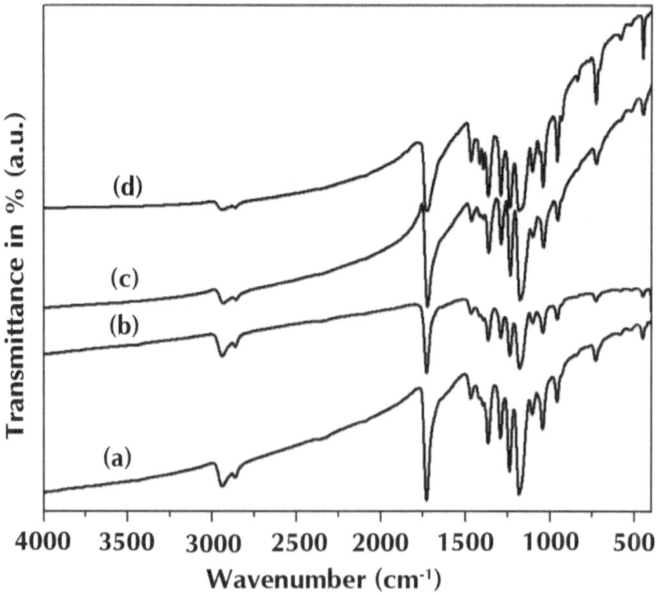

FIGURE 2.8 FTIR spectra of (a) 10% PCLCS, (b) 15% PCLCS, and (c) 20% PCLCS nanofibers stored at room temperature for 2 years. (d) 10% PCLCS nanofibers stored at room temperature for above 3 years.

(Fig. 2.9b and c). 10% PCLCS became almost amorphous on storage beyond 3 years, as indicated by the XRD pattern in Figure 2.9d. These results suggested some crystalline modification of the components of the composite resulting from the thermal variations encountered during prolonged storage. But, reasonable stability of the espun PCLCS fiber mats without any obvious chemical degradation as evidenced from FT-IR results proved their room temperature storage stability for about 2 years. Such a demonstrated storage stability would be significant in exploiting the electrospun mats of PCLCS nanofibers for long-time tissue engineering applications.

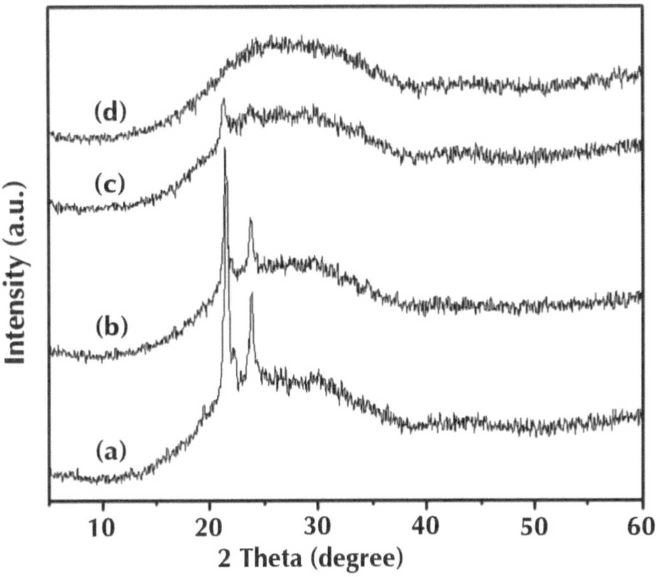

FIGURE 2.9 XRD patterns of (a) 10% PCLCS, (b) 15% PCLCS, and (c) 20% PCLCS nanofibers stored at room temperature for 2 years. (d) 10% PCLCS nanofibers stored at room temperature for above 3 years.

2.4 CONCLUSIONS

The introduction of nanoparticles into biodegradable polymer matrices to produce nanocomposites is one of the most effective approaches to enhance the properties of pristine polymers. In the present study, a simple method of electrospinning of nanofibers was employed to fabricate porous scaffolds using newly developed nanoclay reinforced PCL/starch hybrid solutions. The functional groups and crystallinity of PCL, PCLS, nanoclay, and PCLCS were identified from the XRD and FTIR spectral studies. Nanostructured PCLCS fibers prepared by this method were confirmed from SEM images and their histograms. The introduction of starch and small amounts of clay particles to the PCL marginally increased the thermal degradation temperature of the PCL and PCL–starch blended fibers. Contact angle measurements revealed that introduction of nanoclay compensated the loss of hydrophobicity induced by starch in the hybrid nanofiber and also significantly improved the water repellence. The biocompatibility of the nanocomposite fibers was confirmed from their ability to support the growth of hydroxyl apatite layer on the surface when immersed in simulated body fluid for a period of 1 week. The PCLCS hybrid nanofibers also showed room temperature storage stability for over 2 years. Though some crystallinity loss had been indicated by the XRD patterns of stored samples, no significant chemical change was noticed during this period. The present study established the very desirable combination of biocompatibility together with a reasonable storage stability for the nanoclay reinforced PCL–starch fibers prepared by electrospinning method. The prepared nanohybrid fibrous mat can be further exploited and developed as a potential bone and tissue engineering scaffold material for orthopedic implants.

ACKNOWLEDGMENT

The authors gratefully acknowledge the UGC and DST, Government of India for the analytical instrumentation facilities provided at Department of Chemistry, Anna University, Chennai and Prof. Won-Ki Lee (Department of Polymer Engineering, Pukyong National University, Busan 608-739, Republic of Korea), for his support to measure the contact angles of nanofibers.

KEYWORDS

- **nanofiber**
- **poly(ε-caprolactone)**
- **nanoclay**
- **starch**
- **thermal stability**
- **electrospinning**

BIBLIOGRAPHY

1. Lowery, J.L.; Datta, N.; Rutledge, G.C. *Biomaterials.* **2010,** *31*, 491–504.
2. Swamy, B.Y.; Prasad, C.V.; Rao, K.C.; Subha, M.C.S. *Int. J. Polym. Mater. Polym. Biomater.* **2013,** *62*, 700–705.
3. Elakkiya, T.; Malarvizhi, G.; Rajiv, S.; Natarajan, T.S. *Polym. Int.* **2014,** *63*, 100–105.
4. Armentano, I.; Dottori, M.; Fortunati, E.; Mattioli, S.; Kenny, J.M. *Polym. Degrad. Stability.* **2010,** *95*, 2126–2146.
5. Lu, X.F.; Wang, C.; Wei, Y. *Small.* **2009,** *5*, 2349–2370.
6. Wu, H.; Zhang, R.; Sun, Y.; Lin, D.; Sun, Z.Q.; Pan, W.; Downs, P. *Soft Matter.* **2008,** *4*, 2429–2433.
7. Pucciariello, R.; Villani, V.; Gorrasi, G.; Vittoria, V. *J. Macromol. Sci. B Phys.* **2005,** *44*, 79–92.
8. Nair, L.S.; Laurencin, C.T. *Prog. Polym. Sci.* **2007,** *32*, 762–798.
9. Khatri, Z.; Nakashima, R.; Mayakrishnan, G.; Lee, K.H.; Park, Y.H.; Wei; Ick-Soo Kim. *J. Mater. Sci.* **2013,** *48*, 3659–3664.
10. Link, D.P.; Gardel, L.S.; Correlo, V.M.; Gomes, M.E.; Reis, R.L. *J. Biomed. Mater. Res. A.* **2013,** *101*, 3059–3065.
11. Haque, Md. M.U.; Errico, M.E.; Gentile, G.; Avella, M.; Pracella, M. *Macromol. Mater. Eng.* 2012 *297*, 985–993.
12. Li, W.J.; Danielson, K.G.; Alexander, P.G.; Tuan, R.S. *J. Biomed. Mater. Res. A.* **2003,** *67*, 1105–1114.
13. Chen, B.; Sun, K.; Ren, T. *Eur. Polym. J.* **2005,** *41*, 453–457.
14. Jia, Y.; Liang, K.; Shen, X.Y.; Bowlin, G.L. *Carbohydrate Polym.* **2014,** *101*, 68–74.
15. Zhang, Y.Z.; Ouyang, H.; Lim, C.T.; Ramakrishna, S.; Huang, Z. *J. Biomed. Mater. Res. Part B: Appl. Biomater.* **2005,** *72B*, 156–165.
16. Prasad, C.V.; Swamy, B.Y.; Reddy, C.L.N.; Prasad, K.V.; Sudhakara, P.; Subha, M.C.S.; Il S.J.; Rao, K.C. *J. Polym. Environ.* **2012,** *20*, 344–352.
17. Rhim, J.W.; Park, H.M.; Ha, C.S. *Prog. Polym. Sci.* **2013,** *38*, 1629–1652.
18. Leonor, I.B.; Kim, H.M.; Balas, F.; Kawashita, M.; Reis, R.L.; Kokubo, T. Nakamura, T. *J. Mater. Chem.* **2007,** *17*, 4057–4063.
19. Chen, B.; Evans, J.R.G. *Macromolecules.* **2006,** *39*, 747–754.
20. Chung, Y.L.; Ansari, S.; Estevez, L.; Hayrapetyan, S.; Giannelis, E.P.; Lai, H.M. *Carbohydrate Polym.* **2010,** *79*, 391–396.
21. Choi, E.J.; Kim, C.H.; Park, J.K. *J. Polym. Sci. Part B Polym. Phys.* **1999,** *37*, 2430–2438.
22. Lepoittevin, B.; Pantoustier, N.; Devalckenaere, M.; Alexandre, M.; Calberg, C.; Jerome, R.; Henrist, C.; Rulmont, A.; Dubois, P. *Polymer.* **2003,** 44, 2033–2040.
23. Pannirselvam, M.; Gupta, R.K.; Bhattacharya, S.N.; Shanks, R.A. *Adv. Mater. Res.* **2007,** *29–30*, 295–298.
24. Ikeo, Y.; Aoki, K.; Kishi, H.; Matsuda, S.; Murakami, A. *Polym. Adv. Technol.* **2006,** *17*, 940.
25. Yang, K.K.; Wang, X.L.; Wang, Y.Z. *J. Ind. Eng. Chem.* **2007,** *13*, 485–500.
26. Kokubo, T.; Ito, S.; Huang, Z.; Hayashi, T.; Sakka, S.; Kitsugi, T. *J. Biomed. Mater. Res.* **1990,** *24*, 331–343.
27. Liu, H.Q.; Hsieh, Y.L. *J. Polym. Sci. Part B Polym. Phys.* **2002,** *40*, 2119–2129.
28. Doshi, J.; Reneker, D.H. *J. Electrostatics.* **1995,** *35*, 151–160.

29. Sundaray, B.; Subramanian, V.; Natarajan, T.S.; Xiang, R.Z.; Chang, C.C.; Fann, W.S. *Appl. Phys. Lett.* **2004,** *84*, 1222–1224.
30. Bhattarai, S.R.; Bhattarai, N.; Yi, H.K.; Hwang, P.H.; Cha, D.; Kim, H.Y. *Biomaterials.* **2004,** *25*, 2595–2602.
31. Rajesh, K.P.; Natarajan, T.S.; *J. Nanosci. Nanotechnol.* **2009,** *9*, 5402–5405.
32. Nagappan, S.; Choi, M.C.; Sung, G.; Park, S.S.; Moorthy, M.S.; Chu, S.W.; Lee, W.K.; Ha, C.S. *Macromol. Res.* **2013,** *21*, 669–680.
33. Nagappan, S.; Park, J.J.; Park, S.S.; Lee, W.K.; Ha, C.S. *J. Mater. Chem. A.* **2013,** *1*, 6791–6769.
34. Nagappan, S.; Park, S.S.; Yu, E.J.; Cho, H.J.; Park, J.J.; Lee, W.K.; Ha, C.S. *J. Mater. Chem. A.* **2013,** *1*, 12144–12153.
35. Unger, M.; Vogel, C.; Siesler, H.W. *Appl. Spectrosc.* **2010,** *64*, 805–809.
36. Bikiaris, D.; Panayiotou, C. *J. Appl. Polym. Sci.* **1998,** 70, 1503–1521.
37. Chandra, R.; Rustgi, R. *Polym. Degrad. Stability.* **1997,** *56*, 185–202.
38. Fang, J.M.; Fowler, P.A.; Sayers, C.; Williams, P.A. *Carbohydrate Polym.* **2004,** *55*, 283–289.
39. Wang, J.; Cheung, M.K.; Mi, Y. *Polymer.* **2002,** 43, 1357–1364.
40. Kizil, R.; Irudayaraj, J.; Seetharaman, K. *J. Agric. Food. Chem.* **2002,** *50*, 3912–3918.
41. Lee, K.H.; Kim, H.Y.; Khil, M.S.; Ra, Y.M.; Lee, D.R. *Polymer.* **2003,** *44*, 1287–1294.
42. Machado, A.V.; Botelho, G.; Silva, M.M.; Neves, I.C.; Fonseca, A.M. *J. Polym. Res.* **2011,** 18, 1743–1749.
43. Zanetti, M.; Camino, G.; Reichert, P.; Mulhaupt, R. *Macromol. Rapi. Commun.* **2001,** *22*, 176–180.
44. Dee, K.C.; Bizios, R. *Biotechnol. Bioeng.* **2000,** *50*, **438–442.**
45. Stevens, M.M. *Mater. Today.* **2008,** *11*, **18–25.**
46. Oyane, A.; Kim, H.M.; Furuya, T.; Kukubo, T.; Miyazaki, T.; Nagamura, T. *J. Biomed. Mater. Res. A.* **2003,** *65*, 188–195.
47. Yang, F.; Wolke, J.G.C.; Jansen, J.A. *Chem. Eng. J.* **2008,** *137*, 154–161.

PART II

Nanomaterials for Energy Harvesting and Drinking Water Purification

CHAPTER 3

ROLE OF ECO-FRIENDLY ADSORBENTS IN DEFLUORIDATION OF WATER

NATRAYASAMY VISWANATHAN* and KALIMUTHU PANDI

Department of Chemistry, Anna University, University College of Engineering – Dindigul, Reddiarchatram, Dindigul 624622, India.
**E-mail: drnviswanathan@gmail.com*

CONTENTS

ABSTRACT

Drinking water is the major source of fluoride intake. The fluoride content in drinking water may be beneficial or detrimental depending on its concentration. The presence of excess fluoride (>1.5 mg/L) level in drinking water leads to fluorosis. The impact of fluoride in drinking water is a major concern in developing countries like India where around 17 states are affected by fluorosis. As there is no remedy for fluorosis, defluoridation of drinking water is one of the options to reduce the fluoride concentration in water. The creation of awareness among the people in rural areas is the best possible way to tackle this problem. Though various treatment techniques like precipitation, ion-exchange, adsorption, reverse osmosis, electrodialysis, Donnan dialysis, and nanofilteration have been suggested for fluoride removal, adsorption seems to be effective, eco-friendly, and economical one. Various adsorbents have been reported for fluoride removal, but each of them possesses own advantages and limitations. Most of the reports are academic oriented, but only few reports are available for technology development. Polymeric materials are widely used for water treatment, and they possess various advantages over the other adsorbent materials. This review focuses exclusively on the eco-friendly materials like low-cost inorganic materials, biopolymers, and biopolymeric composites for the defluoridation of water. The fluoride selectivity of the biopolymers/composites will be enhanced by chemical modifications like functionalization, metal ion incorporation, etc. A comparison of the adsorption capacity of various low-cost inorganic adsorbents, biopolymers, and biopolymeric composites were made.

3.1 INTRODUCTION

Water is one of the most vital resources of our life, and it is also the resource most adversely affected, both qualitatively and quantitatively. Water is a unique and universal solvent because it has the ability to dissolve almost all substances that come within its contact. The pollution of water is responsible for large number of mortalities and incapacitations in the world. The potability of water depends upon the quality and quantity of the material dispersed and dissolved in it. Some ions in water are essential in trace amount for human being, while higher concentration of the same can cause toxic effects. Fluoride is one such ion.

3.2 OCCURRENCE AND GEOCHEMISTRY OF FLUORIDE

Fluorine is the 13th most abundant element available in the earth crust. Fluorine in the free state is a pale yellow gas with a pungent and irritating odor. On cooling, it condenses to liquid boiling at −188°C and on further cooling it freezes to a solid melting at −220°C. This fluorine exists as a diatomic molecule with remarkable low dissociation energy (38 kcal/mole). Fluorine is the most electronegative and reactive element in the periodic table. As a result, it possesses strong affinity to combine with other elements to produce compounds known as fluoride. In environment, it is not found as fluorine, but as fluoride ions in solution.[1] It exists as inorganic fluorides or as organic fluoride compounds. Inorganic fluorides are much more abundant than organic fluorides. Since fluoride in drinking water does not change its color, smell, or taste, normally there is no way to detect it unless tested.

3.3 SOURCES OF FLUORIDE

Fluoride is released into the environment mainly through natural sources and by anthropogenic activities.

3.3.1 NATURAL SOURCES

Fluoride is widely dispersed in the geological environment and it can be leached out by rain water, thereby contaminating the ground and surface water. Usually, the surface water is not contaminated with high fluoride, whereas ground water may be contaminated with high fluoride because the usual source of fluoride is fluoride-rich rocks. The minerals and rocks rich in fluoride are fluorspar, cryolite, fluorapatite, topaz, sellaite, basalt, villiamite, and bastanaesite.[2]

3.3.2 ANTHROPOGENIC ACTIVITIES

Several fluoride compounds have been extensively used in various industries and consequently they also contribute to fluoride pollution. For example, the effluents from the industries include glass and ceramic production, manufacturing semiconductors and integrated circuits, aluminum smelters, etc., possess higher fluoride concentrations. The fluoride containing industrial effluents are finally discharged into the nearby water sources.

3.4 FLUORIDE-DUAL NATURE

Drinking water is the single major source of daily fluoride intake.[1] So, World Health Organization (WHO) has specified the tolerance limit of fluoride content in drinking water as 1.5 mg/L. Indian standards for fluoride in the drinking water is 1 mg/L. Fluoride in drinking water resembles a two-edged sword (i.e., dual nature) depending on its concentration and the total amount ingested. Fluoride ion, within the tolerance limit (<1.5 mg/L) is essential for the formation of caries-resistant dental enamel and for the normal mineralization in hard tissues. Because of this beneficial role of fluoride, WHO included fluoride in its list as one of the trace elements essential for human health.[3] But the presence of a large amount of fluoride (>1.5 mg/L) in water leads to fluorosis (cf. Fig. 3.1).

3.5 FLUOROSIS

The problem of excess fluoride in drinking water is mounting day by day. Fluoride, being a highly electronegative ion, has a strong tendency to attract the positively charged ions like calcium.[4,5] The mineralized tissues like bone and teeth are affected due to the deposition of calcium fluorapatite crystals. Tooth enamel is made of crystalline hydroxyapatite. During fluoride intake, the hydroxyl ions present in the hydroxyapatite gets substituted by fluoride ion because fluorapatite is more stable than hydroxyapatite. Fluorosis is a crippling disorder known to occur due to the entry of fluoride into the body. The disease manifestations occur over a period of time. Depending upon the quantum of ingestion of fluoride, three forms of fluorosis viz., dental, skeletal, and non-skeletal fluorosis can occur. Dental fluorosis causes yellowing of teeth, white spots, and pitting or mottling of enamel. It usually affects the permanent teeth and occasionally the primary teeth. Dental fluorosis occurs when the fluoride content of the water is between 1.5 and 3 mg/L. Skeletal fluorosis generally occurs when the fluoride level of water between 3 to 6 mg/L. Fluoride toxicity affects children more severely than it does adults, due to the greater accumulation of fluoride in the metabolically more active growing bones of the children. Skeletal fluorosis affects bones causing stiffness of the spine, limitations of movements, and difficulty in walking.[6] Fluoride when consumed in excess, can also affect non-calcified tissues besides bone and teeth.

3.5.1 INTERNATIONAL STATUS

Fluorosis is a global health problem endemic in many parts of the world. The high fluoride concentration (>1.5 mg/L) in ground water was found in Africa, China, India, Sri Lanka, United States, Mexico, Chile, and Argentina.[7,8] In 1980, it was estimated that ~260 million people in 30 countries were drinking water with >1 mg/L of fluoride.

3.5.2 INDIAN SCENARIO

In India, fluorosis was first detected in Nellore district of Andhra Pradesh in 1937.[9] Since then considerable work has been done in different parts of India to explore the fluoride rich water sources and their impact on human as well as animals.[10] At present, 17 states of India are affected by fluorosis out of which Rajasthan, Tamilnadu, Andhra Pradesh, and Gujarat are highly fluoride endemic. Karthikeyan et al.[11] reported the fluoride endemic villages in Dindigul and Dharmapuri districts of Tamilnadu. Current status indicates that ~1 million people of India are affected by fluorosis.

3.6 NEED FOR DEFLUORIDATION

Fluorosis is an incurable and irreversible disease. As there are no proper treatment procedures available for this disease, the only possible alternative is to prevent fluorosis. Fluorosis can be controlled and prevented through early detection and practice of appropriate interventions. Fluoride poisoning can be prevented by

1. Alternative water resources
2. Improving nutritional level
3. Defluoridation of water

3.6.1 ALTERNATIVE WATER RESOURCES

Other alternative water sources include surface water, low-fluoride ground water, rain water, and transporting water through pipelines from nearby areas, which is free from the risk of fluorosis. Surface water in most areas is highly contaminated with biological and chemical pollutants and it cannot be used for drinking purposes without proper treatment. Utilizing rain water also does not seem to be feasible as this source cannot be relied upon. Utilizing ground water from nearby nonfluoride endemic area also will not be cost-effective technology, as the concentration of fluoride in ground water seems to be changing according to the geographical location of the ground water source and also has to be monitored regularly. Thus, the option of utilizing this method has its own limitations.

3.6.2 IMPROVING NUTRITIONAL LEVEL

The reported clinical data indicate that the adequate calcium intake is directly associated with the reduced risk of dental fluorosis.[12] Intake of vitamin C also safeguards the risk. However, this may not be the best substitute as there is no clear evidence to utilize this method as the viable alternative, as the results indicate that it can only reduce the chances of risk.

3.6.3 DEFLUORIDATION OF WATER

The process of fluoride removal from water is termed as defluoridation. Removing of excess fluoride from drinking water before being supplied to the community is one of the best practically possible options to overcome the problem of fluorosis (cf. Fig. 3.1).

Good characteristics of an ideal defluoridation process

- Independent of input fluoride concentration, alkalinity, pH, and temperature.
- Easy to handle/operate by the rural people.
- Taste of water should not be affected.
- Treated water should not contain any undesirable substances.
- Cost-effective.

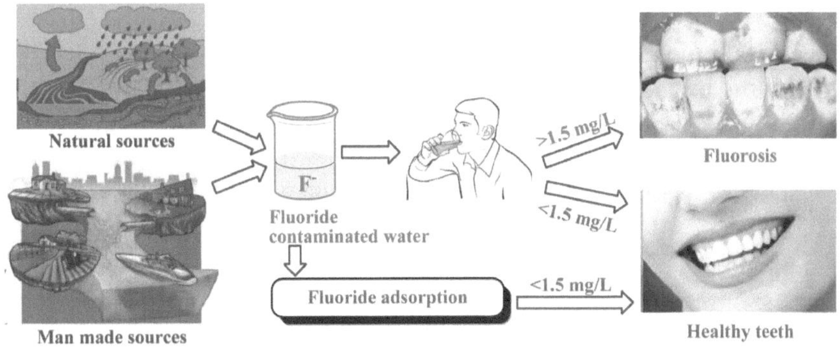

FIGURE 3.1 The dual nature of fluoride in drinking water and defluoridation of water.

3.7 DEFLUORIDATION TECHNIQUES

Several methods have been suggested from time to time for removing the excess fluoride in water since 1930. Researchers have developed a large number of creative techniques viz., adsorption, ion-exchange, precipitation, reverse osmosis, nanofilteration, electrodialysis, Donnan dialysis, etc., have been successfully used to remove fluoride content in water.[2,13–20] Among various methods reported for the defluoridation, the adsorption process seems to be a promising and attractive method for the removal of fluoride in terms of cost, simplicity of design, and operation. Adsorption is one of the potential processes for the removal of excess fluoride from water. The fluoride may adsorb to the adsorbent either by the physical or chemical forces. Various adsorbents have been successfully tested for the removal of fluoride from drinking water.

3.8 ADSORPTION TECHNIQUE

Adsorption process is widely used as remarkable method for defluoridation studies. Researchers have tried numerous efforts in introducing and developing a variety of adsorbents starting from natural to synthetic materials for fluoride removal. Each adsorbent material has its own advantages and limitations. A brief review of numerous eco-friendly adsorbents with their adsorption performance were discussed below.

3.8.1 ALUMINA-BASED MATERIALS

Activated alumina (AA) or calcined alumina, is aluminum oxide (Al_2O_3) prepared by low-temperature dehydration (300–600°C) of aluminum hydroxides. AA has been used for defluoridation of drinking water since 1934. The fluoride uptake capacity of AA depends on the specific grade of AA, particle size, pH, alkalinity, and fluoride concentrations. The fluoride exhausted AA has to be regenerated using suitable regenerant. AA has been the method of choice for defluoridation of drinking water in the developed countries. In recent years, this technology has received more attention even in the developing countries.

Meenakshi et al.[2] carried out the defluoridation studies of AA in batch and column modes. Various fluoride adsorption influencing parameters like contact time, dosage, pH of the medium, co-ions, initial fluoride concentration, and temperature were optimized for maximum defluoridation. The suitability of AA at field conditions was checked by collecting fluoride field water in nearby fluoride endemic areas of Dindigul district in Tamilnadu. Based on the results, domestic defluoridation units have developed using AA. Karthikeyan and co-workers[21] of Gandhigram Rural Institute (India) have installed community defluoridation units in nearby fluoride endemic areas of Dindigul district of Tamilnadu to ensure the supply of water with safe fluoride levels to the poor rural people.

Ghorai and Pant[22] have studied the defluoridation of drinking water with AA in a continuous-flow fluidized system. It is an efficient and economical method with an adsorption capacity of 1.45 mg/g at pH 7. Ku and Chiou[23] have also studied the effects of some operational factors on fluoride removal from aqueous solution by alumina. The removal efficiency of fluoride was influenced significantly by solution pH and the optimum operating pH for maximum fluoride removal (16.3 mg/g) was found to be in the range of 5–7. The authors have observed that the adsorption of fluoride was retarded in acidic solutions because of the electrostatic repulsion, and when the equilibrium solution pH was greater than 7, the fluoride adsorption by alumina reduced, which was attributed to the electrostatic repulsion of fluoride ions to the negatively charged surface of alumina; and the competition for active sites by the excessive amounts of hydroxide ions. For neutral and acidic solutions, the fluoride adsorption capacity of alumina was found to be interfered by the other anions, which are normally present in drinking water.

Wasay et al.[24] have modified the alumina surface by impregnation of La(III) and Y(III) to improve the adsorption efficiency of AA. The selectivity of the developed hybrid adsorbent was compared with alumina for adsorption of fluoride with other co-ions. The selectivity of La(III) and Y(III)-impregnated alumina was found to be in the order: fluoride > phosphate > arsenate > selenite. Puri and Balani[25] have reported the adsorption capacity of the lanthanum hydroxide impregnated on alumina for fluoride adsorption. The lanthanum hydroxide-impregnated alumina possesses more defluoridation capacity (DC) of 0.350 mM/g than alumina, which possesses the fluoride adsorption capacity in the range of 0.170–0.190 mM/g.

Tripathy and Raichur[26] have synthesized manganese dioxide-coated AA for defluoridation of water. The prepared manganese dioxide-coated AA could bring fluoride concentration down to 0.2 mg/L when the initial concentration of fluoride in water is 10 mg/L. The authors concluded that the fluoride removal was occurred by physical adsorption at the porous material surface. Magnesia-amended activated alumina (MAAA) was prepared by impregnating alumina with magnesium hydroxide and calcining the product at 450°C by Maliyekkal et al.[27] for fluoride removal. The synthesized MAAA possess remarkable fluoride adsorption capacity than AA.

The surface functionality of AA was improved by coating of calcium oxide (CaO) or manganese oxide (MnO_2) by sol–gel technique. The CaO-modified AA (CaO-AA) and manganese oxide-coated AA (MnO_2-AA) showed 5–10 times higher fluoride removal capacity of 101.01 and 10.18 mg/g, respectively than AA.[28] The copper oxide-coated mesoporous alumina (COCA) was prepared by Bansiwal et al.[29] The modified mesoporous alumina having enhanced fluoride removal capacity of 7.22 mg/g when compared to AA, which possesses DC of 2.232 mg/g at below pH 8. There is no leaching of copper ions after treatment with COCA was also confirmed.

Tomar et al.[30] reported the synthesis of a hybrid adsorbent namely hydroxyapatite-modified activated alumina (HMAA) by dispersing nanoparticles of hydroxyapatite inside AA granules. The composite adsorbent provided a synergy toward fluoride removal from contaminated drinking water. The hybrid adsorbent possesses a maximum adsorption capacity of 14.4 mg/g, which is at least five times higher than the virgin-activated alumina. A La^{3+}-modified activated alumina (La-AA) adsorbent was prepared for the effective removal of fluoride from water by Cheng et al.[31] Column filtration results shows that La-AA and AA treated 270 and 170 bed volumes of the fluoride spiked tap water, respectively, before fluoride breakthrough occurred. The results demonstrated that La-AA act as a promising adsorbent for effective removal of fluoride from water.

3.8.2 CARBON-BASED MATERIALS

Carbon-based adsorbents notche the interest of the researchers toward defluoridation process. A variety of carbon-based sorbents was used effectively for the fluoride sorption due to its high surface area. Activated carbon with high porosity, selectivity, and high surface area are used efficiently due to its 0% sludge production and easy design in methodology.[32]

Bhargava and Killedar[33] have studied the adsorption of fluoride on fishbone charcoal in a moving media adsorption system. The ratio of attained equilibrium sorbate concentration to the initial sorbate concentration and the fluoride removal capacity of the sorbent were found to vary inversely with the sorbent mass input rate, and varied directly with the sorbate flow rate and initial sorbate concentration. The ratio of attained equilibrium sorbate concentration to the initial sorbate concentration was found to be a function of the sorbent–sorbate mass input rate ratio. Leyya et al.[34] have reported aluminum-impregnated activated carbon was prepared by impregnation activated carbon with aluminum nitrate solution at a fixed pH, followed by calcination under nitrogen at temperatures above 300°C. The adsorption of fluoride on aluminum-impregnated carbon was shown to be dependent upon both the pH of the impregnating solution and the temperature of calcination. Aluminum-impregnated carbon was shown to have a fluoride adsorption capacity of 1.07 mg/g than the plain activated carbon (0.49 mg/g).

Tripathy et al.[35] synthesized alum-impregnated activated carbon for defluoridation of water. The prepared alum-impregnated activated carbon can remove fluoride effectively (up to 0.2 mg/l) from water containing 20 mg/L fluoride. Alagumuthu and Rajan[36] have prepared zirconium-impregnated cashew nut shell carbon (ZICNSC) to assess its capacity for the adsorption of fluoride from aqueous solution. The method is simple and shows great potential for fluoride removal. The various treatment conditions were optimized. The fluoride adsorption capacity of ZICNSC possesses 90.4%, which is higher than activated carbon, which is 27.8% only.

Maa et al.[37] have utilized granular activated carbon (GAC) and synthesized manganese oxide-coated GAC for defluoridation studies. With a notable increase in the surface area of manganese oxide-coated GAC [914.17 m^2/g] than non-coated GAC [850.60 m^2/g], the sorption capacity found to be threefold increase in lower pH range. Increase in the porosity of the manganese oxide-coated GAC also contributes for increase in the sorption capacity. Tchomgui-Kamga et al.[38] have synthesized charcoal modified with calcium compounds, characterized and subjected for defluoridation studies. Charcoal was permeated with calcium chloride by heating it at three different temperatures viz., 500, 650, and 900°C. Charcoal heated with calcium chloride at 650°C was found to possess high sorption capacity (19.5 mg/g) due to its less chloride and more hydroxide groups, which in turn contributes for fluoride sorption.

3.8.3 CALCIUM-BASED MATERIALS

Fluoride is a highly electronegative ion, which has the tendency to get attracted by the positively charged calcium ion. Hence, the researchers have made valuable efforts toward the development of calcium-based adsorbents for fluoride removal. Turner et al.[39] have studied the fluoride removal by crushed limestone (99% pure calcite) in batch studies. They reported that fluoride adsorption occurs immediately over the entire calcite surface with fluorite precipitating at step edges and kinks, where dissolved Ca^{2+} concentration is highest. Islam and Patel[40] have removed fluoride by using activated and ordinary quick lime. The removal of fluoride was 80.6% and the final concentration was 9.7 mg/L at optimum condition from the synthetic solution having initial fluoride concentration of 50 mg/L. The removal process followed Langmuir adsorption isotherm.

Sairam Sundaram et al.[41] have studied the advantages of nano-hydroxyapatite (n-HAp), a cost-effective sorbent for fluoride removal. n-HAp possesses a maximum DC of 1.30 mg/g. The various adsorption influencing parameters were optimized for maximum defluoridation. The equilibrium data were fitted with Freundlich, Langmuir, and Redlich–Peterson isotherms. The nature of fluoride removal is spontaneous and endothermic. The fluoride removal by n-HAp was governed by both ion-exchange and adsorption process. The suitability of n-HAp at field conditions was tested with field water sample.

Jiménez-Reyes and Solache-Ríos[42] reported that hydroxyapatite (HAP) is a potential material that could be used for the treatment of water contaminated with fluoride ions. The effect on the sorbent dosage ($mass_{HAP}/volume_{solution}$) in the sorption system was significant. The higher efficiency was found by using 0.01 g of HAP and 25 mL of solution (4.7 mg/g HAP); whereas with 0.1 g of HAP and 25 mL of solution, 96 ± 1% of fluoride was removed. The maximum sorption of fluoride ions was in the pH range between 5 to 7.3, when the insoluble $Ca_5(PO_4)_3OH$ exchanged OH^- for F^-. Freundlich model described the sorption isotherm process.

An examination of defluoridation by industrial grade limestone indicated that a combination of precipitation and adsorption of fluoride can be more effective for defluoridation of water, which was studied by Nath and Dutta.[43] Preacidified fluoride water using two edible acids, *viz.*, acetic acid and citric acid (CA) have been used for treatment by crushed limestone of diameter 3–4 mm to precipitate fluoride as CaF_2 in addition to adsorption of fluoride on limestone. Addition of the acids to the water before treatment with the crushed limestone in batch tests significantly improved the fluoride removal and this increased with the increase in the concentrations of the acids. The concentration of citric acid and acetic acid required bringing down the fluoride concentration from 10 to 1.5 mg/L, are 0.05 and 0.033 M, respectively, when the crushed limestone chips sample was used for the first time with contact time of 12 h. The acids are neutralized by limestone during the defluoridation process and the resulting final pH of the treated water was found to be in the ranges of 6.2 and 7.0 for citric acid and 5.7 and 7.0 for acetic acid. Jain and Jayaram[44] have reported the adsorption of fluoride on lime stone (LS) and aluminum hydroxide-impregnated lime stone (AlLS) in batch mode. The removal of fluoride was observed to be the most effective at pH 8. Freundlich equation fits better than Langmuir equation. The maximum sorption capacities for LS and AlLS adsorbents were found to be 43.10 and 84.03 mg/g, respectively.

3.8.4 MIXED OXIDE-BASED MATERIALS

Sujana et al.[45] evaluated the effectiveness of amorphous iron and aluminum mixed hydroxides in removing fluoride from aqueous solution. A series of mixed Fe/Al samples were prepared at room temperature by coprecipitating Fe and Al mixed salt solutions at pH 7.5. The compositions (Fe:Al molar ratio) of the oxides were varied as 1:0, 3:1, 2:1, 1:1, and 0:1. The fluoride adsorption capacities of the materials were highly influenced by solution pH. All samples exhibited very high Langmuir adsorption capacities; the sample with molar ratio 1 has shown maximum adsorption capacity of 91.7 mg/g.

The synthetic hydrated iron(III)–tin(IV) mixed oxide (HITMO) has been prepared by Biswas et al.[46] The mixed oxide is found hydrated and amorphous with irregular surface morphology. The fluoride adsorption capacity is nearly constant in the pH range 5.0–7.5. Langmuir isotherm describes the equilibrium data well and gives high Langmuir capacity (~10.50 mg/g) value. The mean adsorption energy (9.05 kJ mol^{-1}) computed from Dubinin–Redushkevich isotherm suggests the ion-exchange mechanism for fluoride adsorption. Biswas et al.[47] have also synthesized and characterized hydrated iron(III)–aluminum(III)–chromium(III) ternary mixed oxide (HIACMO) for fluoride adsorption. The fluoride removal efficiency at varied conditions showed that the reaction was pH sensitive, and optimum pH (initial) was between 4.0 to 7.0. Langmuir isotherm equation described the equilibrium well. The regeneration of fluoride-adsorbed material could be possible up to 90% with 0.5 M NaOH. A total of 0.2 g of HIACMO reduced fluoride level well below the maximum permissible value from 50 mL of fluoride spiked tap water (10 mgF^{-} dm^{-3}) sample.

Deng et al.[48] have reported a novel Mn–Ce oxide adsorbent with high sorption capacity for fluoride was prepared via coprecipitation method. The granular adsorbent was successfully prepared by calcining the mixture of the Mn–Ce powder and pseudoboehmite. Sorption isotherms showed that the sorption capacities of fluoride on the powdered and granular adsorbent were 79.5 and 45.5 mg/g, respectively at the equilibrium fluoride concentration of 1 mg/L. Both anion exchange and electrostatic interaction were involved in the sorption of fluoride onto Mn–Ce oxide adsorbent. Dou et al.[49] carried out the defluoridation performance and mechanism of hydrous zirconium oxide adsorbent. The maximum adsorption capacities of 124 and 68 mgF^{-}/g adsorbent were obtained in batch studies at pH 4 and 7, respectively. The adsorption capacity showed a continuously increasing trend with decreasing pH.

Chen et al.[50] have synthesized a novel bimetallic oxide adsorbent by the coprecipitation of Fe(II) and Ti(IV) sulfate solution using ammonia titration at room temperature. An optimized Fe–Ti adsorbent had Langmuir adsorption capacity of 47.0 mg/g, which was much higher than that of either a pure Fe oxide or Ti oxide adsorbent. There was a synergistic interaction between Fe and Ti, in which Fe—O—Ti bonds on the adsorbent surface and hydroxyl groups provide the active sites for adsorption, and Fe—O—Ti—F bonds were formed by fluoride adsorption. Wang et al.[51] have prepared CeO_2–ZrO_2 nanocages and their fluoride removal performance was investigated in batch studies. The adsorption isotherm could be better described by Langmuir model than Freundlich model. The maximum adsorption capacity of CeO_2–ZrO_2 was calculated to be 175 mg/g at pH 4.0.

A novel adsorbent of sulfate-doped Fe_3O_4/Al_2O_3 nanoparticles with magnetic separability was developed for fluoride removal from drinking water by Chai et al.[52] The fluoride adsorption isotherm was well described by Elovich model. The calculated adsorption capacity of this nanoadsorbent for fluoride by two-site Langmuir model was 70.4 mg/g at pH 7.0. Moreover, this nanoadsorbent performed well over a considerable wide pH range of 4–10, and the fluoride removal efficiencies reached upto 90 and 70% throughout the pH range of 4–10 with initial fluoride concentrations of 10 and 50 mg/L, respectively.

Ghosh et al.[53] have prepared hydrous Ce(IV)–Zr(IV) oxide (Ce/Zr ~ 1:1, mol/mol) (HCZMO) by simple chemical precipitation method. The investigation of fluoride adsorption over HCZMO from its aqueous phase was optimized at pH ~ 6. More than 95% fluoride was released from F^- HCZMO (24.8 mgF^-/g) by 1.0 M NaOH, confirming the ion-exchange adsorption mechanism inclining to chemisorption. In total, 1 g HCZMO per liter of a groundwater (F–: 4.40 mg/L) can reduce fluoride level below 1.5 mg/L in batch treatment. Yu and Chen[54] have reported a new manganese–lanthanum bimetal composite, which was developed by a coprecipitation method. The adsorption was highly pH dependent, and the optimal adsorption was obtained at pH 5.0. Langmuir equation fits better with the experimental data of the adsorption isotherm. The composite possess a maximum adsorption capacity of 292.9 mg/g.

Wang et al.[55] have synthesized Mg–Fe–La trimetal composite by a facile coprecipitation method for fluoride removal. The experimental results demonstrated that the adsorbent with a Mg/Fe/La molar ratio of 25:1:4, synthesized at room temperature (25°C) and calcined at 400°C, obtained the largest adsorption capacity of 112.17 mg/g for fluoride at near neutral pH. The adsorption process was fitted well with Langmuir isotherm. The mechanism for fluoride removal was governed by adsorption/complexation and ion-exchange. Yu et al.[56] have prepared a novel iron–magnesium–lanthanum trimetallic composite by a coprecipitation approach for efficient removal of fluoride in water. The adsorption process was highly pH dependent, and the optimal adsorption was obtained at pH 4.0. The adsorption isotherm could be well described by Langmuir equation and possess the maximum adsorption capacity of 270.3 mg/g.

3.8.5 CLAY-BASED MATERIALS

The low-cost naturally available inorganic clay materials have been tried for fluoride removal. To enhance the adsorption capacity, clay materials have been modified by increasing surface area, metal ion incorporation, etc. Meenakshi et al.[57] have processed kaolinite clay obtained from the mines and studied for fluoride uptake capacity. The surface area of the clay mineral was increased from 15.11 m^2/g (raw) to 32.43 m^2/g (activated) by mechanochemical activation. Batch adsorption studies were conducted to optimize various equilibrating conditions like contact time, dosage, pH, and other interfering anions for both raw and micronized kaolinites (RK and MK). The studies revealed that there is an enhanced fluoride sorption on MK. The sorption data obtained at optimized conditions were subjected to Freundlich and Langmuir isotherms. The values of sorption capacity *Qo (0.609, 0.714,* and 0.782 mg/g) at three different temperatures have been estimated using Langmuir isotherm. Thermodynamic studies revealed that the sorption of fluoride on MK is endothermic and spontaneous process.

Tor[58] utilized montmorillonite as an adsorbent for the removal of fluoride from aqueous solution. The equilibrium time for the removal of fluoride was determined to be 180 min. The maximum removal of fluoride was obtained at pH 6. The best-fitting adsorption isotherm was Freundlich isotherm. The fluoride saturation capacity of montmorillonite is 0.263 mg/g at room temperature. Desorption studies showed that the fluoride can be easily desorbed at pH 12 indicating that the adsorbent can be reused.

Adsorption potential of metal oxide (lanthanum, magnesium, and manganese) incorporated bentonite clay was investigated for defluoridation of drinking water using batch equilibrium by Kamble et al.[59] 10% La-bentonite shows higher fluoride uptake capacity for defluoridation of drinking water as compared to Mg-bentonite, Mn-bentonite, and bare bentonite clay. The uptake of fluoride in acidic pH was higher as compared to alkaline pH. Langmuir isotherm describes the equilibrium data well and gives high Langmuir capacity of 4.24 mg/g.

The low-cost bentonite clay was chemically modified using magnesium chloride in order to enhance its fluoride removal capacity by Takre et al.[60] The magnesium incorporated bentonite (MB) works effectively over wide range of pH and showed a maximum fluoride removal capacity of 2.26 mg/g. Desorption study of MB suggest that almost all the loaded fluoride was desorbed (~97%) using 1 M NaOH solution; however, the maximum fluoride removal decreases from 95.47 to 73% after regeneration. Sujana and Anand[61] reported the feasibility of utilizing bauxite for fluoride removal from synthetic and natural fluoride bearing groundwater samples of Orissa, India. The system followed Langmuir adsorption isotherm model with adsorption capacity of 5.16 mg/g. The competition of CO_3^{2-} ions for surface sites is comparatively less as compared to NO_3^-, SO_4^{2-}, and PO_4^{3-} ions in the concentration range of 5–20 mg/g at solution pH of 6 during fluoride adsorption.

Defluoridation using Fe(III)-modified montmorillonite was investigated using batch experiments by Bia et al.[62] The effect of reaction time, pH, ionic strength, and phosphate, as a competitive anion, was evaluated. The adsorption rate increases by increasing the fluoride concentration and by decreasing pH. The presence of phosphate reduces fluoride adsorption and reveals that both the ions are in competition for surface sites. Fe(III)-modified montmorillonite clay for fluoride removal and 95% of fluoride adsorption was achieved in 3 h.

Das et al.[63] have reported a Zn/Al hydrotalcite-like compound (HTlc) was prepared by coprecipitation (at constant pH) method. The ability of Zn/Al oxide to remove F^- from aqueous solution was investigated. It was found that the maximum adsorption takes place within 4 h at pH 6.0. The percentage of adsorption increases with increase in the adsorbent dose, but decreases with increase in the adsorbate concentration. The adsorption data fitted well into the linearly transformed Langmuir equation. With 0.01 M NaOH solution, the adsorbed F^- could be completely desorbed from Zn/Al oxide in 6 h. The study of fluoride removal from aqueous solution by calcined Mg–Al–CO_3-LDH has been carried out by Lv et al.[64] The layered doubled hydroxide (LDH) calcined at 500°C possess the highest capacity of fluoride removal, because of retention of its intrinsic structure. The adsorption loading is higher for the calcined Mg-Al-layered double hydroxide (Mg–Al-LDH) than for calcined Zn–Al and Ni–Al-LDH. It was found that maximum removal of fluoride from aqueous solutions was obtained in 6 h at pH 6.0 with an initial concentration of 50 mg/L, and that the retention of fluoride ions by calcined layered double hydroxide (CLDH) material was 98% or higher. The residual fluoride concentration was found to be 0.4 mg/L with an initial concentration of 20 mg/L, which meets the national standard for drinking water quality. Freundlich and Langmuir isotherms were used to fit the data of equilibrium experiments.

Wang et al.[65] have studied fluoride adsorption by synthetic Mg/Al–CO_3 hydrotalcite-like compounds (HT) and their calcined products (HTC). The results from their adsorption experiments indicated that fluoride uptake by HTC was much stronger than their precursors HT. The removal efficiency is inversely related to the initial fluoride concentration for HTC obtained from HT synthesized at 130°C (HTC130). The adsorption isotherm at pH 7 for HTC130 was linear and did not follow Langmuir equation. HTC130 was shown to be able to bring fluoride concentration from 5 mg/L down to less than 1 mg/L in aqueous solution, suggesting that this material may be a possible candidate for fluoride removal.

A simple ultrasound-assisted coprecipitation method in combination with a calcination treatment was developed to prepare magnetic Mg–Al-layered double hydroxide composite as an adsorbent material to remove fluoride ions from aqueous solution by Chang et al.[66] The composite prepared under the ultrasound irradiation have exhibited fairly high maximum adsorption capacity of fluoride (47.7 mg/g), which was 60% higher than that of the composite prepared without the ultrasound irradiation assistance with the same aging time. Ma et al.[67] have utilized the calcination product of Mg–Al–Fe hydrotalcite compound at 500°C (HTlc500) as the adsorbent to remove fluoride ions from aqueous solution. Mg–Al–Fe hydrotalcite compound was synthesized by coprecipitation method. It was found that HTlc500 have the largest adsorption capacity of 14 mg/g at pH 6 with adsorbent dose of 0.2 g/L.

A novel fast and efficient adsorbent based on lamellar compound, namely CeO_2/Mg–Fe layered double hydroxide composite has been designed for fluoride removal from water by Zhang et al.[68] In order to improve fluoride removal efficiency, nonthermal plasma (NTP) was used to modify the surface state of composites. The experimental results indicated that the adsorption capacity was enhanced with NTP surface modification. The maximum adsorption capacity was found to be 38.7–60.4 mg/g. Mandal et al.[69] have synthesized Mg–Cr–Cl layered double hydroxide for fluoride removal from aqueous solution. The fluoride removal was 88.5% and 77.4% at pH 7 with an adsorbent dose of 0.6 g/100 mL solution and initial fluoride concentration of 10 and 100 mg/L, respectively. The equilibrium was established at 40 min.

Facile microemulsion methods have been developed to synthesize mesoporous Co–Al hydroxide carbonates with rod-like and hexagonal sheet-like morphologies for defluoridation of water by Zhao et al.[70] A series of samples with different molar ratio of cobalt/aluminum were prepared to investigate the impact of cobalt content. The specific surface area of the calcined S-10 sample with the rod-like morphology is 379 m^2/g; it provides porosity to the material and increases the fluoride removal efficiency to 95.6%.

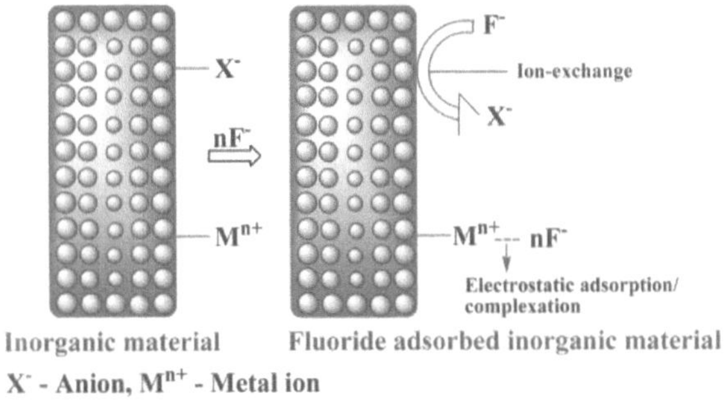

FIGURE 3.2 The feasible mechanism of fluoride removal using inorganic materials.

The anions present in the inorganic materials get exchanged for fluoride by ion-exchange. Due to high electronegativity and small ionic size, fluoride is classified as hard base. The multivalent metal ions (hard acids) prefer to bind with fluoride ion by electrostatic adsorption/complexation. In general, the inorganic materials will remove fluoride by ion-exchange and/or adsorption as shown in Figure 3.2.

3.8.6 BIOPOLYMERS

The typical adsorbent materials (e.g., activated carbon) are produced by means of processes, which consume great quantities of energy and/or chemical substances involving high environmental costs. Therefore, there is a need of cheaper and more sustainable materials. This need has promoted the increased scientific study of biosorbents during the last decades. Biosorption is an emerging and environmentally sound technique for the removal of toxic ions from water.[71–80] Biosorbents are adsorbent materials obtained from biomass by relatively simple process that use considerably lower quantities of chemical substances and energy, compared to the production of the typical adsorbents. Besides, most biosorbents are obtained from materials typically considered like byproducts or wastes. The biomass is naturally composed by very diverse substances, such as proteins, polysaccharides, pigments, carboxylic acids, etc., which have many functional groups, such as hydroxyl, carboxyl, carbonyl, thiol, sulfhydryl, sulfonate, phosphonate, amine, amide, etc. Under determined conditions, such groups can interact physically or chemically with the contaminants present in water, which causes their retention on the surface of the biosorbent. Only certain types of biomass (cellulose, chitin, chitosan, alginate, gelatin, etc.) have the capacity and selectivity to remove an appreciable quantity of a contaminant from an aqueous medium and, thus, can be considered for use in high-scale biosorption process.

Biosorbents have been studied as ions removers, being heavy metal (cations) or anions. The removal of cations by biosorbents from water has been considered as more viable than the removal of anions, since most chemical groups of biomass act as ligands, that is, as electron donors, which enable them to form coordination complexes with heavy metals. Only some chemical groups, specifically amine groups (primary, secondary, and tertiary) are capable to acquire positive charge at acid pH values, because the protonation phenomenon takes place. Protonation occurs due to the chemical nature of the amine groups, which act as Bronsted–Lowry bases regarding that these accept hydrogen ions, or as Lewis bases regarding they donate their nitrogen free electron pair to the hydrogen ions. Hence, nitrogen acquires positive charge and the amine group becomes an anion adsorption site.

Chitin, a biopolymer consisting of (1,4)-2-acetamido-2-deoxy-d-glucose units, is the second most abundant polymer next to cellulose. Chitosan is a derivative of chitin obtained from deacetylation of chitin. Chitosan is a linear polysaccharide of ß-1,4-D-glycosyl-linked glucosamine residue, which is a major component of the shells of crustacean shells and fungal biomass.[72] Gelatin is a biopolymer obtained by physicochemical or thermal denaturation of collagen. The industries like food, pharmaceuticals, cosmetic, and photography are used as a commercial gelling agent in the manufacturing process. The gelatin contains many reactive functional groups like amine, carbonyl group in the polypeptide chain, and the carboxyl group in the side chain. In gelatin, the molecular structure exists in the form of continuous repetition of Gly–proline–hydroxyproline and it is stabilized by both intra- and inter-molecular hydrogen bonding. Cellulose is the most abundant renewable biopolymer on earth, and is a very promising raw material available at low cost to synthesize an adsorbent for defluoridation. Alginate is a biopolymer isolated from brown seaweeds, which is composed of two monomers viz., (1→4) β-D-mannuronate and (1→4) α-L-guluronate. These biopolymers possess advantages like biocompatible, biofunctional, nontoxic, nonimmunogenic, and biodegradable. Biopolymers like cellulose, lignin, starch, alginate, gelatin, chitin, and chitosan are promising a new class of low-cost adsorbents for the removal of toxic ions from aqueous solution.

In adsorption process, chitosan is often used in the form of flakes or powder, which is less stable and causes a significant pressure drop, which would affect filtration during field applications and is difficult to regenerate. A major material limitation of chitosan is poor chemical resistance and mechanical strength. The adsorption capacity of the unmodified chitosan was found to be minimum. These disadvantages outweigh its advantages of biodegradability, biocompatibility, and indigenous. If chitosan has been modified into a stable form, which could overcome the abovementioned challenges, then definitely it would throw more light on the field of defluoridation. Focuses were made on the development of cross-linked chitosan beads (CB), which are stable, regenerated and reused in subsequent operations and also they help to avoid pressure drop during field applications. However, the fluoride selectivity of CB was observed only after chemical modifications.[72–74]

Viswanathan and coworkers[72] have prepared chitosan in its more usable bead form by dissolving chitosan in acetic acid solution and dropped to NaOH solution to get CB. To increase the stability, CB were cross-linked with glutaraldehyde. But the raw cross-linked chitosan beads (CB) possess low DC. In order to effectively utilize the reactive amino groups of CB, protonation was carried out using hydrochloric acid (HCl) and employed as a most promising defluoridating agent. Protonated chitosan beads showed a maximum DC of 1.66 mg/g whereas raw CB possess only 0.052 mg/g. Viswanathan et al.[73] try to effectively utilize the hydroxyl group of CB, carboxylation was carried out using chloroacetic acid to give carboxylated CCB and utilized for defluoridation of water. CCB shows a significant DC of 1.39 mg/g than the raw CB, which displayed only 0.052 mg/g. Viswanathan and coworkers[74] chemically modified CB by introducing multifunctional groups, viz., NH_3^+ and COOH groups by means of protonation and carboxylation in order to utilize both amine and hydroxyl groups of chitosan for fluoride removal. The synthesized protonated cum carboxylated chitosan beads (PCCB) showed a maximum DC of 1.80 mg/g whereas raw CB displayed only 0.052 mg/g. Sorption process was found to be independent of pH and slightly influenced in the presence of other common anions. The equilibrium sorption data were fitted to Freundlich and Langmuir isotherms. The values of thermodynamic parameters indicate the nature of fluoride removal is spontaneous and endothermic. The suitability of the modified CB was also tested under field conditions.

Recently, the higher valence metal ions incorporated materials are focused to develop new adsorbents with good performance for the selective removal of fluoride. Due to its high electronegativity and small ionic size, fluoride is classified as hard base. Electropositive multivalent metal ions viz., Al^{3+}, Fe^{3+}, La^{3+} Ce^{3+}, Zr^{4+}, Ti^{4+}, etc., are classified as hard acids. According to hard–soft-acid base (HSAB) principle hard acids prefers to bind with hard base and soft acids prefers to bind with soft base. Since F^- is a hard base, it has strong affinity toward multivalent metal ions. Hence, the higher valence metal ions viz., Al^{3+}, Fe^{3+}, Ce^{3+}, La^{3+}, Ti^{4+}, and Zr^{4+} ions have been utilized for enabling metal coordination with biopolymers, which helps to increase its strength and fluoride uptake capacity.

A novel adsorbent, La^{3+}-impregnated cross-linked gelatin, was prepared for the removal of fluoride from drinking water by Zhou et al.[75] The experimental results showed that this adsorbent exhibited a maximum adsorption efficiency of 98.8% at pH of 5–7, contact time for 40 min, and dose of 0.2 g when used to treat 50 ml of F^- containing aqueous solution.

The applicability of chitin, chitosan, and 20% lanthanum-incorporated chitosan (20% La-chitosan) as adsorbents for the removal of excess fluoride from drinking water was studied by Kamble et al.[76] Lanthanum chitosan adsorbents show excellent removal of fluoride from water, which is much better than bare chitosan and chitin. The mechanism of adsorption of fluoride on lanthanum-modified chitosan was explained in terms of the ligand exchange mechanism between fluoride ion and hydroxide ion coordinated to La^{3+} ion immobilized on the chitosan. 20% La-chitosan possess the maximum adsorption capacity of 3.1 mg/g.

Viswanathan and Meenakshi[77–80] have synthesized metal ion incorporated carboxylated chitosan beads (M-CCB) for fluoride removal. The metal ions chosen are La^{3+}, Fe^{3+}, Ce^{3+}, and Zr^{4+} ions, which have high affinity toward fluoride. The different M-CCB viz., Fe(III)-incorporated carboxylated chitosan beads (Fe-CCB), La(III)-incorporated carboxylated chitosan beads (La-CCB), Ce(III)-incorporated carboxylated chitosan beads (Ce-CCB), and Zr(IV)-incorporated carboxylated chitosan beads (Zr-CCB) were synthesized by incorporating CB in the respective metal ions solutions for 24 h. Then, M-CCB was washed with distilled water to neutral pH. The dried M-CCB was utilized for defluoridation experiments in batch mode. The DC of Fe-CCB, La-CCB, Ce-CCB, and Zr-CCB was found to be 4.23, 4.71, 4.80, and 4.85 mg/g respectively, whereas carboxylated chitosan beads (CCB) possess a DC of 1.39 mg/g. In general, M-CCB possess higher DC than CCB, which indicates their high selectivity toward fluoride. M-CCB showed higher selectivity toward fluoride in the presence of co-ions. The equilibrium data were fitted to Freundlich and Langmuir isotherms. The values of thermodynamic parameters indicate that the nature of fluoride removal is spontaneous and endothermic. The suitability of the modified CB was also tested under field conditions. The fluoride removal by M-CCB is governed by both adsorption and complexation mechanism.

The applicability of neodymium-modified chitosan as F^- adsorbent from aqueous solution was studied by Yao et al.[81] The effect of temperature, adsorbent dose, particle size, and the presence of co-ions on the adsorption process was investigated. The experimental data were fitted to Langmuir and Freundlich isotherms. The maximum adsorption capacity was 11.41 mg/g. The adsorbent was regenerated in 24 h by 4 g/L NaOH solution. Lanthanum incorporated chitosan beads were prepared using precipitation method by Bansiwal et al.[82] The synthesis was optimized by varying different synthesis parameters namely lanthanum loading, complexation and precipitation time, strength of ammonia solution used for precipitation, drying time, etc. The equilibrium adsorption data fitted well to Langmuir adsorption isotherm and showing maximum fluoride adsorption capacity of 4.7 mg/g with negligible lanthanum release. The effect of pH was also studied and the best efficiency was observed at pH 5.

The kinetics and thermodynamics of fluoride adsorption on cerium-impregnated chitosan (CIC) have been studied by Swain et al.[83] They are observed that the percentage removal of F^- increased from 63 to 93% with increase in adsorbent dose and reached a maximum value of 93% at 8.0 g/L of CIC. Maximum fluoride could be removed at solution pH 6.5–7; therefore, CIC can be conveniently utilized for the removal of fluoride from drinking water sources without undertaking any further treatment facilities. Vijaya et al.[84] have developed a novel biosorbent by the cross-linking of an anionic biopolymer, calcium alginate (CA) with glutaraldehyde. The glutaraldehyde cross-linked calcium alginate(GCA) has maximum monolayer adsorption capacity of 73.5 mg/g for fluoride. Huo et al.[85] have developed lanthanum alginate beads for fluoride sorption. They investigated its adsorption performance and adsorption mechanism. The adsorption isotherm for fluoride onto lanthanum alginate bead fits the Langmuir model well, and the maximum adsorption capacity is 197.2 mg/g. The adsorption mechanism of lanthanum alginate bead is considered as an ion-exchange between F^- and Cl^- or OH^-, as verified from the adsorbent and the solution by pH effect, energy dispersive X-ray, and ion chromatography.

The green seaweed, *Ulva japonica*, modified by loading multivalent metal ions, such as Zr(IV) and La(III) after $CaCl_2$ cross-linking to produce metal-loaded cross-linked seaweed (M-CSW) adsorbents by Paudyal et al.[86] The maximum sorption potential for fluoride was drastically increased after La(III) and Zr(IV) loading, which were evaluated as 0.58 and 0.95 mmol/g, respectively. Adsorbed fluoride was quantitatively desorbed by using dilute alkaline solution for its regeneration. The mechanism of fluoride adsorption was inferred in terms of ligand exchange reaction between the hydroxyl ion and fluoride ion in aqueous solution.

Liang et al.[87] described the fluoride removal from water using a new adsorbent namely mixed rare earths modified chitosan (CR). Mixed rare earths mainly contained La followed by Ce, which was analyzed by inductively coupled plasma mass spectrometry. La(III)-modified chitosan (CL) was also prepared as control. It was observed that the fluoride adsorption capacity of CR (3.72 mg/g) was higher than CL (3.16 mg/g) at 2 h. The presence of coions, such as bicarbonate and carbonate greatly affected the fluoride adsorption from water.

A novel sorbent for the removal of fluoride was prepared by immobilizing Zr(IV) on carboxymethyl cellulose sodium (CMC-Zr) by Wang et al.[88] The results show that the saturation adsorption capacity is 47 mg/g at pH 4. The fluoride removal of Zr(IV) is loaded on CMC through ion-exchange of sodium of CMC and surface adsorption, coordination reaction happened during the adsorption process between CMC-Zr and fluoride. A novel zirconium-impregnated cellulose (ZrIC) sorbent was prepared using ultrasonication by Barathi et al.[89] Fluoride from aqueous solution interacts with the cellulose hydroxyl groups and the cationic zirconium hydroxide. The prepared ZrIC has an adsorption capacity of 4.95 mg/g for fluoride.

Pandi and Viswanathan[90] have synthesized carboxylated alginic acid (CAA) and metal ions coordinated CAA (M-CAA) for defluoridation studies in batch mode. The oxidation of alginic acid (AA) with $KMnO_4$ gives CAA and the metal coordination was enabled in CAA by using high valence metal ions viz., La^{3+}(La-CAA) and Zr^{4+}(Zr-CAA). The synthesized materials

Zr-CAA, La-CAA, and CAA possess the DCs of 4.06, 3.14, and 0.88 mg/g, respectively. An enhanced DC was observed for M-CAA than CAA. The defluoridation mechanism of M-CAA was governed by electrostatic adsorption and complexation. The equilibrium data were fitted to Freundlich, Langmuir, Dubinin–Radushkevich, and Temkin isotherms. The values of thermodynamic parameters indicate that the nature of fluoride removal is spontaneous and endothermic. At field conditions, M-CAA reduce the fluoride concentration below the tolerance limit.

Qiusheng et al.[91] synthesized porous zirconium alginate beads as a novel adsorbent for fluoride removal. The isotherm data were well fitted to Langmuir isotherm model and the maximum adsorption capacity was 32.80 mg/g. The effect of solution pH, adsorbent dose, initial fluoride concentration, contact time, medium temperature, and the coexisting ions on adsorption capacity of fluoride ion was studied. The presence of HCO_3^-, SO_4^{2-} and PO_4^{3-} ions had a large negative impact on fluoride removal.

Muthu Prabhu and Meenakshi[92] explains the fluoride removal from aqueous solution using alginate-zirconium complex prepared with respective dicarboxylic acids like oxalic acid (Ox), malonic acid (MA), and succinic acid (SA) as a medium. The effects of various operating parameters were optimized. The result showed that the maximum removal of fluoride 9.65 mg/g was achieved by Alg–Ox–Zr complex at acidic pH in an ambient atmospheric condition.

3.8.7 POLYMERIC COMPOSITES

The technology is dedicated to the development of new materials, which are able to gratify the specific requirements in terms of both the structural and functional properties. The fashionable development of polymeric materials have introduced a new generation of composites. Polymeric composites are new classes of materials with ultrafine dimensions, and have received considerable attention in recent years due to its better handling properties. The main advantage of the polymeric composites resides in the possibility of combining the physical properties of the constituents to obtain new structural and functional properties. Composites can be made into any desired forms viz., beads, candles, and membranes, and hence environmentally sound. Due to the possibility of crafty properties, polymeric composites have been used for field applications.[93–98]

Most of the inorganic materials like hydroxyapatite, alumina, magnesia, etc., possess high DC, but cannot be directly used in fixed bed columns or any other flow-through systems because it cause enormous pressure drops. To avoid such technological bottlenecks, it is intended to synthesis organic–inorganic hybrid polymeric composites by dispersing inorganic materials in organic polymeric matrix, which results in the generation of new hybrid materials.

HAp powder is an excellent defluoridating material, but it suffers from significant pressure drop during filtration. To overcome such technology bottle necks, nano-hydroxyapatite/chitosan (n-HApC) and nano-hydroxyapatite/chitin (n-HApCh) composites have been prepared by Sairam Sundaram and co-workers.[93,94] The DC of n-HApC composite (1.56 mg/g) and n-HApCh composite (2.84 mg/g) is higher than n-HAp, which possesses DC of 1.30 mg/g. The fluoride sorption was explained with Freundlich and Langmuir isotherms. Thermodynamic parameters indicate that the nature of fluoride sorption is spontaneous and endothermic. Both the biocomposites remove fluoride by adsorption and ion exchange mechanism. The suitability of the composites was also tested under field conditions.

Magnesia (MgO) is a well-known adsorbent showing extremely high DC. In order to overcome the limitations of MgO for field applications, Sairam Sundaram et al.[95] have made an attempt to modify magnesia with abundant biomaterial chitosan to form magnesia/chitosan (MgOC) composite and its merits over conventional magnesia and raw chitosan is established. At equilibrium, MgOC composite posses the DC of 4.44 mg/g, while for magnesia it is only 2.18 mg/g.

Viswanathan and Meenakshi[96] developed a new biocomposite by incorporating inorganic ion-exchanger, namely zirconium(IV) tungstophosphate (ZrWP) into the chitosan biopolymeric matrix. The sorption behavior of fluoride from aqueous solutions by ZrWP/chitosan (ZrWPCs) composite has been investigated by batch technique. The DC of the adsorbent was found to be 2.03 mg/g. The equilibrium sorption data were fitted to Freundlich and Langmuir isotherms. The values of thermodynamic parameters indicate that the nature of fluoride removal is spontaneous and endothermic. The suitability of the biocomposite was also tested under field conditions. ZrWPCs composite removes fluoride by adsorption/complexation and ion exchange mechanism.

Hydrotalcite (HT) is a LDH used as catalyst, ion-exchanger, adsorbent, and as fillers of polymers. HT possesses an appreciable DC. To enhance the adsorption capacity of HT and to make into a usable form hydrotalcite/chitosan (HTCs) composite was prepared and studied for defluoridation of water by Viswanathan and Meenakshi[97]. HTCs composite showed a DC of 1.26 mg/g compared to 1.03 and 0.05 mg/g for HT and chitosan, respectively. The equilibrium adsorption data was fitted with the Langmuir model. It removes fluoride by ion-exchange and adsorption mechanism. The equilibrium data were fitted to Freundlich and Langmuir isotherms. The values of thermodynamic parameters indicate that the nature of fluoride removal is spontaneous and endothermic. The suitability of HTCs composite was also tested under field conditions.

Alumina possesses an appreciable DC of 1.57 mg/g. In order to improve its DC, it was aimed to prepare alumina polymeric composites using the chitosan by Viswanathan and Meenakshi.[98] Alumina/chitosan (AlCs) composite was prepared by incorporating alumina particles in the chitosan polymeric matrix, which can be made into any desired forms viz., beads, candles, and membranes. AlCs composite displayed a maximum DC of 3.81 mg/g than alumina and chitosan (0.05 mg/g). The equilibrium data were fitted to Freundlich and Langmuir isotherms. The values of thermodynamic parameters indicate that the fluoride sorption is spontaneous and endothermic in nature. The fluoride removal mechanism of AlCs composite was governed by electrostatic adsorption and complexation. At field conditions, AlCs composite reduce fluoride concentration below the tolerance limit. This work provides a potential platform for the development of defluoridation technology.

Mandal et al.[99] prepared hydrotalcite-like anionic clays (Zn/Al and Mg/Al), which were intercalated with sodium alginate to form organic–inorganic composite adsorbents for water treatment applications. The adsorption potential of the alginate–clay composites was examined for the removal of fluoride ions from water by adsorption. The adsorption capacity of the composites possesses higher fluoride removal capacity of 5.1 mg/g than that of either alginate or clay, used individually. Sujana et al.[100] have prepared the composite of hydrous ferric oxide (HFO) alginate beads and studied for its fluoride efficiency from water. The modified beads demonstrated Langmuir fluoride adsorption capacity of 8.90 mg/g at pH 7. It was found that about 80% of the adsorbed fluoride could be desorbed by using 0.05 M HCl.

Liu et al.[101] prepared enhanced chitosan/bentonite composite by treating chitosan/bentonite composite with concentrated HCl. The optimum operating conditions for fluoride removal by the enhanced chitosan/bentonite composite were pH is 7 and adsorbent dosage is 1.2 g. Increasing initial fluoride concentration reduced the adsorption of fluoride onto the enhanced chitosan/bentonite composite. The maximum monolayer adsorption capacity was found to be 2.95 mg/g at 293 K.

Yu et al.[102] have synthesized cellulose@hydroxyapatite nanocomposites in NaOH/thiourea/urea/H_2O solution via *situ* hybridization. The composite materials combine the advantages of cellulose and hydroxyapatite with the high specific surface area and the strong affinity toward fluoride. At the initial fluoride concentration of 10 mg/L, the residual concentration using above 3 g/L adsorbent dose could meet the drinking water standard of WHO norms. Furthermore, the coexisting anions had no significant effect on fluoride adsorption.

Swain et al.[103] have developed a new hybrid material of (Fe/Zr)–alginate (FZCA) microparticles. The average particle size of Fe–Zr particle was found to vary between 70.89 and 477.7 nm. The result indicated that the maximum sorption capacity (q_m, mg/g), 0.981, is related to Langmuir adsorption model, for which both experimental and calculated values are quite close to each other. The desorption characteristic of the hybrid material shows that nearly 89% of fluoride could be leached out at pH 12. Further, the reusable properties of the material support further development for commercial application purpose.

Basu et al.[104] have studied the feasibility of alumina-impregnated calcium alginate beads to sorb the excess fluoride ions from the potable water. The optimal condition for synthesis of calcium alginate alumina (Cal-Alg-Alu) beads is 2% (wt/vol) having 22% (wt/vol) alumina loading, stirring time: 1 h, drying temperature: 60°C for 8 h. The results of batch sorption experiments suggest that Cal-Alg-Alu beads is very effective for defluoridation in the pH range of 3.5–9.0 and sorption is more than 99.9% in the concentration range of 1–100 mg/L. The equilibrium sorption follows Langmuir isotherm well and the maximum fluoride uptake calculated is 17.0 mg/g. Swain et al.[105] developed a new hybrid material of (Al/Ce)-alginate micro particles for fluoride adsorption. The average particle size of Al/Ce mixed metal oxide was found to vary between 29.39 and 553.2 nm. The influence of pH upon sorption–desorption characteristic of the hybrid material was quite prominent as evident from leaching of 89% of fluoride at pH 12.

Barathi et al.[106] proposed a method, which involves the impregnation of Al–Zr in cellulose matrix wherein fluoride ion from aqueous medium interacts with the cellulose hydroxyl groups, as well as cationic Zr, and Al hydroxides. The sorbent has an adsorption capacity of 5.76 mg/g. The electrostatic and hydrogen bonding interactions support the mechanism of fluoride removal by Al–Zr in cellulose matrix. Ma et al.[107] reported aluminum and doping chitosan–Fe(III) hydrogel (Al–CS–Fe) as a possible adsorbent for the removal of fluoride from aqueous solutions. Besides, a series of single metal-impregnated chitosan were also synthesized to compare their defluoridation ability with Al–CS–Fe. The results indicate that Al–CS–Fe has a good ability to resist coexisting anions and the change of pH value, and owns a relatively high adsorption capacity of 31.16 mg/g. After regeneration, this adsorbent performs higher defluoridation efficiency than before. The mechanism of adsorption may be described by the ligand and ions-exchange that occurred on the active sites.

Adsorbents, lanthanum(III)-zirconium(IV) mixed oxide (LZMO), and chitosan supported lanthanum(III)-zirconium(IV) mixed oxides beads (CLZMOB), were prepared and utilized for fluoride removal by Muthu Prabhu and Meenakshi[108]. The DC at 50 min contact time for LZMO and CLZMOB was found to be 3.59 and 4.97 mg/g, respectively at pH 7. The results showed that CLZMOB could be an effective sorbent for the removal of fluoride from aqueous solutions.

Pandi and Viswanathan[109] have developed an eco-friendly adsorbent by alginate (Alg) bioencapsulating *n*-HAp, namely *n*-HApAlg composite for defluoridation studies in batch mode. The DC of synthesized n-HApAlg composite possesses an enhanced DC of 3.87 mg/g when compared to n-HAp and CaAlg composite, which possess DC of 1.30 and 0.68 mg/g, respectively. SEM images of n-HApAlg composite and fluoride sorbed n-HApAlg composite are presented in Figure 3.3(a) and (b), respectively. The surface of the sorbent is rough with abundant bump and many pores are present on the surface, which facilitates the diffusion of fluoride ions into n-HApAlg composite for its sorption. The change in the SEM micrographs of the sorbent before and after fluoride treatment facilitates the structural changes in the sorbent. This is further supported by Energy Dispersive X-ray

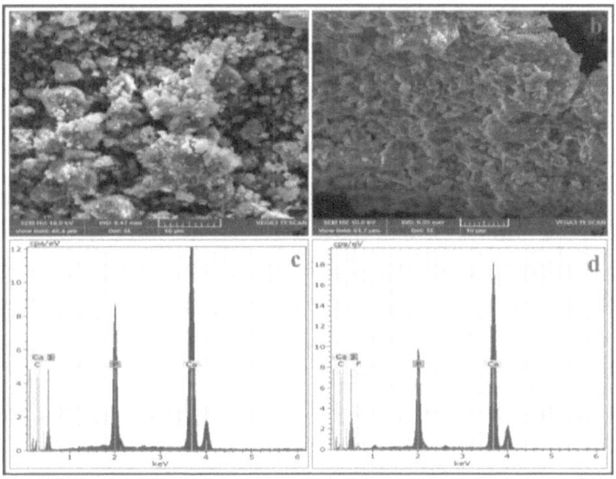

FIGURE 3.3 SEM images of (a) n-HApAlg composite and (b) fluoride sorbed n-HApAlg composite. EDAX spectra of (c) n-HApAlg composite and (d) fluoride sorbed n-HApAlg composite.

Analyzer (EDAX) analysis, which provides the direct confirmation for the sorption of fluoride ions onto n-HApAlg composite. The EDAX spectra of n-HApAlg composite confirm the presence of respective ions in the composite (Fig. 3.3c). The fluoride sorption, which occurred onto n-HApAlg composite was confirmed by the presence of fluoride peaks in the EDAX spectra of fluoride treated n-HApAlg composite (Fig. 3.3d). The sorption of fluoride on n-HApAlg composite follows Langmuir isotherm. The nature of fluoride sorption is spontaneous and endothermic. Field trial results indicate that n-HApAlg composite could be effectively employed as selective and promising defluoridating agent.

Prasad et al.[110] reported a novel approach for synthesis of Zr nanoparticles using aqueous extract of *Aloe vera*. The resulting nanoparticles were embedded into chitosan biopolymer and termed as CNZr composite. The composite was subjected to detailed adsorption studies for removal of fluoride from aqueous solution. The adsorption of fluoride onto CNZr composite worked well at pH 7.0, where ~99% of fluoride was found to be adsorbed on adsorbent. Langmuir isotherm model best fitted the equilibrium data.

A novel alginate–montmorillonite biopolymer–clay composite bead for defluoridation was developed by Kaygusuz et al.[111] Montmorillonite was dispersed in alginate solution, and the mixture was cross-linked in the aqueous solution of aluminum(III). The maximum adsorption capacity was reached as 31.0 mg/g at 25°C. The results show that aluminum alginate–montmorillonite composite beads can be used as effective and natural sorbents for fluoride removal from water. Mg–Al-LDH nanoflake-impregnated magnetic alginate beads (LDH-n-MABs) were successfully synthesized and their defluoridation performance was systematically evaluated by Gao et al.[112] Mg–Al-LDH nanoflakes with a higher adsorption capacity for F^- were immobilized into alginate beads without leaching, taking the safety issues concerning nanomaterials into consideration. The prepared beads had a high magnetic sensitivity to an external magnetic field providing an easy and efficient way to separate them from aqueous solution. LDH-n-MABs possess a maximum adsorption capacity of 32.4 mg/g.

Viswanathan et al.[113] focused on the development of silica gel/chitosan (SGCS) composite for fluoride removal. To enhance the fluoride selectivity of the biocomposite, lanthanum-III (La) was incorporated into SGCS composite, namely LaSGCS composite. A comparative evaluation of DC of LaSGCS composite, SGCS composite, silica gel (SG), and chitosan (CS) was made in batch mode. The results showed that LaSGCS composite possesses an enhanced DC of 4.90 mg/g, whereas SGCS composite, SG and CS possess the DCs of 1.56, 1.30, and 0.052 mg/g, respectively. The sorption of fluoride on LaSGCS composite follows Langmuir isotherm. The nature of fluoride sorption is spontaneous and endothermic.

Chitosan/montmorillonite/zirconium oxide (CTS/MMT/ZrO_2) nanocomposites were made by Teimouria et al.[114] The optimal conditions for the removal of fluoride were found to be: molar ratio of CTS/MMT/ZrO_2, 1:1; pH: 4; temperature: 30°C for 60 min in 25.00 mL of 20 mg/L of fluoride solutions and 0.1 g of adsorbent. The fluoride adsorption capacity of CTS/MMT/ZrO_2 was also found to be 23 mg/g experimentally. The results indicated that the adsorption capacity of CTS/MMT/ZrO_2 nanocomposite was higher than the average values of those of CTS (0.052 mg/g for fluoride removal), MMT, ZrO_2, CTS/ZrO_2, and CTS/MMT.

A novel adsorbent of γ-AlOOH@CS (pseudoboehmite and chitosan shell) magnetic nanoparticles (ACMN) with magnetic separation capabilities was developed to remove fluoride from drinking water by Wan et al.[115] The mechanism for the adsorption involved electrostatic interaction and hydrogen bonding. Moreover, the calculated adsorption capacity of the ACMN for fluoride using Langmuir model was 67.5 mg/g (20°C, pH = 7.0 ± 0.1), higher than other fluoride removal adsorbents. This nanoadsorbent performed well over a pH range of 4–10. The study found that PO_4^{3-} was the coexisting anion most able to hinder the nanoparticle's fluoride adsorption, followed by NO_3^- then Cl^- ions.

Pandi and Viswanathan[116] have studied synthesis of *n*-HAp incorporated gelatin (Gel) biocomposite namely *n*-HAp@Gel for efficient removal of fluoride from aqueous solution. The results demonstrated that, the developed *n*-HAp@Gel biocomposite possess an enhanced DC of 4.16 mg/g. The sorption data were fitted with various isotherm models. The acquired thermodynamic parameters showed that the adsorption of fluoride onto the sorbent was endothermic and spontaneous in nature. At field conditions, *n*-HAp@Gel composite reduce the fluoride concentration below the tolerance limit. A regeneration technique was proposed in order to reuse the composite.

The synthesized polymeric composites possess enhanced fluoride uptake capacity than the individual components. The feasible mechanism of fluoride removal by the polymeric composites is shown in Figure 3.4. The anions present in the polymeric composite get exchanged for fluoride by ion-exchange. Due to high electronegativity and small ionic size, fluoride is classified as hard base and electropositive multivalent metal ions like Al^{3+}, Fe^{3+}, La^{3+}, Ce^{3+}, Zr^{4+}, Ti^{4+}, etc., are classified as hard acids. According to HSAB principle, hard acids prefer to bind with hard base and soft acids prefer to bind with soft base. The multivalent metal ions (hard acids) prefer to bind with fluoride ion by electrostatic adsorption/complexation. In general, the polymeric composites will remove fluoride by ion-exchange and adsorption/complexation.

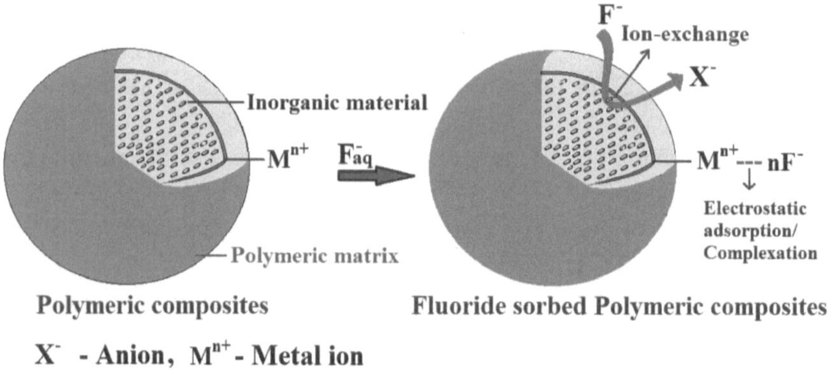

FIGURE 3.4 Feasible mechanism of defluoridation using polymeric composites.

3.9 COMPARISON OF VARIOUS ADSORBENTS WITH THEIR FLUORIDE UPTAKE CAPACITIES

The adsorption capacity of various low-cost inorganic adsorbents, biopolymers, and biopolymeric composites were shown in Tables 3.1, 3.2, and 3.3, respectively. The effectiveness of the adsorbent depends on various influencing parameters, like contact time, pH, dosage, particle size, initial fluoride concentration, ionic strength, temperature, presence of competing ions, etc. In general, the fluoride adsorption capacity of an adsorbent increases with increasing initial fluoride concentration until saturation conditions.

TABLE 3.1 The Adsorption Capacity of Various Low-Cost Inorganic Adsorbents.

S. No.	Adsorbent	Adsorption Capacity (mg/g)	Reference
1	Activated alumina (AA)	1.57	2
2	AA	1.45	22
3	Alumina	16.30	23
4	Lanthanum hydroxide-impregnated alumina	0.350 mM/g	25
5	Magnesia-amended AA	10.12	27
6	Calcium oxide-modified AA	101.01	28
7	Manganese oxide-coated AA	10.18	28

TABLE 3.1 *(Continued)*

S. No.	Adsorbent	Adsorption Capacity (mg/g)	Reference
8	Copper oxide-coated mesoporous alumina	7.22	29
9	Hydroxyapatite-modified AA	14.40	30
10	Aluminum-impregnated activated carbon	1.07	34
11	Charcoal was permeated with calcium	19.50	38
12	Nano-hydroxyapatite	1.30	41
13	Hydroxyapatite	4.70	42
14	Lime stone	43.10	44
15	Aluminum hydroxide-impregnated lime stone	84.03	44
16	Iron and aluminum mixed hydroxides	91.70	45
17	Iron(III)–tin(IV) mixed oxide	10.50	46
18	Mn–Ce oxide powder	79.50	48
19	Mn–Ce oxide granular	45.50	48
20	Hydrous zirconium oxide @ pH 4	124.00	49
21	Hydrous zirconium oxide @ pH 7	68.00	49
22	Fe(II)-Ti(IV) metal oxides	47.00	50
23	CeO_2–ZrO_2 nanocages	175.00	51
24	Sulfate-doped Fe_3O_4/Al_2O_3 nanoparticles	70.40	52
25	Hydrous Ce(IV)–Zr(IV) oxides	26.10	53
26	Manganese–lanthanum bimetal composite	292.9	54
27	Mg–Fe–La trimetal composite	112.17	55
28	Iron–magnesium–lanthanum trimetallic composite	270.30	56
29	Micronized kaolinites	0.78	57
30	Montmorillonite clay	0.26	58
31	10% La-bentonite clay	4.24	59
32	Magnesium-incorporated bentonite	2.26	60
33	Bauxite clay	5.16	61
34	Mg–Al-layered double hydroxides	47.70	66
35	Mg–Al–Fe hydrotalcite	14.00	67
36	CeO_2/Mg–Fe-layered double hydroxide composite	38.70–60.40	68

TABLE 3.2 The Adsorption Capacity of Various Biopolymers and Their Chemical Modifications.

S. No.	Adsorbent	Adsorption Capacity (mg/g)	Reference
1	Raw CB	0.05	72
2	Protonated CB	1.66	72
3	Carboxylated CB	1.39	73
4	Protonated cum carboxylated CB	1.80	74
5	20% lanthanum-incorporated chitosan	3.10	76
6	La(III)-incorporated carboxylated CB	4.71	77
7	Fe(III)-incorporated carboxylated CB	4.23	78
8	Ce(III)-incorporated carboxylated CB	4.80	79
9	Zr(IV)-loaded carboxylated CB	4.85	80
10	Neodymium-modified chitosan	11.41	81
11	Lanthanum-incorporated CB	4.70	82
12	Glutaraldehyde cross-linked calcium alginate	73.50	84
13	Lanthanum alginate bead	197.20	85
14	La(III)-loaded *Ulva japonica*	0.58 mmol/g	86
15	Zr(IV)-loaded *U. japonica*	0.95 mmol/g	86
16	Cerium-modified chitosan	3.72	87
17	Lanthanum-modified chitosan	3.16	87
18	Immobilizing Zr(IV) on carboxymethyl cellulose	47.00	88
19	Zirconium-impregnated cellulose	4.95	89
20	La^{3+}-coordinated carboxylated alginic acid	4.06	90
21	Zr^{4+}-coordinated carboxylated alginic acid	3.14	90
22	Porous zirconium alginate beads	32.80	91
23	Alginate–zirconium complex	9.65	92

TABLE 3.3 The Adsorption Capacity of Various Biopolymeric Composite Adsorbents and Their Chemical Modifications.

S. No.	Adsorbent	Adsorption Capacity (mg/g)	Reference
1	Nano-hydroxyapatite/chitosan	1.56	93
2	Nano-hydroxyapatite/chitin composite	2.84	94
3	Magnesia/chitosan composite	4.44	95
4	Zirconium(IV) tungstophosphate composite	2.03	96
5	Hydrotalcite/chitosan composite	1.26	97
6	Alumina/chitosan composite	3.81	98
7	Zn/Al and Mg/Al intercalated in sodium alginate	5.10	99

TABLE 3.3 *(Continued)*

S. No.	Adsorbent	Adsorption Capacity (mg/g)	Reference
8	HFO alginate beads	8.90	100
9	Chitosan/bentonite composite	2.95	101
10	Fe/Zr—alginate microparticles	0.98	103
11	Alumina-impregnated calcium alginate beads	17.00	104
12	Al–Zr in cellulose matrix	5.76	106
13	Aluminum and doping chitosan–Fe(III) hydrogel	31.16	107
14	Lanthanum(III) and zirconium(IV) mixed oxides beads	4.97	108
15	Nano-hydroxyapatite@alginate composite	3.87	109
16	Alginate–montmorillonite composite beads	31.00	111
17	Mg–Al-LDH nanoflake-impregnated magnetic alginate beads	32.40	112
18	Silica gel chitosan composite	1.56	113
19	La^{3+}-incorporated silica gel chitosan composite	4.90	113
20	Chitosan/montmorillonite/zirconium oxide nanocomposites	23.00	114
21	γ-AlOOH@CS magnetic nanocomposite	67.50	115
22	Nano-hydroxyapatite-incorporated gelatin biocomposite	4.16	116

3.10 CONCLUSIONS

The review focuses attention towards the eco-friendly adsorbents for defluoridation of drinking water. Based on the literature reviewed, the following conclusions were arrived:

- The adsorption technique is the most suitable technique for removing the excess fluoride content in water because the concentration of fluoride in field water is low when compared to the concentrations of other ions. In order to remove fluoride selectively, suitable chemical modifications were carried out.
- Rare earth metals and transition multivalent metal ions like Al^{3+}, Fe^{3+}, La^{3+}, Ce^{3+}, Zr^{4+}, Ti^{4+}, etc., possess higher selectivity toward fluoride. Hence, these metal ions have been loaded into the adsorbents to enhance the fluoride selectivity.

- The low-cost carbon, inorganic, mixed oxide, and clay-based materials possess appreciable DC, but it suffers from drawbacks like pressure drop during field applications.
- Polymeric materials possess advantages over other materials because it provides good quality of water, easily to handle, acceptable by the users, and reuse of polymeric materials makes cost effective.
- Biopolymeric materials are interesting, attractive, and more effective sorbents because of their unique characteristics like nontoxicity, hydrophilicity, biocompatibility, biodegradability, and biofunctionality.
- But some of the biopolymeric materials possess low fluoride selectivity, and hence chemical modifications like functionalization, metal ion incorporation were carried out to increase their adsorption capacity.
- Chemically modified biopolymeric materials possess enhanced fluoride uptake capacity than the raw biopolymeric materials.
- Polymeric composites are new class of hybrid materials, which could overcome the pressure drop during field applications. The polymeric composites can be made into any desirable form like beads, candles, membranes, etc.
- The polymeric composites possess enhanced fluoride adsorption capacity than the individual components.
- Metal ion-loaded polymeric composites possess high DC than polymeric composites.
- The effectiveness of the adsorbent depends on various influencing parameters like contact time, pH, dosage, particle size, initial fluoride concentration, ionic strength, temperature, presence of competing ions, etc.
- The regeneration and residual toxicity studies should be carried out in the adsorbent-treated water.
- The suitability of polymeric materials at field conditions should be checked before developing the technology because the field water contains other competing ions in addition to fluoride.
- The people who are living in rural areas are severely affected by fluorosis because they have no protected water supply. Providing defluoridated water will improve their health and social status.
- Polymeric materials/composites would pave the way for the development of defluoridation technology.

KEYWORDS

- **fluoride**
- **fluorosis**
- **defluoridation**
- **biopolymers**
- **polymeric composites**

BIBLIOGRAPHY

1. Hem, J.D. *Study and Interpretation of the Chemical Characteristics of Natural Water, 3rd ed.;* U.S. Geological Survey Water-Supply Paper 2254, Washington, D.C., 1985.
2. Meenakshi, S. *Ph.D Thesis*, Gandhigram Rural University, Gandhigram, Tamilnadu, India, 1992.
3. WHO Report. Fluoride and Fluorides, Environmental Health Criteria; World Health Organization: Geneva, Switzerland, 1984.
4. Jagtap, S.; Yenkie, M.K.; Labhsetwar,N.; Rayalu, S. Fluoride in Drinking Water and Defluoridation of Water. *Chem. Rev.* **2012**, *112*, 2454–2466.
5. Bhatnagar, A.; Kumar, E.; Sillanpaa, M. Fluoride Removal from Water by Adsorption—A Review. *Chem. Eng. J.* **2011**, *171*, 811–840.
6. Gupta, S.K.; Gambhir, S.; Mithal, A; Das, B.K. Skeletal Scintigraphic Findings in Endemic Skeletal Fluorosis. *Nuclear Medicine Communications*, **1993**, *14*, 384–390.
7. Susheela, A.K. Fluorosis Management Programme in India. *Curr. Sci.* 1999, *77,* 1250–1256.
8. Agrawal, V.; Vaish, A.K.; Vaish, P. Groundwater Quality: Focus on Fluoride and Fluorosis in Rajasthan. *Curr. Sci.* **1997**, *73*, 743–746.
9. Shortt, W.E. Endemic Fluorosis in Nellore District, South India. *Indian Med. Gaz.* **1937**, *72*, 396.
10. Pandit, C.G.; Raghavachari, T.; Rao, D.S.; Murthy, K. Endemic Fluorosis in South India. *Indian J. Med. Res.* **1940**, *28*, 533–558.
11. Karthikeyan, G.; Shanmugasundaraj, A. Isopleth Marking and *In Situ* fluoride Dependence on Water Quality in the Krishnagiri Block of Tamil Nadu in South India. *Fluoride* **2000**, *33*, 121–127.
12. Dinesh, C. Fluoride and Human Health Cause for Concern. *Indian J. Envir. Protect.* **1998**, *19*, 81–89.
13. Meenakshi, S.; Viswanathan, N. Identification of Selective Ion-Exchange Resin for Fluoride Sorption, *J. Colloid Interface Sci.* **2007**, *308*, 438–450.
14. Viswanathan, N.; Meenakshi, S. Effect of Metal Ion Loaded in a Resin Towards Fluoride Retention, *J Fluorine Chem.* **2008**, *129*, 645–653.
15. Kunin, R.; McGarvey, F.X. Oxidative Stability of Cellulose Derivatives—strong base anion exchange resin. *Ind. Eng. Chem.* **1949,** *41*, 1265–1268.

16. Boruff, C.S. Removal of Fluoride from Drinking Waters. *Ind. Eng. Chem.* **1934**, *27*, 69–71.
17. Joshi, S.V.; Mehta, S.H.; Rao, A.P.; Rao, A.V. Estimation of Sodium Fluoride Using HPLC in Reverse Osmosis Experiments. *Water Treatment,* **1992**, *7*, 207–211.
18. Simons, R. Trace Element Removal from Ash Dam Waters by Nanofiltration and Diffusion Dialysis. *Desalination.* **1993**, *89*, 325–341.
19. Adihikary, S.K.; Tipnis, U.K.; Harkare, W.P.; Govindan, K.P. Defluoridation during Desalination of Brackish Water by Electro Dialysis. *Desalination.* **1989**, *71*, 301–312.
20. Hichour, M.; Persin, F.; Sandeaux, J.; Gavach, C. Fluoride Removal from Waters by Donnan Dialysis. *Sep. Purif. Technol.* **2000**, *18*, 1–11.
21. Karthikeyan, G.; Shanmugasundarraj, A.; Meenakshi, S.; Elango, K.P. Adsorption Dynamics and the Effect of Temperature of Fluoride at Alumina-Solution Interface. *J. Indian Chem. Soc* **2004**, *81*, 461–466.
22. Ghorai, S.; Pant, K.K. Investigations on the Column Performance of Fluoride Adsorption by Activated Alumina in a Fixed-Bed. *Chem. Eng. J.*, **2004**, *98*, 165–173.
23. Ku, Y.; Chiou, H.M. The Adsorption of Fluoride Ion from Aqueous Solution by Activated Alumina. *Water Air Soil Pollut.* **2002**, *133*, 349–361.
24. Wasay, S.A.; Tokunaga, S.; Park, S.W. Removal of Hazardous Anions from Aqueous Solutions by La(III)- and Y(lll)-Impregnated Alumina. *Sep. Sci. Technol.* **1996**, *31*, 1501–1514.
25. Puri, B.K.; Balani, S. Trace Determination of Fluoride Using Lanthanum Hydroxide Supported on Alumina. *J. Environ. Sci. Health, A Tox./Hazard. Subst. Environ. Eng.* **2000**, *35*, 109–121.
26. Tripathy, S.S.; Raichur, A. Abatement of Fluoride from Water Using Manganese Dioxide Coated Activated Alumina. *J. Hazard. Mater.* **2008**, *153* (3), 1043–1051.
27. Maliyekkal, S.M.; Shukla, S.; Philip, L.; Indumathi, M.N. Enhanced Fluoride Removal from Drinking Water by Magnesia-Amended Activated Alumina Granules *Chem. Eng. J.* **2008**, *140*, 183–192.
28. Camacho, L.M.; Torres, A.; Saha, D.; Deng, S. Adsorption Equilibrium and Kinetics of Fluoride on Sol–Gel Derived Activated Alumina Adsorbents. *J. Colloid Interface Science*, **2010**, *349*, 307–313.
29. Bansiwal, A.; Pillewan, P.; Biniwale, R.B.; Rayalu, S.S.; Copper Oxide Incorporated Mesoporous Alumina for Defluoridation of Drinking Water. *Microporous Mesoporous Mater.* **2010**, *129*, 54–61.
30. Tomar, G.; Thareja, A.; Sarkar, S. Enhanced Fluoride Removal by Hydroxyapatite-Modified Activated Alumina. *Int. J. Environ. Sci. Technol.* 2014, doi. 10.1007/s13762-014-0653-5.
31. Cheng, J.; Meng, X.; Jing, C.; Hao, J. La^{3+}-Modified Activated Alumina for Fluoride Removal from Water. *J. Hazard. Mater.*, **2014**, *278*, 343–349.
32. Li, L.; Quinlivan, P.A. Effects of Activated Carbon Surface Chemistry and Pore Structure on the Adsorption of Organic Contaminants from Aqueous Solution. *Carbon* **2002**, *40*, 2085–2100.
33. Bhargava, D.S.; Killedar, D.J. Fluoride Adsorption on Fishbone Charcoal through a Moving Media Adsorbent. *Water Res.* **1992**, *26*, 781–788.
34. Leyya, R.; Ovalle-Turrubartes, J.; Sanchez-Castillo, M.A. Adsorption of Fluoride from Aqueous Solution on Aluminum-Impregnated Carbon. *Carbon* **1999**, *37*, 609–617.

35. Tripathy, S.S.; Bersillon, J.; Gopal, K. Removal of Fluoride from Drinking Water by Adsorption onto Alum-Impregnated Activated Alumina. *Sep. Purif. Technol.*, **2006**, *50*, 310–317.
36. Alagumuthu, G.; Rajan, M. Equilibrium and Kinetics of Adsorption of Fluoride onto Zirconium Impregnated Cashew Nut Shell Carbon. *Chem. Eng. J.* **2010**, *158*, 451–457.
37. Maa, Y.; Wanga, S.G.; Fanb, M.; Gonga, W.X.; Gaoa, B.Y. Characteristics and Defluoridation Performance of Granular Activated Carbons Coated with Manganese Oxides. *J. Hazard. Mater.* **2009**, *168*, 1140–1146.
38. Tchomgui-Kamga, E.; Ngameni, E.; Darchen, A. Evaluation of Removal Efficiency of Fluoride from Aqueous Solution Using New Charcoals that Contain Calcium Compounds. *J. Colloid Interface Sci.* **2010**, *346*, 494–499.
39. Turner. B.; Binning, P.; Stipp, S.L.S. Fluoride Removal by Calcite: Evidence for Fluorite Precipitation and Surface Adsorption. *Environ. Sci. Technol.* **2005**, *39*, 9561–9568.
40. Islam, M.; Patel, R.K. Evaluation of Removal Efficiency of Fluoride from Aqueous Solution Using Quick Lime. *J. Hazard. Mater.* **2007**, *143*, 303–310.
41. Sairam Sundaram, C.; Viswanathan, N.; Meenakshi, S. Defluoridation Chemistry of Synthetic Hydroxyapatite at Nano Scale: Equilibrium And Kinetic Studies. *J. Hazard. Mater.* **2008**, *155*, 206–215.
42. Jiménez-Reyes, M.; Solache-Ríos, M. Sorption Behavior of Fluoride Ions from Aqueous Solutions by Hydroxyapatite. *J. Hazard. Mater.* **2010**, *180*, 297–302.
43. Nath, S.K.; Dutta, R.K. Fluoride Removal from Water Using Crushed Limestone, *Indian J. Chem. Technol.* **2010**, *17*, 120–125.
44. Jain, S.; Jayaram, R.V. Removal of Fluoride from Contaminated Drinking Water Using Unmodified and Aluminium Hydroxide Impregnated Blue Lime Stone. *Waste Separation Sci. Technol.* **2009**, *44*, 1436–1451.
45. Sujana, M.G.; Soma, G.; Vasumathi, N.; Anand, S. Studies on Fluoride Adsorption Capacities of Amorphous Fe/Al Mixed Hydroxides from Aqueous Solutions. *J. Fluorine Chem.* **2009**, *130*,749–754.
46. Biswas, K.; Gupta, K.; Ghosh, U.C. Adsorption of Fluoride by Hydrous Iron(III)–Tin(IV) Bimetal Mixed Oxide from the Aqueous Solutions. *Chem. Eng. J.* **2009**, *149*, 196–206.
47. Biswas, K.; Gupta, K.; Goswami, A.; Ghosh, U.C. Fluoride Removal Efficiency from Aqueous Solution by Synthetic Iron(III)–Aluminum(III)–Chromium(III) Ternary Mixed Oxide, *Desalination* **2010**, *255*, 44–51.
48. Deng, S.; Liu, H.; Zhou, W.; Huang, J.; Yu, G. Mn–Ce Oxide as a High-Capacity Adsorbent For Fluoride Removal from Water. *J. Hazard. Mater*. **2011**, *186*, 1360–1366.
49. Dou, X.; Mohan, D.; Pittman, C.U.; Yang, S. Remediating Fluoride from Water Using Hydrous Zirconium Oxide. *Chem. Eng. J.* **2012**, *198–199*, 236–245.
50. Chen, L.; He, B.Y.; He, S.; Wang, T.J.; Su, C.L.; Jin, Y. Fe—Ti Oxide Nano-Adsorbent Synthesized by Co-Precipitation for Fluoride Removal from Drinking Water and Its Adsorption Mechanism. *Powder Technol.* **2012**, *227*, 3–8.
51. Wang, J.; Xu, W.; Chen, L.; Jia, Y.; Wang, L.; Huang, X.J.; Liu, J. Excellent Fluoride Removal Performance by CeO_2–ZrO_2 Nanocages in Water Environment. *Chem. Eng. J.* **2013**, *231*, 198–205.
52. Chai L.; Y. Wang; Zhao, N.; Yang, W.; You, X. Sulfate-Doped Fe_3O_4/Al_2O_3 Nanoparticles as a Novel Adsorbent for Fluoride Removal from Drinking Water. *Water Res.* **2013**, *47*, 4040–4049.

53. Ghosh, A.; Chakrabarti, S.; Biswas, K.; Ghosh, U.C. Agglomerated Nanoparticles of Hydrous Ce(IV) + Zr(IV) Mixed Oxide: Preparation, Characterization and Physicochemical Aspects on Fluoride Adsorption. *Appl. Surf. Sci.* **2014**, *307*, 665–676.
54. Yu, Y.; Chen, J.P. Fabrication and Performance of a Mn–La Metal Composite for Remarkable Decontamination of Fluoride. *J. Mater. Chem. A* **2014**, *2*, 8086–8093.
55. Wang, J.; Kang, D.; Ge, M.; Chen,Y. Synthesis and Characterization of Mg–Fe–La Trimetal Composite as an Adsorbent for Fluoride Removal. *Chem. Eng. J.* **2015**, *264*, 506–513.
56. Yu, Y.; Yu, L.; Chen, J.P. Adsorption of Fluoride by Fe–Mg–La Triple-Metal Composite: Adsorbent Preparation, Illustration of Performance and Study of Mechanisms. *Chem. Eng. J.* **2015**, *262*, 839–846.
57. Meenakshi, S.; Sundaram, C.S; Sukumar, R. Enhanced Fluoride Sorption by Mechanochemically Activated Kaolinites. *J. Hazard. Mater.* **2008**, *153*, 164–172.
58. Tor, A. Removal of Fluoride from an Aqueous Solution by Using Montmorillonite. *Desalination* **2006**, *201*, 267–276.
59. Kamble, S.P.; Dixit, P.; Rayalu, S.S.; Labhsetwar, N.K. Defluoridation of Drinking Water Using Chemically Modified Bentonite Clay. *Desalination* **2009**, *249*, 687–693.
60. Takre, D.; Rayalu, S.; Kawade, R.; Meshram, S.; Subrt, J.; Labhsetwar, N. Magnesium Incorporated Bentonite Clay for Defluoridation of Drinking Water. *J. Hazard. Mater.* **2010**, *180*, 122–130.
61. Sujana, M.G.; Anand, S. Fluoride Removal Studies from Contaminated Ground Water by Using Bauxite. *Desalination* **2011**, *267*, 222–227.
62. Bia, G.; De Pauli, C.P.; Borgnino, L. The Role of Fe(III) Modified Montmorillonite On Fluoride Mobility: Adsorption Experiments and Competition with Phosphate. *J. Environ. Manage.* **2012**, *100*, 1–9.
63. Das, D.P.; Das, J.; Parida, K. Physicochemical Characterization and Adsorption Behavior of Calcined Zn/Al Hydrotalcite-Like Compound (HTlc) Towards Removal of Fluoride from Aqueous Solution. *J. Colloid Interface Sci.* **2003**, *261*, 213–220.
64. Lv, L.; He, J.; Wei, M.; Evans, D.G.; Duan, X. Factors Influencing the Removal of Fluoride from Aqueous Solution by Calcined Mg–Al–CO_3 Layered Double Hydroxides. *J. Hazard. Mater.* **2006**, *133*, 119–128.
65. Wang, H.; Chen, J.; Cai, Y.; Ji, J.; Liu, L.; Teng, H.H. Defluoridation of Drinking Water by Mg/Al Hydrotalcite-Like Compounds and Their Calcined Products. *Appl. Clay Sci.* **2007**, *35*, 59–66.
66. Chang, Q.; Zhu, L.; Luo, Z.; Lei, M.; Zhang, S.; Tang, H. Sono-Assisted Preparation of Magnetic Magnesium–Aluminum Layered Double Hydroxides and Their Application for Removing Fluoride. *Ultrason. Sonochem.* **2011**, *18*, 553–561.
67. Ma, W.; Zhao, N.; Yang, G.; Tian, L.; Wang, R. Removal of Fluoride Ions From Aqueous Solution by the Calcination Product of Mg–Al–Fe Hydrotalcite-Like Compound. *Desalination* **2011**, *268*, 20-26.
68. Zhang, T.; Li, Q.; Xiao, H.; Mei, Z.; Lu, H.; Zhou, Y. Enhanced Fluoride Removal from Water by Non-Thermal Plasma Modified CeO_2/Mg–Fe layered double hydroxides. *App. Clay Sci.* **2013**, *72*, 117–123.
69. Mandal, S.; Tripathy, S.; Padhi, T.; Sahu, M.K.; Patel, R.K. Removal Efficiency of Fluoride by Novel Mg-Cr-Cl Layered Double Hydroxide by Batch Process from Water. *J. Environmental Sci.***2013,** *25*, 993–1000.
70. Zhao, X.; L. Zhang, Pan Xiong,Wenjing Ma, Na Qian, Wencong Lu, et al., A Novel Method for Synthesis of Co–Al Layered Double Hydroxides and Their Conversions

to Mesoporous $Coal_2o_4$ Nanostructures for Applications in Adsorption Removal of Fluoride Ions. *Microporous Mesoporous Mater.* **2015**, *201*, 91–98.
71. Miretzky, P.; Cirelli, A.F. Fluoride Removal from Water by Chitosan Derivatives and Composites. *J. Fluorine Chem.* **2011**, *132*, 231–240.
72. Viswanathan, N.; Sairam Sundaram, C.; Meenakshi, S. Removal of Fluoride from Aqueous Solution using Protonated Chitosan Beads. *J. Hazard. Mater.***2009,** *161*, 423–430.
73. Viswanathan, N.; Sairam Sundaram, C., Meenakshi, S. Sorption Behaviour of Fluoride on Carboxylated Cross-Linked Chitosan Beads. *Colloids Surf. B: Biointerfaces* **2009**, *68*, 48–54.
74. Viswanathan, N.; Sairam Sundaram, C.; Meenakshi, S. Development of Multifunctional Chitosan Beads for Fluoride Removal. *J. Hazard. Mater.* **2009**, *167*, 325–331.
75. Zhou, Y.; Yu, C.; Shan, Y. Adsorption of Fluoride from Aqueous Solution on La^{3+}-Impregnated Cross-Linked Gelatin. *Sep. Purif. Technol.***2004**, *36*, 89–94.
76. Kamble, S.P.; Jagtap, S.; Labhsetwar, N.K.; Thakare, D.; Godfrey, S.; Devotta, S.; Rayalu, D.S. Defluoridation of Drinking Water Using Chitin, Chitosan and Lanthanum-Modified Chitosan. *Chem. Eng. J.* **2007**, *129*, 173–180.
77. Viswanthan, N.; Meenakshi, S. Enhanced Fluoride Sorption Using La(III) Incorporated Carboxylated Chitosan Beads. *J. Colloid Interface Sci.* **2008**, *322*, 375–383.
78. Viswanthan, N.; Meenakshi, S. Selective Sorption of Fluoride Using Fe(III) Loaded Carboxylated Chitosan Beads. *J. Fluorine Chem.* **2008**, *129*, 503–509.
79. Viswanthan, N.; Meenakshi, S. Synthesis of Zr(IV) Entrapped Chitosan Polymeric Matrix for Selective Fluoride Sorption. *Colloids Surf. B: Biointerfaces* **2009**, *72*, 88–93.
80. Viswanthan, N.; Meenakshi, S. Enhanced and Selective Fluoride Sorption on Ce(III) Encapsulated Chitosan Polymeric Matrix. *J. Appl. Polym. Sci.* **2009**, *112*, 1114–1121.
81. Yao, R.; Meng, F.; Zhang, L.; Ma, D.; Wang, M. Defluoridation of Water Using Neodymium-Modified Chitosan. *J. Hazard. Mater.* **2009**, *165*, 454–460.
82. Bansiwal, A.; Takre, D.; Labhshetwar, N.; Meshram, S.; Rayalu, S. Fluoride Removal Using Lanthanum Incorporated Chitosan Beads. *Colloids Surf. B: Biointerfaces*, **2009**, *74*, 216–224.
83. Swain, S.K.; Padhi, T.; Patnaik, T.; Patel, R.K.; Jha, U.; Dey, R.K. Kinetics and Thermodynamics of Fluoride Removal Using Cerium-Impregnated Chitosan. *Desalin. Water Treat.* **2010**, *13*, 369–381.
84. Vijaya, Y.; Popuri, S.R.; Reddy, A.S.; Krishnaiah, A. Synthesis and Characterization of Glutaraldehyde-Crosslinked Calcium Alginate for Fluoride Removal from Aqueous Solutions. *J. Appl. Polym. Sci.* **2011**, *120*, 3443–3452.
85. Huo, Y.; Ding, W.; Huang, X.; Xu, J.; Zhao, M. Fluoride Removal by Lanthanum Alginate Bead: Adsorbent Characterization and Adsorption Mechanism. *Chin. J. Chem. Eng.* **2011**, *19*, 365–370
86. Paudyal, H.; Pangeni, B.; Inoue, K.; Kawakita, H.; Ohto, K.; Ghimire, K.N.; Alam, S. Preparation of Novel Alginate Based Anion Exchanger from *Ulva japonica* and its Application for the Removal of Trace Concentrations of Fluoride from Water. *Bioresour. Technol.* **2013**, *148*, 221–227.
87. Liang, P.; Wang, D.; Wang, D.; Xu, Y; Luo, L. Preparation of Mixed Rare Earths Modified Chitosan for Fluoride Adsorption. *J. Rare Earth.* **2013**, *31* (8), 817–822.
88. Wang, J.; Lin, S.; Luo, S.; Long, Y. A Sorbent of Carboxymethyl Cellulose Loaded with Zirconium for the Removal of Fluoride from Aqueous Solution. *Chem. Eng. J.* **2014**, *252*, 415–422.

89. Barathi, M.; Kumar, A.S.K.; Rajesh, N. A Novel Ultrasonication Method in the Preparation of Zirconium Impregnated Cellulose for Effective Fluoride Adsorption. *Ultrason. Sonochem.* **2014**, *21*, 1090–1099.
90. Pandi, K.; Viswanathan, N. A Novel Metal Coordination Enabled in Carboxylated Alginic Acid for Effective Fluoride Removal. *Carbohydr. Polym.* **2015**, *118*, 242–249.
91. Qiusheng, Z.; Jin, L.X.Q.; Jing, W.; Xuegang, L. Porous Zirconium Alginate Beads Adsorbent for Fluoride Adsorption from Aqueous Solutions. *RSC Adv.* **2015**, *5*, 2100–2112.
92. Muthu Prabu, S.; Meenakshi, S. Novel One-Pot Synthesis of Dicarboxylic Acids Mediated Alginate-Zirconium Biopolymeric Complex for Defluoridation of Water. *Carbohydr. Polym.* **2015**, *120*, 60–68.
93. Sairam Sundaram, C.; Viswanathan, N.; Meenakshi, S. Uptake of Fluoride by Nano-Hydroxyapatite/Chitosan, a Bioinorganic Composite. *Bioresour. Technol.* **2008**, *99*, 8226–8230.
94. Sairam Sundaram, C.; Viswanathan, N.; Meenakshi, S. Fluoride Sorption by Nano-Hydroxyapatite/Chitin Composite. *J. Hazard. Mater.* **2009**, *172*, 147–151.
95. Sairam Sundaram, C., Viswanathan, N., and Meenakshi, S. Defluoridation of water using magnesia/chitosan composite. *J. Hazard. Mater.* **2009**, *163*, 618–624.
96. Viswanathan, N.; Meenakshi, S. Development of Chitosan Supported Zirconium(IV) Tungstophosphate Composite for Fluoride Removal. *J Hazard. Mater.* **2010**, *176*, 459–465.
97. Viswanathan, N.; Meenakshi, S. Selective Fluoride Adsorption by a Hydrotalcite/Chitosan Composite. *Appl. Clay Sci.* **2010**, *48*, 607–611.
98. Viswanathan, N.; Meenakshi, S. Enriched Fluoride Sorption using Alumina/Chitosan Composite. *J. Hazard. Mater.* **2010**, *178*, 226–232.
99. Mandal, S.; Patil, V.S; Mayadevi, S. Alginate and Hydrotalcite-Like Anionic Clay Composite Systems: Synthesis, Characterization and Application Studies. *Microporous Mesoporous Mater.* **2011**, *158*, 241–246
100. Sujana, M.G.; Mishra, A.; Acharya, B.C. Hydrous Ferric Oxide Doped Alginate Beads for Fluoride Removal: Adsorption Kinetics and Equilibrium Studies. *Appl. Surf. Sci.* **2013**, *270*, 767–776.
101. Liu, Q.; Huang, R.; Yang. B.; Liu, Y. Adsorption of Fluoride from Aqueous Solution by Enhanced Chitosan/Bentonite Composite. *Water Sci. Technol.* **2013**, *68*, 2074–2081.
102. Yu, X.; Tonog, S.; Ge, M.; Zuo, J. Removal of Fluoride from Drinking Water by cellulose@hydroxyapatite nanocomposites. *Carbohydr. Polym.* **2013**, *92*, 269–275.
103. Swain, S.K.; Patnaik, T.; Patnaik, P.C.; Jha, U.; Dey, R.K. Development of New Alginate Entrapped Fe(III)–Zr(IV) Binary Mixed Oxide for Removal of Fluoride from Water Bodies. *Chem. Eng. J.* **2013**, *215–216*, 763–771.
104. Basu, H.; Singhal, R.K.; Pimple, M.V.; Reddy, A.V.R. Synthesis and Characterization of Alumina Impregnated Alginate Beads for Fluoride Removal from Potable Water. *Water Air Soil Poll.* **2013**, *224*, 1572–1577.
105. Swain, S.K.; Patnaik, T.; Dey, R.K. Efficient Removal of Fluoride using New Composite Material of Biopolymer Alginate Entrapped Mixed Metal Oxide Nanomaterials. *Desalin. Water Treat.* **2013**, *22–24*, 4368–4378.
106. Barathi, M.; Santhana Krishna Kumar, A.; Rajesh, N. Efficacy of Novel Al–Zr Impregnated Cellulose Adsorbent Prepared Using Microwave Irradiation for the Facile Defluoridation of Water. *J. Environ. Chem. Eng.* **2013,** *1*, 325–1335.

107. Ma, J., Shen, Y., Shen, C., Wen, Y., and Liu, W. Al-Doping Chitosan-Fe(III) Hydrogel for the Removal of Fluoride from Aqueous Solutions. *Chem. Eng. J.* **2014**, *248*, 98–106.
108. Muthu Prabhu, S.; Meenakshi, S. Enriched Fluoride Sorption Using Chitosan Supported Mixed Metal Oxides Beads: Synthesis, Characterization and Mechanism. *J. Water Process Eng.* **2014**, *2*, 96–104.
109. Pandi, K.; Viswanathan, N. Synthesis of Alginate Bioencapsulated Nano-Hydroxyapatite Composite for Selective Fluoride Sorption. *Carbohydr. Polym.* **2014**, *112*, 662–667.
110. Prasad, K.S.; Amin, Y.; and Selvaraj, K. Defluoridation Using Biomimetically Synthesized Nano Zirconium Chitosan Composite: Kinetic and Equilibrium Studies. *J. Hazard. Mater.* **2014,** *276*, 232–240.
111. Kaygusuz, H.; Çoşkunırmak, M.H.; Kahya, N.; Erim, F.B. Aluminum Alginate–Montmorillonite Composite Beads for Defluoridation of Water. *Water Air Soil Poll.* **2014**, *226*, 2257.
112. Gao, C.; Yu, X.Y.; Luo, T.; Jia, Y.; Sun, B.; Liu, J.H.; Huang, X.J. Millimeter-Sized Mg–Al-LDH Nano Flake Impregnated Magnetic Alginate Beads (LDH-n-MABs): A Novel Bio-Based Sorbent for the Removal of Fluoride in Water. *J. Mater. Chem. A* **2014**, *2*, 2119–2128.
113. Viswanathan, N.; Pandi, K.; Meenakshi, S. Synthesis of Metal Ion Entrapped Silica Gel/ Chitosan Biocomposite for Defluoridation Studies. *Int. J. Biol. Macromol.* **2014**, *70*, 347–353.
114. Teimouria, A.; Nasaba, A.J.; Habibollahia, S.; Najafabadib, M.F.; Chermahinic, A.N. Synthesis and Characterization of Chitosan/Montmorillonite/ZrO_2 Nanocomposite and its Application as Adsorbent for Removal of Fluoride, *RSC Adv.* **2015**, *5*, 6771–6781.
115. Wan, Z.; Chen, W.; Liu C.; Liu, Y.; Dong, C. Preparation and Characterization of γ-AlOOH @CS Magnetic Nanoparticle as a Novel Adsorbent for Removing Fluoride from Drinking Water. *J. Colloid Interface Sci.* **2015**, *443*, 115–124.
116. Pandi, K.; Viswanathan, N. *In Situ* Precipitation of Nano-Hydroxyapatite in Gelatin Polymatrix toward Specific Fluoride Sorption. *Int. J. Biol. Macromol.* **2015**, *74*, 351–359.

CHAPTER 4

ADVANCED MATERIALS FOR SUPERCAPACITORS

A. NIRMALA GRACE* and R. RAMACHANDRAN

*Center for Nanotechnology Research, VIT University, Vellore, Tamil Nadu, India.*E-mail: anirmalagladys@gmail.com*

CONTENTS

ABSTRACT

The exciting development of advanced nanostructured materials has driven the rapid growth of research in the field of electrochemical energy storage (EES) systems, which are critical to a variety of applications ranging from portable consumer electronics, hybrid electric vehicles, to large industrial scale power and energy management. In this view, this chapter deals about the advanced nanostructures materials development toward the electrodes for supercapacitors. After a short introduction of energy storages systems, the various types of supercapacitor electrode materials and its properties are discussed and explained.

4.1 INTRODUCTION TO ENERGY STORAGE SYSTEMS

With a fast and tremendous industrial development and escalating human population along with an increase in energy demand, the global energy consumption has been accelerating at a startling rate. The global energy will get exhausted at this present consumption rate. To address this issue caused by energy exhaustion, urgency in the use of renewable energy is needed to suffice the demand. To facilitate the effective usage of renewable energy, it is imperative to develop high-performance, low-cost, and eco-friendly energy conversion and storage systems. Among the various developed protocols, fuel cells and supercapacitors are effective and simple systems that have been adopted for ages toward electrochemical energy conversion and storage. The performance of these systems is dependent on the intrinsic properties of the materials utilized for building the same. Hence, to cater the development of electrochemical energy conversion and storage systems, material technology plays a pivotal and supporting role.

4.2 BATTERIES

Electric batteries are the most common energy storage devices for portable applications, wherein the energy is stored in chemical form and gets reconverted into electrical form. The components of it are two electrodes (anode and cathode) and one electrolyte, which can be either solid or liquid. In the redox reaction that powers the battery, reduction occurs at the cathode, while oxidation occurs at the anode.[1] This energy storage form has changed noticeably over years, even though the basic principles have been known since the invention by the Italian physicist Alessandro Volta in year 1800. The first form was a stack of zinc and copper disk separated by an acid electrolyte. Nickel cadmium (Ni–Cd) and Nickel metal hydrate (Ni–Mh) was dominating the market until the developments of Lithium batteries (Fig. 4.1). This latter category of batteries has captured the market finally due to the higher specific energy [150–500 vs. 50–150 Wh/kg of Ni–Mh, Ni–Ca (Fig. 4.1)], and are nowadays one of the most common batteries available. These batteries have two subclasses viz. lithium-ion (Li-ion) batteries and lithium polymer (Li-Po) batteries (which basically are an evolution of the Li-ion).

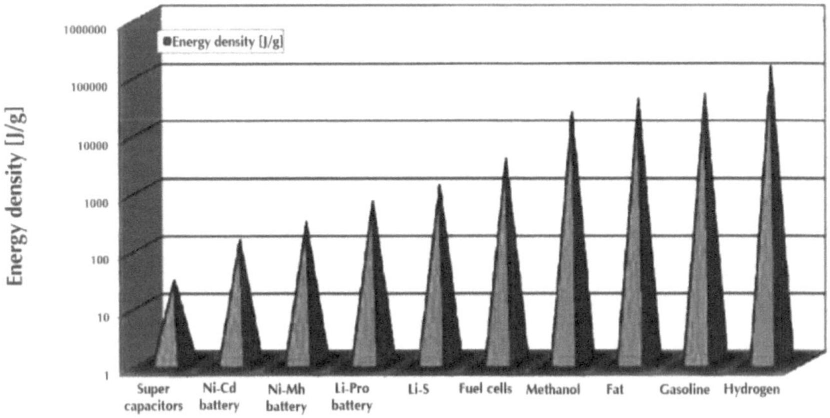

FIGURE 4.1 Energy density comparison graph of different energy storage sources.[2]

The high volume of the market (around 50 billion dollar market in 2006[3] is expected to grow even more in the coming years, and is expected to grow to 85.76 billion dollars by 2016 with a compounded annual growth rate of 7% over the next 5 years.[4] These figures suggests the enormous volume of growth, generating huge revenues and part of it is reinvested in battery research, which is focused to improve energy density, life time, and cycling stability, without using dangerous materials that can create health hazards. It is difficult to find all these properties optimized in one material combination. One way for enhancement of the battery capacity is using nanotechnologies for increasing and structuring the surface area of the electrodes (e.g., by depositing nanomaterials or by growing nanostructures, such as nanotubes).[1] But despite all these efforts, the technological improvements of batteries are still much, much slower when compared with the evolutionary progress of electronics. For these reasons, many researchers are trying to include smart circuitry inside the batteries for optimizing the discharge curve by optimizing the load. They call it intelligent batteries and they exploit some battery-related characteristic, such as charge recovery effect, for improving their lifetime.

4.3 FUEL CELLS

A next alternative to battery is fuel cell and are already in the market, with many vehicles running on fuel cell-based engines.[5] They are electrochemical devices wherein the chemical energy stored in the fuels is converted into electrical energy. Such devices are made of two electrodes and a membrane, which forma reaction chamber and have external stored reactants. The working principle is similar to batteries, but here the species near the vicinity of the electrode surface are continuously replenished and they can be refilled, thus enabling a continuous electrical supply for a long period. There are varieties of fuel cells available, and they mainly differ from the species used as fuel, but all the types use one element as fuel and a second element as oxidizer (commonly air).[6] In fuel cells, it is possible to generate any power or current by changing the physical dimension of the cell and the flow rate of the fuel. But, the voltage across the single-cell electrode is fixed and it is not possible to change it. In general, this voltage is very low (less than 1 V for realistic operating condition[6] and thus multiple cell stacks connected in series are needed to achieve larger potentials. To aid the same, mixed series and parallel connections between different cells can also be used for increasing the voltage and the maximum current supplied. Among all the fuel cell types, the most appealing and promising are the proton exchange membrane (PEM) and the direct methanol fuel cells (DMFC), a subset of the former. A schematic of PEM fuel cell is given in Figure 4.2, where the cell uses hydrogen and oxygen as species. The membrane sandwiched between the two electrodes allows the passage of the ions only, whereas the electrons are forced to travel through the electric circuit.

The mechanism can be detailed using chemical reactions as below. The reaction occurring at the anode is:

$$H_2 \rightarrow 2H^+ + 2e^-$$

On the cathode side, the electrons recombine with the ions and they react with the cathode species (oxygen in this case) forming water through the following reaction:

$$4H^+ + O_2 + 4e^- \rightarrow 2H_2O$$

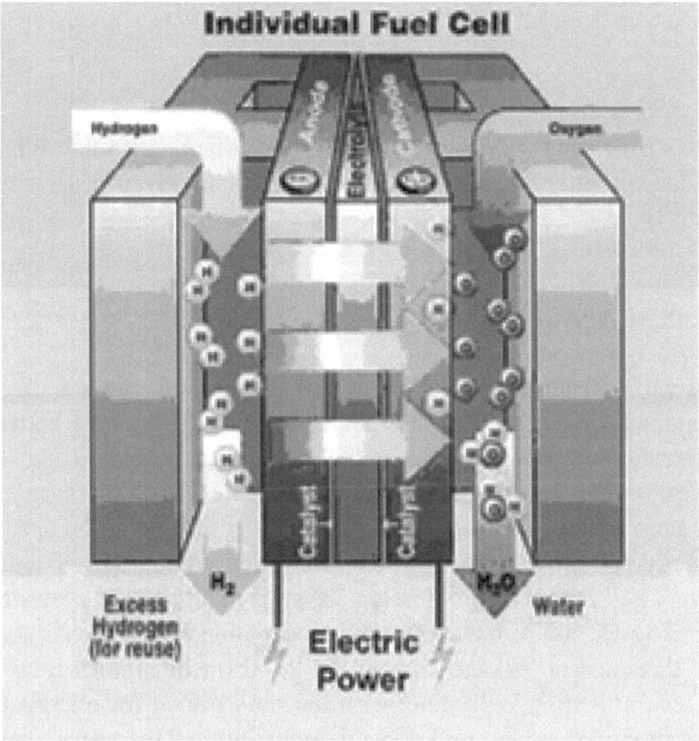

FIGURE 4.2 Schematic diagram of fuel cell.

The main problems associated with PEM are the impurity present in the hydrogen fuel, such as CO, H_2S, NH_3, organic sulfur carbon, and carbon hydrogen compounds, and in air, such as NO_x, SO_x, and small organics, which are brought in air feed and fuel streams into the electrodes of a proton exchange membrane fuel cells (PEMFC) stack, causing degradation of performance or damages of membrane.[7] It is reported that a small amount of these impurity materials could cause poisoning of the anode, cathode, or the membrane of the cell, in turn depreciating the performance. To overcome these issues (especially the one associated with the hydrogen production and storage), a new type of fuel cell, which uses a direct oxidation of methanol instead of hydrogen was developed by a team from University of Southern California's Lokerhydrocarbon Institute. Methanol is also advantageous being liquid at room temperature, thus making storage and transportation issues easier. Using methanol in fuel cells is called DMFC and the working principle is similar to the PEM, but with more complicated reactions at the anode and at the cathode[8]

$$CH_3OH + H_2O \rightarrow CO_2 + 6e^- + 6H$$

$$12H^+ + 6O_2 + 12e^- \rightarrow 6H_2O$$

Methanol has mass densities and energy volume of 5600 Wh/kg and 4380 Wh/l, which are about 11 times higher than current Li-ion batteries (≈500 Wh/l).

4.4 SUPERCAPACITORS

Batteries suffer from a lot number of limitations, like limited life cycles, high manufacturing cost, and low power density. Also, large batteries require several hours for charging. On the other hand, standard capacitors offer high-power density, almost unlimited life cycles, and fast charging rates. With all such advantages, their energy density is too low to be used as primary energy storage systems. Supercapacitors combine the advantages of both battery and capacitor for creating a system that has high-power density, long cyclic time, and at the same time, acceptable energy density. In such systems, the internal leakage current (in the form of dipoles relax and/or charges recombination) will determine the time period for energy storage, while the maximum power will depend on the internal resistance equivalent series resistance (ESR). Table 4.1 shows the advantage and disadvantage of supercapacitors.

TABLE 4.1 Advances and Disadvantages of Supercapacitors.

Advantage	Disadvantage
High-power density	Lower energy density
Long shell life	Higher cost
Fast charge	High self-discharge rate
High efficiency	Low cell voltage
Wide temperature operation range	Linear voltage drop

Figure 4.3 is Ragone plot, which details the specific energy and power capabilities of several energy storage and conversion systems (conventional capacitors, supercapacitors, batteries, and fuel cells). As noticed, no single energy source can match the entire power and energy region. Supercapacitors and batteries, filling up the gap between conventional capacitors and fuel cells, are ideal electrochemical energy-storage systems.[9,10]

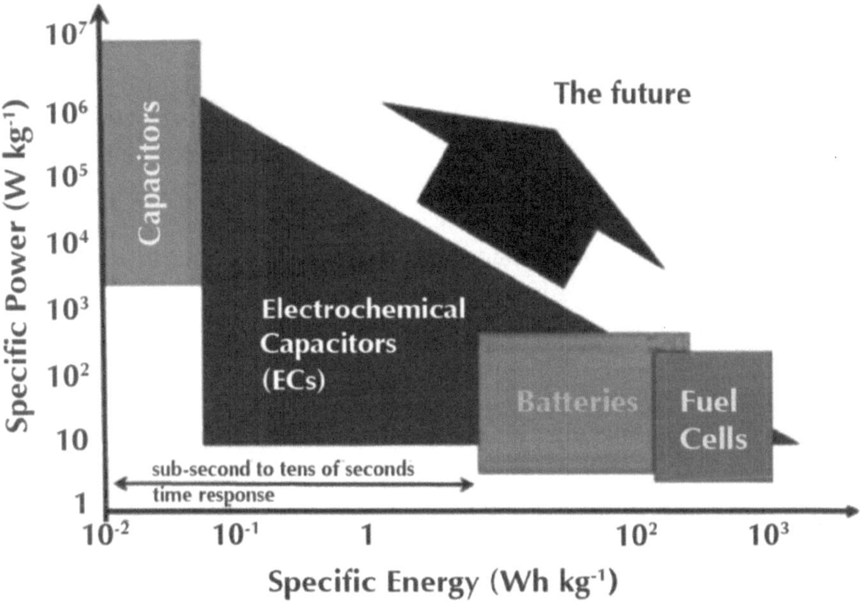

FIGURE 4.3 Ragone plot of specific energy and power capabilities for various energy storage and conversion devices.[11]

A commonality observed in both the systems is that the energy release takes place at the interface of electrode and electrolyte, and that electron and ion transports are separated[12] due to the characteristic differences between batteries and supercapacitors with respect to energy storage mechanism and electrode materials, the characteristic performance of supercapacitors make it unique from batteries. Table 4.2 gives a summary of such differences between batteries and supercapacitors as well as the conventional capacitors (electrolytic capacitors).[13]

In a battery, energy is stored in chemical form, whereas energy is released in an electrical form by connecting a load across the terminals of a battery. The electrochemical reactions of electrode materials with ions in an electrolyte occur, leading to the conversion of chemical energy to electrical energy.[14] Lithium-ion batteries (LIBs) are the most popular rechargeable batteries. A battery mainly consists of an anode, a cathode, an electrolyte, and a separator. When an LIB is cycled, Li-ions exchange between the anode and cathode. The discharge rate and power performance are governed by kinetics and mass transport of active materials. In this aspect, batteries yield high energy densities (150 Wh/kg for LIBs), low power rates, and limited life cycles. The ever-increasing demand for power requirements is a great challenge to the capability of battery design.[15] Conventional capacitors (namely electrolytic capacitors) store energy physically as positive and negative charges on two parallel conductive plates, offering a high-power density but a low energy density (Fig. 4.4). A supercapacitor is nothing but a conventional capacitor with high surface area (A) in its electrodes and less distance (D) between the electrodes with a thinner dielectric. The configuration of a typical supercapacitor consists of a pair of polarizable electrodes with current collectors, a separator, and an electrolyte, similar to that of a battery. The fundamental difference between supercapacitors and batteries lies in the fact that energy is physically stored in a supercapacitor by means of ion adsorption at the electrode/electrolyte interface [namely, electrical double-layer capacitors (EDLCs)]. As a result, the supercapacitor offers the ability to store/release energy in timescales of a few seconds with extended cycle life.

TABLE 4.2 Difference between Batteries, Conventional Capacitors, and Supercapacitors.

Parameters	Electrolytic Capacitor	Supercapacitor	Battery
Storage mechanism	Physical	Physical	Chemical
Charge time	10^{-6}–10^{-3} sec	1–30 sec	1–5 h
Discharge time	10^{-6}–10^{-3} sec	1–30 sec	0.3–3 h
Energy density (Wh/kg)	<0.1	1–10	20–100
Power density (kW/kg)	~10	5–10	0.5–1
Cycle life	Infinite	>500,000	500–2000
Charge stored determinants	Electrode area and dielectric	Microstructure and electrolyte	Active mass and thermodynamics

Supercapacitors offer a higher specific power density than most batteries and a higher energy density than conventional capacitors. Capacitor, known as condenser, is an energy storage device with two electrodes separated by a dielectric material (Fig. 4.5). When a voltage is applied between the two electrodes, static electric field is formed between the dielectric material causing positive and negative charges to get separated between two electrodes thus storing energy. A schematic diagram for a conventional capacitor is given below.

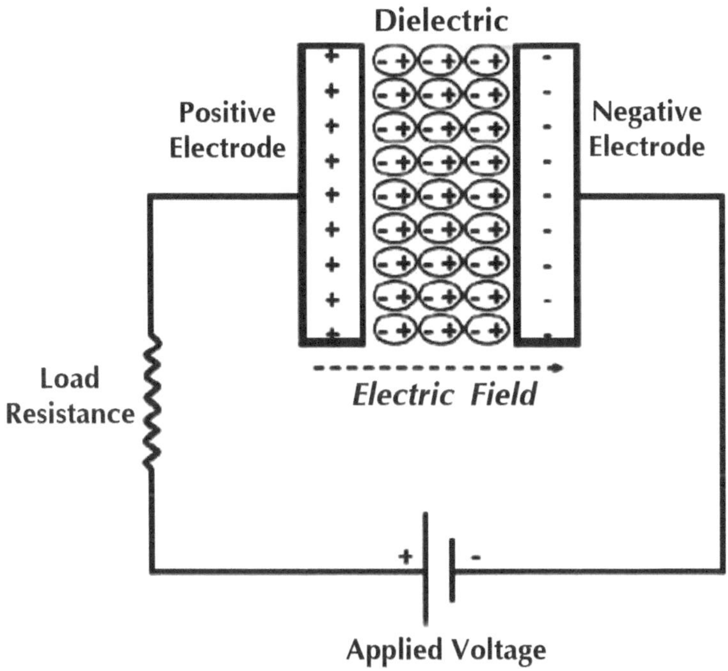

FIGURE 4.4 Schematic diagram of conventional capacitor.

Capacitance for a capacitor is denoted by C and is the amount of charge (Q) stored to the applied voltage (V).

$$C = Q/V$$

This capacitance C is directly proportional to the area (A) of the electrodes and inversely proportional to the distance (D) between the electrodes.

Therefore, $C = \mathcal{E}0\mathcal{E}r\frac{A}{D}$

Here, $\mathcal{E}0$ and $\mathcal{E}r$ are constants of proportionality. $\mathcal{E}0$ is the permittivity or dielectric constant of free space and $\mathcal{E}\mathcal{E}0$ the dielectric constant of the insulating material between the two electrodes. Energy and power are two imperative factors related to batteries and capacitors or any other energy storage devices. Energy is the total amount of work that can be done or the total amount of charge that can be stored by an energy storage device. The amount of charge that can be delivered per unit time is known as power. In other words, work done per unit time is known as power. Energy and power per unit mass of the energy storage device are known as energy density and power density, respectively. Energy (E) stored in a capacitor is directly proportional to its capacitance.

$$E = \tfrac{1}{2}CV^2$$

For calculating the power of a capacitor, it is considered with a series resistance in a circuit. The internal components (current collector, dielectric material, electrodes, etc.) of a capacitor collectively form a series resistance inside a capacitor known as ESR, which has impact on the power of a capacitor. When the capacitor is discharged, its voltage is determined by the "ESR." The relationship between the maximum power (P_{max}) of a capacitor and ESR is as follows:

$$P_{max} = \frac{V^2}{4 \times \text{ESR}}$$

Thus, it could be seen that a greater value of ESR will decrease the maximum power of a capacitor.

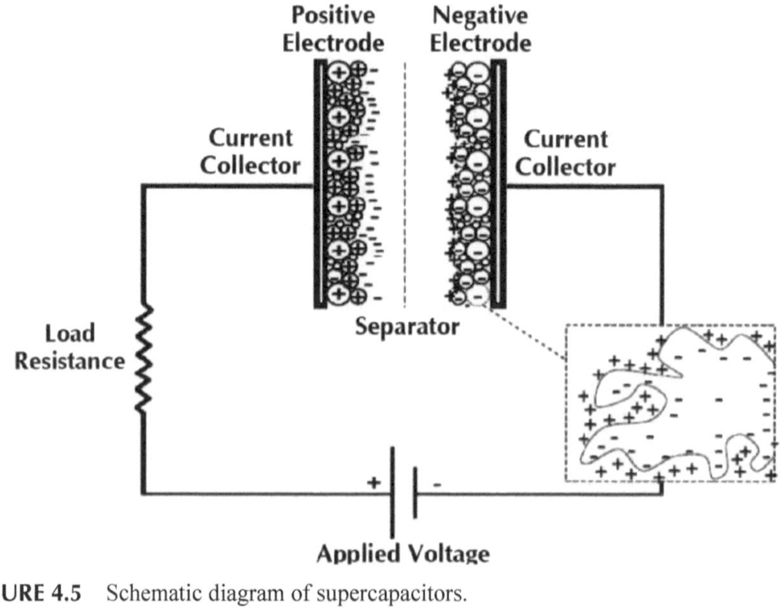

FIGURE 4.5 Schematic diagram of supercapacitors.

Based on the energy storage mechanism, supercapacitors are categorized into two categories. One is the EDLC, where the capacitance is originated from the pure electrostatic charge accumulated at the electrode/electrolyte interface. Hence, it is strongly dependent on the surface area of the electrode materials that is accessible to the electrolyte ions. The second type is called as the pseudocapacitor, in which fast and reversible faradic processes take place due to electroactive species. The former is also called as non-Faradaic supercapacitors because there is no charge transfer occurring between the electrode and electrolyte. The energy storage mechanism is thus similar to standard capacitor, where the area is much larger and the distance is in the atomic range of charges.[16]

4.5 ELECTRIC DOUBLE LAYER CAPACITOR (EDLC)

EDLC consists of two electrodes, one membrane between the two electrodes, which separates the electrodes and electrolyte that can be either aqueous or nonaqueous depending on EDLC.[17] The material of the electrode plays a dominant role for the final supercapacitor performances. For the supercapacitors of modern world, the most common material of the electrodes are activated carbons as it is cheap, has large surface area, and is easy to process. The basic structure of a carbon activated EDLC is shown in Figure 4.6 and the graphical representation of EDLC given in Figure 4.7.

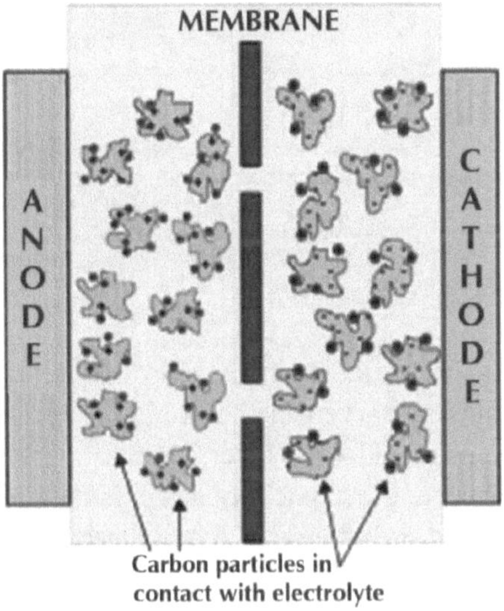

FIGURE 4.6 Graphical representation of active carbon EDLC.

In the case of carbon-based electrodes, the other alternative to activated carbon is carbon nanotube (CNT), which consists of carbon atoms organized in cylindrical nanostructures and can be considered as rolled-up graphene sheets. The roll-up orientation is given by two indexes (*n* and *m*) and is very important in CNTs since different directions render different properties. Among CNTs, both single-walled (SWNTs) and multi-walled (MWNTs) were investigated as EDLC electrodes. Due to their high conductivity and their open shape both SWNTs and MWNTs are particularly suitable for high-power density capacitors.

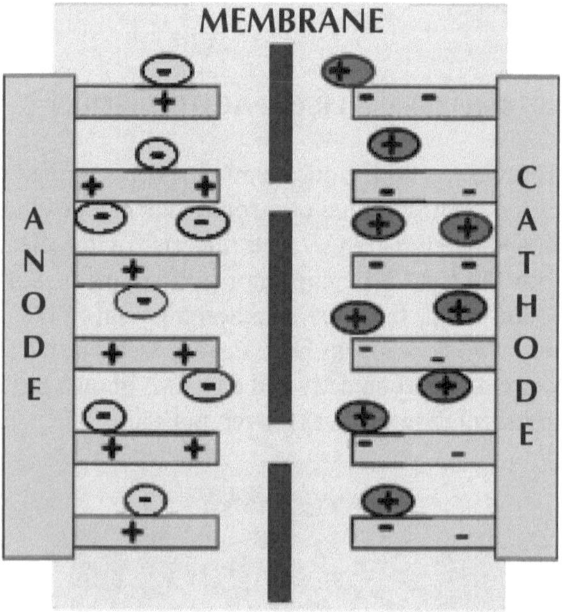

FIGURE 4.7 Graphic representation of carbon nanotube-based EDLC.

4.6 PSEUDOCAPACITOR

Pseudocapacitance is Faradaic in origin, and different from the classical electrostatic capacitance observed in the double layer. In this case, charge will be transferred across the double layer, similar to discharging and charging in a battery, and hence, the capacitance is calculated by using the extent of charge stored (Δq) and the change of the potential (ΔV). The relationship between them can be described by equation:

$$\frac{\partial(\quad)}{\partial(\quad)}$$

Some of the causes of Faradic process in pseudocapacitors are:

i. Rapid and reversible redox reactions between the electrodes and the electrolyte
ii. Surface adsorption of ions from the electrolyte
iii. Doping and undoping of active conducting polymer material of the electrode

Among the above causes, the first two processes are surface mechanism, and hence the capacitance depends on the surface materials of the electrodes, while the last one is more bulk-based technique and hence the capacitance will feebly depend on the surface materials of the electrodes. As these processes are more battery-like rather than capacitor-like, the capacitors are also called as electrochemical pseudo supercapacitors (EPC). When compared with EDLC, EPC have smaller power density since faradic processes are normally slower than that of non-Faradic reactions (Table 4.3). However, EPCs can offer much higher capacitances and thus they store much higher energy density.

TABLE 4.3 Comparison of Double-Layer Capacitance and Pseudocapacitance.

Double-Layer Capacitance	Pseudocapacitance
NonFaradaic	Involves Faradaic process
20–50 F cm^{-2}	2000 F cm^{-2} for single-state process; 200–500 F cm^{-2} for multistate, overlapping Processes
Highly reversible charging/discharging	Quite reversible but has intrinsic electrode-kinetic rate limitation
Has restricted voltage range	Has restricted voltage range

In the supercapacitors research going on, an emphasis is given on investigating two types of materials for achieving high pseudocapacitance viz. metal oxides (MO) and conductive polymers. Among all the MO, ruthenium oxide (RuO_2) has been a subject of investigation due to its intrinsic reversibility for various surface redox couples and high conductivity. In particular, research is focused to explore the chemical reactions of RuO_2 in acid electrolytes. Results showed that the pseudo capacitance of EPC is due to the adsorption of protons at ruthenium oxide surface, combined with a quick and reversible electron transfer,

$$RuO_2 + xH^+ + xe^- \leftrightarrow RuO_{2-x}(OH)_x,$$

where $0 < x < 2$. With such type of electrode materials, a specific capacitance of 700 F/g was achieved. Though high capacitance, the commercial applications of RuO_2-based supercapacitors are limited due to its rarity and high cost. Hence for these reasons, research was deviated to probe other oxides that may provide the same performances with lower costs.

4.7 MATERIALS FOR ELECTRODES OF SUPERCAPACITORS

Supercapacitors are made from a variety of materials, whose selection depends largely on the type of capacitance to be utilized, such as materials utilizing double-layer capacitance, pseudocapacitance, and combination of double-layer and pseudocapacitance. With respect to the electrode materials, there are three main categories: carbon-based, transition MO, and conductive polymers.

4.8 CARBON-BASED MATERIALS

Carbon-based materials in various modifications (carbon aerogels, graphites, CNTs, carbon nanofibers, and nanosized carbon) exhibit a large and a stable double-layer capacitance due to their good conductivity, large specific area, high chemical stability, easy processability, relatively low cost, and wide temperature range of stability. Among the various forms of carbon, activated carbons are the most widely used active materials for EDLC applications, because of their high surface area and relatively low cost. Activated carbons are derived from carbon-rich organic precursors by heat treatment process in an inert atmosphere followed by selective oxidation process in water, CO_2, or KOH, to obtain the high surface area and pore volumes. Specific surface area is a deciding factor for EDLCs in accordance with the storage mechanism. Theoretically, the specific capacitance of carbon materials is proportional to their specific area. Though activated carbons are popular for EDLCs, however, the main issue to ponder is that, not all the surface area is electrochemically accessible when in contact with electrolyte. It is more difficult to adsorb large solvated ions in micropores than in mesopores, and that is more difficult with a higher rate of charge–discharge. In this view, it has become the most imperative task to investigate meso- and macroporous carbon materials to develop high performance EDLC electrode materials. Due to large specific surface area and high electrical conductivity of activated carbon fibers, researchers focused this material for the electrodes of EDLC. The main advantage of activated carbon fiber than activated carbon powders is that they can directly use as active material films in device and do not require any binder addition. Activated carbon fibers can be obtained from polymeric fibers, such as rayon and polyacrylonitrile. Once activated, the surface area is comparable to activated carbons and most of the porosity is developed in the mesoporous range. Kim et al., 2004 developed activated carbon fibers with exposed graphite edges as electrode materials where a large capacitance of 149 F/g was achieved. Xu et al. prepared activated carbon fiber cloth electrodes with high double-layer capacitance and good rate capability from polyacrylonitrile fabrics. However, their use was limited as the costs of these activated carbon fibers are high. Recently, CNTs and graphene have attracted a great zeal because of its good mechanical strength, high electric conductivity, good electrolyte accessibility, and good chemical and thermal stability. CNTs can be conceptualized as seamless hollow tubes rolled up from two-dimensional graphene sheets with diameters in the nanometer range and lengths usually on the micron scale. For purified CNT powders, specific capacitance is not high, typically in the range of 20–80 F/g. This has been mainly attributed to the hydrophobic property of CNT surface. Surface functionalization introduced pseudocapacitance through oxidation treatments, thus leading to significant improvement of the specific capacitive behavior. In a work on the use of acid treated CNT, a specific capacitance of 130 F/g was achieved by Frackowiak.[18] The next popular material is graphene, a two-dimensional carbon material with high surface area (2675 m^2/g), unique mechanical and electronic properties, and high thermal conductivity, and has been regarded as a promising candidate for next-generation supercapacitors.[19] Pristine graphene sheets are grown by methods like mechanical exfoliation of graphite, epitaxial growth, and chemical vapor deposition (CVD).[20] Zhao and coworkers employed a CVD method to prepare carbon nanosheets consisting of 1–7 layers of graphene sheet on conventional carbon fibers and carbon papers.[21] When utilized as supercapacitor electrodes, the outer surfaces of extended graphene sheets were exposed to the electrolyte and thus were available for forming electrical double layers. It was found that such graphene sheets possessed a volume capacitance of 0.076 F/cm^2, which was measured according to the geometric testing area in a H_2SO_4 electrolyte. Despite wide interests and continuously escalating number of publications, applications of pristine graphene sheets have yet to be realized. This is attributed to the problems and difficulties faced in reliable production of high-quality graphene from a scalable approach. To harness the excellent properties of graphene for macroscopic applications, both large-scale synthesis and integration of graphene sheets with single- and few-layer into advanced multifunctional structures are required. Reduced graphene oxide (RGO) can be prepared in large-scale and at a relatively low cost through chemical conversion route. Also, RGO sheets are accessible to different types of electrolyte ions, which enable their wide applications in energy conversion and storage devices.[22] Early studies on the capacitive performance using RGO as electrode materials in supercapacitors rendered specific capacitances of 135, 99, and 75 F/g in aqueous, organic, and ionic liquid electrolytes, respectively.[23] If the whole surface area of 2675 m^2/g is utilized, graphene sheet is capable of storing EDL capacitance of up to 550 F/g. Thus, the reported capacitances are limited by the agglomeration of graphene sheets and do not reflect the intrinsic capacitance of an individual graphene sheet. Recently, a vast improvement has been seen on the use of chemically activated RGO sheets.[24] The three-dimensional (3D) porous carbon network structure had a BET surface area up to 3100 m^2/g, even higher than the theoretical value. A packaged supercapacitor device using such activated RGO sheets as both electrodes displayed an energy density of above 20 Wh/kg in an organic electrolyte, which is much higher than that of AC-based supercapacitors, and nearly equal to that of lead acid batteries. After 10,000 GCD cycles, 97% of its initial specific capacitance was retained. A comparison of various carbon electrode materials for supercapacitors is given in Table 4.4.

4.9 TRANSITION MO

A number of transition MO with variable oxidation states are attractive components as supercapacitor electrode materials due to its exemplary structural flexibility and high specific capacitance. As far as transition MO as electrode materials are considered, energy storage is based on the reversible redox reaction in addition to the electric double-layer storage and thus showing the predominance of pseudocapacitance in charge storage process. Ruthenium oxides, manganese oxides, vanadium pentoxide, nickel oxides, cobalt oxides, etc, have been investigated as electrode materials for supercapacitor applications.[25]

TABLE 4.4 Comparison of Various Carbon Electrode Materials for Supercapacitors.

Carbon	Specific Surface Area (m^2g^{-1})	Density ($g\ cm^{-2}$)	Electrical Conductivity ($S\ cm^{-1}$)	Cost	Aqueous Electrolyte		Organic Electrolyte	
					($F\ g^{-1}$)	($F\ cm^{-3}$)	($F\ g^{-1}$)	($F\ cm^{-3}$)
Fullerene	1100–1400	1.72	10^{-8}–10^{-14}[18]	Medium				
CNTs	120–500	0.6	10^4–10^5[19]	High	50–100	<60	<60	<30
Graphene	2630	>1	10^6	High	100–205	>100–205	80–110	>80–110
Graphite	10	2.26	10	Low				
ACs	1000–3500	0.4–0.7	0.1–1	Low	<200	<80	<100	<50
Templated porous carbon	500–3000	0.5–1	0.3–10[24]	High	120–350	<200	60–140	<100
Activated carbon fiber	1000–3000	0.3–0.8	5–10	Medium	120–370	<150	80–200	<120
Carbon aerogels	400–1000	0.5–0.7	1–10	Low	100–125	<80	<80	<40

a. *Ruthenium oxide-based nanostructures*

The most active form is hydrous ruthenium oxide due to its intrinsic reversibility for various surface redox couples and high conductivity. Ruthenium oxide electrodes with definitely built structure, crystalline phase, particle size, and morphology have been shown to display good capacitive properties.[15] The specific capacitances for hydrous ruthenium oxides in the range of 200–1300 Fg^{-1} have been reported.[26] Previous studies showed that the electrical and capacitive properties of ruthenium oxides were linked to the amount of water in the materials and the degree of crystallinity.[27] A series of ruthenium oxide nanoparticles with different water contents were investigated.[28] Results demonstrated the importance of hydrous regions (either interparticle or interlayer) to allow appreciable protonic conduction for high energy and high-power supercapacitors. Hydrothermal treatment and heating in air are usually used to regulate water content in ruthenium oxides. Results showed that hydrous ruthenium oxides not only lowers proton diffusion resistance but also enhances the electronic conductivity for the redox transitions of active species. This is because the hydrous regions allow facile proton permeation through the materials.[10] The crystallinity nature affects the capacitance property of RuO_2. For instance, the reaction sputtering or vapor deposition technique usually results in ruthenium oxides with good crystallinity. But, their specific capacitances are drastically less as the crystallized structure leads to difficulty in the proton permeation to the bulk materials. In this regard, the obtained capacitance is only due to the surface reaction of ruthenium oxides with good crystallinity. In contrast, amorphous ruthenium oxides obtained from solution procedures allow fast, continuous, and reversible Faradaic reactions, thus yielding appreciable specific capacitances. Hence, the structure has a direct effect on the capacitance of RuO_2 materials. Recently, mesoporous RuO_2 thin films were found to exhibit a specific capacitance of about 1000 F/g at a scan rate of 10 mV/s. The high capacitance might have originated from the intrinsic nature of hydrous ruthenium oxide and the high mesoporosity. Mesoporous ruthenium oxide films with excellent capacitive properties have also been prepared via the evaporation-induced self-assembled method.[28] As a supercapacitor electrode, a nanotubular arrayed $RuO_2 \cdot xH_2O$ electrode prepared by means of an anodic deposition method exhibited a specific capacitance as high as 1300 F/g.[29] The superior electrochemical performance was ascribed to its tailored nanotubular arrayed porous architecture as well as the hydrous nature. The nanotubular structure allowed to easily control the capacity and rate of charge/discharge by controlling the length and thickness of nanotube walls. In addition, the nanotubular structure provided a high surface area and short pathway for ion transport, which could address the problem of laggard charge transfer. The major drawback of ruthenium oxide as supercapacitor electrode is the high cost due to the limited availability of ruthenium. With an aim to lower the cost but improving the utilization, composite materials have been prepared by combining ruthenium oxide with cheap MO, such as TiO_2, SnO_2, and MnO_2.

b. Manganese oxide-based nanostructures

As compared to the expensive RuO_2, less expensive MO in which the metal ions can reversibly transfer among the multiple valencies have been the subject of much interest. One among it are manganese oxides (MnO_x),which is probed due to its low cost, environmental benign nature, natural abundance, and high energy density.[30] In this view, various preparatory routes have been developed to synthesize manganese oxides as supercapacitor electrodes. The pseudocapacitive nature of amorphous manganese oxides were reported earlier in 1999.[31] The manganese oxide was synthesized by a coprecipitation method, in which the redox reaction between potassium permanganate and manganese acetate occurred in water. It was found that the amorphous manganese oxide with a surface area of 303 m^2/g displayed ideally capacitive behavior with a specific capacitance of about 200 F/g. It was after this report that manganese oxides were used as electrode materials for supercapacitor applications. Apart from materials, manganese oxide porous thin films were also greatly used for capacitive applications. It was found that the specific capacitances of the sol–gel-derived MnO_2 thin films depended on the film thickness and calcination temperature. The specific capacitance of these films quickly decreased from 700 to 200 F/g when the film thickness increased from tens of nanometers to several micrometers, indicating that only a thin layer of MnO_2 was involved in the capacitive reaction process.[32] In yet another report, electrodeposition technique was adopted to grow manganese oxide films by using anodic or cathodic modes[11] and such porous films were investigated for supercapacitance properties. Such films were found favorable for the penetration of electrolyte and reactants into the entire electrode matrix, showing better capacitive performances. It was found that the electrodeposition process could be controlled by a variety of experimental variables like voltage or current density, solution concentration, pH value, and temperature.

c. *Nickel oxide*

Nickel oxide is yet another potential material among the MO-based electrode material in pseudocapacitors. The only set back of $Ni(OH)_2$ cathode is that the specific capacitance decreased dramatically with an increase of current density. Zhao et al. (2007) electrodeposited a hexagonal nanoporous $Ni(OH)_2$ film, and reported a maximum specific capacitance of 578 F/g, but its long-term cyclic stability is poor, which is approximately 4.5% loss of specific capacitance after 400 cycles.[33] Even with electrochemically deposited nickel oxide films, the specific capacitance of the sample significantly depended on the applied potential window viz. when the upper limit potential was higher than 0.35 V, the cyclic voltammetry exhibits rectangular mirror image and the specific capacitance is increased due to the additional redox reaction occurring on the surface layer of the NiO grains. However, its long-term cyclic stability was poor, which is approximately 87.5% of its maximum capacitance after 5000 cycles at a current density of 0.49 mA/cm^2.

d. *Conducting polymers*

Conducting polymers (CPs) are the next set of materials probed widely as it can store charge not only in the electrical double layer, but also through the rapid faradic charge transfer (pseudocapacitance). Because of this dual nature, the specific capacitances of CPs are generally higher than EDL capacitors constructed with carbon materials. There are number of reports on the synthesis and electrochemical properties of CPs and their derivatives.[34] Polyaniline (PANi) nanofibers doped with citric acid were prepared by a surfactant-assisted dilute polymerization technique, which exhibited a specific capacitance of 298 F/g.[35] PANI nanofibres were prepared via an interfacial polymerization in the presence of paraphenylenediamine (PPD). The presence of additives resulted in the formation of longer and less entangled PANI nanofibers than those in the absence of PPD. When tested for its capacitance properties, a high specific capacitance of 548 F/g was obtained at a current density of 0.18 A/g.[36] In another report, polypyrrole (PPy) deposited on a Ti foil via cyclic voltammetry was utilized as an electrode material and results exhibited a high specific capacitance of about 480 ± 50 F/g with a good stability in 1 M KCl electrolyte. Poly(3-methyl thiophene) (PMeT) and poly(3,4-ethylenedioxythiophene) (PEDOT) were also investigated as supercapacitor electrodes with calculated specific capacitances of 165–220 F/g for PMeT and about 110 F/g for PEDOT, respectively.[34]

e. *Composite materials*

Apart from the two categories of pure materials (carbon-based material and redox pseudocapacitive materials), much interest is on the use of composite materials combining the above two pure materials for supercapacitors. The approach for such material combination is to utilize the merits of different materials in the composite. Carbon-based materials will enable easy charge transfer and also allow double-layer charge storage with high surface area, which is able to increase the contacting areas between the capacitive materials and electrolyte. With this manipulation, both the energy and power densities of the composite materials are increased.

(i) Carbon nanotube and metal oxide composites

With the above said benefits of the composite materials, numerous composites have been reported. For instance, the dispersion of RuO_2 on a variety of carbon materials, like mesoporous carbon, carbon nanotubes, and graphene is effective to reduce the mass of noble metals and improve the utilization of ruthenium oxides. Hydrous ruthenium oxide nanodots decorated on the CNTs surface functionalized with poly(sodium 4-styrene sulfonate) (PSS) have been utilized for supercapacitors[28] (Fig. 4.8). As shown in Figure 4.8,

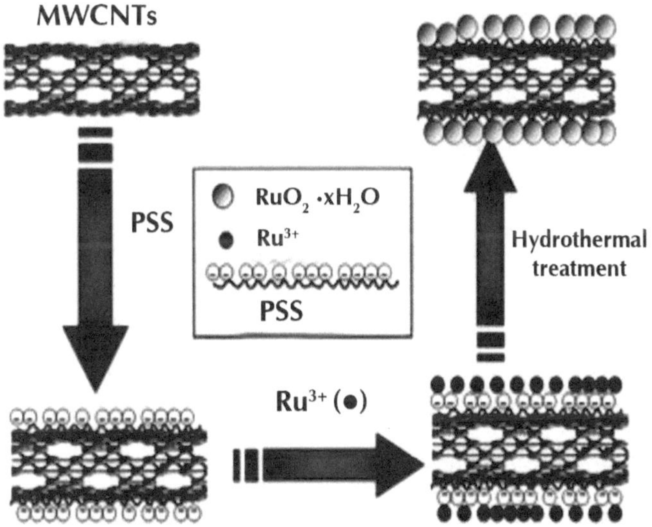

FIGURE 4.8 Formation process of $RuO_2 \cdot xH_2O$/FCNTs nanocomposites—a schematic representation.[28]

the alkyl chain part of PSS molecules lies along the outer surface of the CNTs. The negatively charged sulfonic groups provide electrostatic repulsion to stabilize the aqueous suspension of CNTs, which in turn leads to effective adsorption of metal ion precursors (Ru^{3+}) on the surface of CNTs and facilitates the following deposition of ruthenium species. $RuO_2 \cdot xH_2O$ nucleation takes place on the functionalized carbon nanotubes (FCNTs) surfaces. The as-formed composite gave a specific capacitance as high as 1474 F/g for $RuO_2 \cdot xH_2O$ and electrochemical utilization of 71% for the Ru species were achieved on the composite electrode with 10 wt.% hydrous ruthenium oxide.

Electrical conductivity of manganese oxides is poor (10^{-5}–10^{-6} S cm^{-1}), resulting in less rate capability and reversibility.[37] This possibly explains why the experimentally observed specific capacitance values of MnO_x electrodes are far below the theoretical value. This drawback could be overcome with carbon materials, that is, by making it in the form of a composite. For instance, amorphous MnO_2 on carbon aerogel[38] was developed and such a design exploited the use of a nanoscopic MnO_2–carbon interface to produce a large surface area material with a normalized capacitance of 1.5 F/cm^2, as well as a high volumetric capacitance (90 F/cm^3). CNT was utilized to form a composite material with MnO_2 nanowires leading to a free-standing and flexible electrode.[39] The as-fabricated flexible electrode displayed a specific capacitance of 168 F/g at a current density of 0.077 A/g. It was observed that the fabricated electrode showed good cyclibility even after 3000 cycles with about 88% of its initial capacitance being retained. Manganese oxide nanoflower/carbon nanotube array (CNTA) composite electrodes with a hierarchical porous structure and large surface area were investigated, which showed a high capacitance (305 F/cm^2) and long cycle life (3% capacity loss after 20000 charge/discharge cycles).[22] In yet another study, Au–MnO_2/CNT coaxial arrays were prepared inside porous alumina templates using a combination of electrodeposition, infiltration, and CVD methods (Fig. 4.9). In this porous structure, the conductive and porous CNTA facilitate the electron transport and ion diffusion into the core MnO_2 thus leading to a high utilization of MnO_2. As a result, the hybrid coaxial nanotubes showed good electrochemical performance with a specific capacitance of 68 F/g, a power density of 33 kW/kg, and an energy density of 4.5 Wh/kg.[40]

(ii) Other transition metal oxide and CNT nanostructures

Apart from the above mentioned composite materials, a number of other transition metal oxides like vanadium pentoxide, nickel hydroxide/oxide,

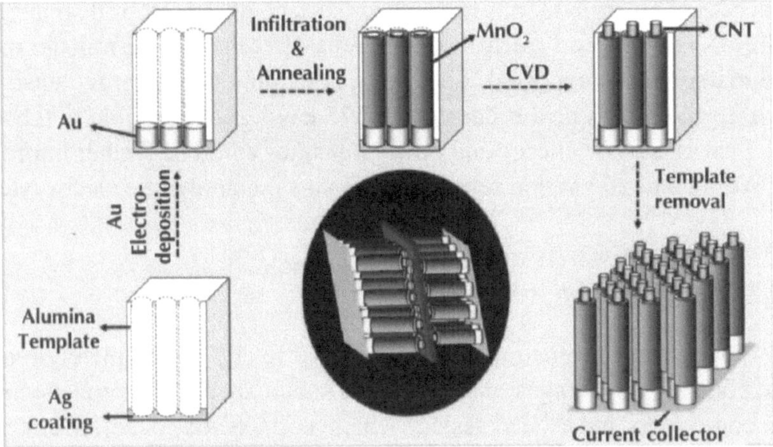

FIGURE 4.9 Schematic diagram showing the fabrication of Au-MnO2/CNT hybrid coaxial nanotube arrays inside an alumina template using a combination of electrodeposition, vacuum infiltration, and CVD techniques.[22]

and cobalt oxides, have been used as supercapacitor electrodes. A layered structure of vanadium pentoxide (V_2O_5) prepared using sol–gel method exhibited a higher specific capacitance of 214 F/g in 2 M KCl electrolyte than in NaCl and LiCl electrolytes.[40] V_2O_5 powders prepared by a coprecipitation method yielded a specific capacitance of 262 F/g.[41] In a recent report, nanocomposites of CNTs and V_2O_5 nanowires were prepared hydrothermally to get a porous and fibrous structure (Fig. 4.10). Such porous structures allowed

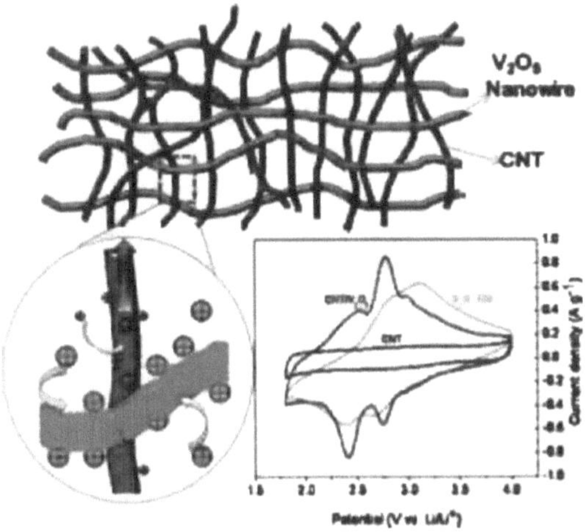

FIGURE 4.10 Hierarchical porous structure based on interpenetrating networks of CNTs and V_2O_5 nanowires, and cyclic voltammograms of CNT, V_2O_5 nanowire, and composite material schematic diagram.

easy access of electrolyte ions to the inner surface of the electrode materials, and the CNTs with good electric conductivity offer fast charge transfer in the active materials. As expected, such devices exhibited an energy density of 16 Wh/kg and a high power density of 3.75 kW/kg, comparable to those of Ni-MH batteries. The energy and power densities could be further improved to 40 Wh/kg and 20 kW/kg, respectively, by using an organic electrolyte.

(iii) Graphene/metal oxide composites

One of the recent intrigued materials is graphene and to completely realize all the potential advantages of graphene in supercapacitors, graphene/metal oxide composites have been tested, which is expected to be an effective material.

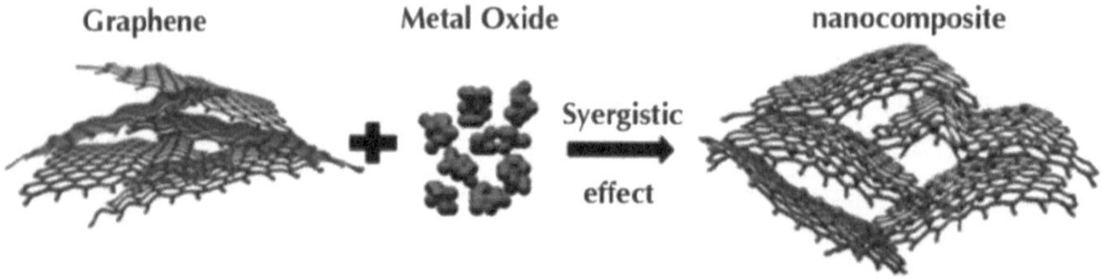

FIGURE 4.11 Schematic of the preparation of graphene/metal oxide composites with synergistic effects between graphene and MO.

The aim is to maximize the practical use of the combined advantages of both graphene and MO as active materials for improving the electrochemical energy storage (Fig. 4.11), and to lower or even solve the current electrode problems of the individual components of graphene or MO as active materials. In the case of graphene composites, graphene provides chemical functionality and compatibility to allow easy processing of MO in the composite. The MO component gives high capacity depending on its structure, size, and crystallinity. The resultant composite from graphene and MO is not just the additive of the individual counterparts, but rather a new material with exciting functionalities and properties. In terms of its structural aspects, MO anchored or dispersed on GNS (Graphene Nanosheets) not only suppress the agglomeration and restacking of GNS, which is shown in Figure 4.12, but also increases the available surface area of the GNS alone, leading to high electrochemical activity. The following are the typical challenges for GNS to be used as electrodes in supercapacitors (Table 4.5).

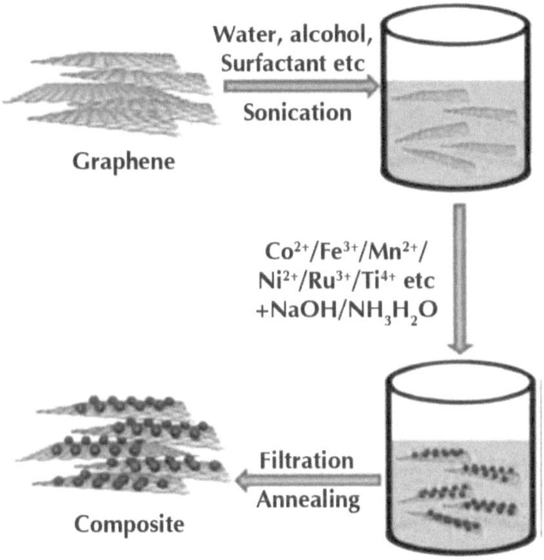

FIGURE 4.12 A general wet chemistry strategy to fabricate graphene/metal oxide composites.

TABLE 4.5 Comparison of Challenges and Advantages of Graphene and Metal Oxides.

Pros of Graphene	Cons of Graphene	Pros of MO	Cons of MO	Pros of Graphene/ MO Composites
Superior electrical conductivity	Serious agglomeration	Very large capacity/ capacitance	Poor electrical conductivity	Synergistic effects
Abundant surface functional groups	Restacking	High packing density	Large volume change	Suppressing the volume change of MO
Thermal and chemical stability	Large irreversible capacity	High energy density	Severe aggregation/ agglomeration	Suppressing agglomeration of MO and restacking of graphene
Large surface area	Low initial coulombic efficiency	Rich resources	Low initial coulombic efficiency	Uniform dispersion of MO and highly conducting and flexible network
High surface-to-volume ratio	Fast capacity fading		Poor rate capability	High capacity/ capacitance, good rate capability

General wet chemistry strategies, like chemical *in situ* deposition, sol–gel processes, and hydrothermal synthesis, are widely used in the fabrication of a broad range of graphene/metal oxide composites starting from a dispersed solution of suspended graphene, which acts as a two-dimensional precursor for an integrated support network for discrete metal nanoparticles.

Ruthenium dioxide (RuO_2) was utilized to synthesize RuO_2/graphene sheet-based composite (ROGSCs) prepared by the sol–gel and low-temperature annealing processes. A maximum SC of 570 Fg^{-1} was achieved for ROGSCs at 38.3 wt.% Ru loading, as compared to 148 F g^{-1} for pure graphene with 97.9% of the initial capacitance being retained and the corresponding energy density and power density was 20.1 Wh kg^{-1} and power density value was 50 Wh kg^{-1} at a current density of 0.1 Ag^{-1}. Graphene composites with ZnO were reported to act as a supercapacitor electrode by Zhang et al, where ZnO was deposited on graphene matrix by ultrasonic spray pyrolysis. Such fabricated electrodes exhibited a specific capacitance of 11.3 Fg^{-1} with excellent charge/discharge stability, wherein the capacitance is due to the contribution of both EDLC from graphene and pseudocapacitance from ZnO.

MnO_2 was also utilized in conjunction with graphene matrix prepared by electro deposition process. The prepared material showed a high specific capacitance of 315 F/g at a scan rate 2 mV/s, high stability upto 95%, power density of 110 kW/kg, and energy density of 12.5 Wh/kg. Cobalt hydroxide-($Co(OH)_2$)based electrode has shown a better performance than the expensive RuO_2 electrode. Hence, graphene oxide/cobalt hydroxide (GO/$Co(OH)_2$) composite was utilized and it showed a very high specific capacitance of 972.5 F/g at a current density of 0.5 A/g. The role of graphene sheets is impeding the agglomeration of $Co(OH)_2$ particles. As a result ,the higher $Co(OH)_2$ feed ratio(below 1/50 for GO/$Co(OH)_2$) results in decreased specific capacitance, because excessive Co^{2+} in solution is aggregated and hinders dispersing $Co(OH)_2$ particles uniformly.

(iv) Graphene/conducting polymer composites

Conducting polymers, such as PANi, PPy, polythiophene, etc., have been widely used as starting materials for the preparation of multifunctional composites, which showed enhanced properties like structural reinforcement and electrical conductivity. Upon the addition of GO into PANi, specific capacitance was found enhanced to 531 F/g. Such enhancement could be attributed to $\pi \rightarrow \pi$ stacking between PANi backbone and GO nanosheets. In another report, on graphene/PANi hybrid material, formed by three steps of in situ polymerization reduction/de-doping–redoping process, a very high specific capacitance of 1126 F/g was observed with a corresponding specific energy density and power density to be 37.9 Wh/kg and 141.1 W/kg, respectively. PPy is yet another interesting polymer and has some advantages for pseudo supercapacitor materials, because such materials could be produced at low cost with high specific capacitance, easy to process in water solution with fast electrochemical switching. GO/PPy composite gave a specific capacitance of 1510 F/g at a scan rate of 10 mV/s in 0.1 M aqueous $LiClO_4$ electrolyte. Graphene-polyethylenedioxythiophene (G-PEDOT) composite was also studied and the material showed a capacitance of 374 F/g at a scan rate of 0.01 A/g.

4.10 CONCLUSION

Overall, this chapter has given an overview of various materials for supercapacitors and among which graphene-based composites are really promising because of its large surface area and the minimal cost of production. Also, with increased development in terms of energy storage limits for supercapacitors in general, graphene-based or hybrid supercapacitors will eventually be utilized in a number of different applications. Such technological growth will come out within the next five to ten years incorporating these supercapacitors. With an escalating growth in this field, there is a possibility for mobile phones and other mobile electronic devices by supercapacitors not because that they can be charged at a higher rate than LIBs, but they can also last longer. Such developments in technology will become more accessible in the future as the efficiency and energy density of supercapacitors increases with a decrease in the cost of manufacture.

KEYWORDS

- **supercapacitors**
- **carbon electrodes**
- **specific capacitance**

BIBLIOGRAPHY

1. Odile Bertoldi, S. B. Report on Energy, Technical Report, European Commission-Observatory Nano. 2009.
2. Petricca, L., Ohlckers, P.; Grinde, C. (2011). Micro-and nano-air vehicles: State of the art. URL: http://dx.doi.org/10.1155/2011/214549
3. Miller, J.R.; Simon, P. Electrochemical Capacitors for Energy Management. *Science* **2008,** *321* (5889), 651–652.
4. Lucintel Growth Opportunities in Global Battery Market 2011–2016: Market Size, Market Share and Forecast Analysis. Lucintel's report.
5. Folkesson, A.; Andersson, C.; Alvfors, M.; Overgaard, L. Real Life Testing of a Hybrid PEM Fuel Cell Bus. *J. Power Sources* **2003,** *118* (1–2), 349–357.
6. Mench, M.M. Fuel Cell Engines, John Wiley and Sons, 2008.
7. Cheng, X.; Shi, Z.; Glass, N.; Zhang, L.; Zhang, J.; Song, D.; Liu, Z.-S.; Wang, H.; Shen, J.A Review of PEM Hydrogen Fuel Cell Contamination: Impacts, Mechanisms, and Mitigation. *J. Power Sources* **2007,** *165* (2), 739–756.
8. Guan, H.; Fan, L.-Z.; Zhang, H.; Qu, X. Polyaniline Nanofibers Obtained by Interfacial Polymerization for High-Rate Supercapacitors. *Electrochim. Acta* **2010,** *56* (2), 964–968.
9. Novak, P.; Muller, K.; Santhanam, K.S.; Hass, O. Electrochemically Active Polymers for Rechargeable Batteries. *Chem. Rev.* **1997,** *97* (1), 207–282.
10. Long, J.W.; Dunn, B.; Rolison, D.R.; White, H.S. Three-Dimensional Battery Architectures. *Chem. Rev.* **2004,** *104* (10), 4463–4492
11. Nakayama, M., Kanaya, T.; Inoue, R. Anodic Deposition of Layered Manganese Oxide into a Colloidal Crystal Template for Electrochemical Supercapacitor. *Electrochem. Commun.* **2007,** *9* (5), 1154–1158.
12. Winter, M.; Brodd, R.J. What Are Batteries, Fuel Cells, and Supercapacitors? *Chem. Rev.* **2004,** *104* (10), 4245–4270.
13. Pandolfo, A.G.; Hollenkamp, A.F. Carbon Properties and their Role in Supercapacitors. *J. Power Sources* **2006,** *157* (1), 11–27.
14. Burke, A. Ultracapacitors: Why, How, and Where is the Technology. *J. Power Sources* **2000,** *91* (1), 37–50
15. Simon, P.; Gogotsi, Y. Materials for Electrochemical Capacitors. *Nat. Mater.* **2008,** *7* (11), 845–854.
16. Zhang, L.L.; Zhao, X.S. Carbon-Based Materials as Supercapacitor Electrodes, *Chem. Soc. Rev.* **2009,** *38* (9), 2520–2531.
17. Jayalakshmi, M.; Balasubramanian, K. Simple Capacitors to Supercapacitors—an Overview. *Int. J. Electrochem. Sci.* **2008,** *3*, 1196–1217.
18. Frackowiak, E.; Jurewicz, K.; Szostak, K.; Delpeux, S.; Béguin, F. Nanotubular Materials as Electrodes for Supercapacitors. *Fuel Process. Technol.* **2002,** *77–78*, 213–219.
19. Geim, A.K.; Novoselov, K.S. The Rise of Graphene. Nat. Mater. **2007,** 6 (3), 183–191.
20. Allen, M.J.; Tung, V.C.; Kaner, R.B. Honeycomb Carbon: A Review of Graphene. *Chem. Rev.* **2010,** *110* (1), 132–145.
21. Zhao, X.; Tian, H.; Zhu, M.; Tian, K.; Wang, J.J. Kang, F.; Outlaw, R.A. Carbon Nanosheets As the Electrode Material in Supercapacitors. *J. Power Sources* **2009,** *194* (2), 1208–1212.

22. Zhang, Y.; Feng, H.; Wu, X.; Wang, L.; Zhang, A; Xia, T.; Dong, H.; Li, X.; Zhang, L. Progress of Electrochemical Capacitor Electrode Materials: A Review. *Int. J. Hydrogen Energ.* **2009,** *34*, (11) 4889–4899.
23. Zhu, Y.; Murali, S.; Stoller, M.D.; Velamakanni, A.; Piner, R.D.; Ruoff, R.S. Microwave Assisted Exfoliation and Reduction of Graphite Oxide for Ultracapacitors. *Carbon* **2010,** *48*, (7) 2118–2122.
24. Zhu, Y., Murali, S.; Stoller, M.D.; Ganesh, K.J.; Cai, W.; Ferreira, P.J.; Pirkle, A.; Wallace, R.M.; Cychosz, K.A.; Thommes, M.; Su, D.; Stach, E.A.; Ruoff, R.S. Carbon-Based Supercapacitors Produced by Activation of Graphene. *Science* **2011,** *332*, (6037) 1537–1541.
25. Xiaohui, S.; Hui, C. Dianzeng, J. Shujuan, B. Wanyong, Z. Meiling, Z. Effective Microwave-Assisted Synthesis of Graphene Nanosheets/NiO Composite for High-Performance Supercapacitors. *New J. Chem.* **2013,** *37*, 439–443.
26. Chang, K.-H.; Hu, C.-C.; Chou, C.-Y. Textural and Pseudocapacitive Characteristics of Sol-Gel derived $RuO_2 xH_2O$: Hydrothermal annealing vs. annealing in air. *Electrochim. Acta* **2009,** *54* (3), 978–983
27. Fu, R.; Ma, Z.; Zheng, J.P. Proton NMR and Dynamic Studies of Hydrous Ruthenium Oxide. *J. Phys. Chem. B* **2002,** *106* (14), 3592–3596.
28. Yuan, C.; Chen, L.; Gao, B. Su, L.; Zhang, X. Synthesis and Utilization of $RuO_2 xH_2O$ Nanodots Well Dispersed on poly(sodium 4-styrene sulfonate) Functionalized Multi-Walled Carbon Nanotubes for Supercapacitors. *J. Mater. Chem.* **2009,** *19* (2), 246–252.
29. Hu, C.C.; Chang, K.H.; Lin, M.C.; Wu, Y.T. Design and Tailoring of the Nanotubular Arrayed Architecture of Hydrous RuO2 for Next Generation Supercapacitors. *Nano Lett.* **2006,** *6* (12), 2690–2695
30. Toupin, M.; Brousse, T.; Bélanger, D. Influence of Microstucture on the Charge Storage Properties of Chemically Synthesized Manganese Dioxide. *Chem. Mater*. **2002,** *14* (9), 3946–3952.
31. Lee, H.Y.; Goodenough, J.B. Supercapacitor Behavior with KCl Electrolyte. *J. Solid State Chem.* **1999,** *144* (1), 220–223.
32. Xu, M.-W.; Zhao, D.-D.; Bao, S.-J.; Li, H.-L. Mesoporous Amorphous MnO_2 as Electrode Material for Supercapacitor. *J. Solid State Electrochem.* **2007,** *11* (8), 1101–1107.
33. Zhao D.D.; Bao, S.J.; Zhou, W.J.; Li, H.L. Preparation of Hexagonal Nanoporous Nickel Hydroxide Film and its Application for Electrochemical Capacitor. *Electrochem. Commun.* **2007,** *9*, 869–874.
34. Snook, G.A.; Kao, P.; Best, A.S. Conducting-Polymer-Based Supercapacitor Devices and Electrodes. *J. Power Sources* **2011,** *196* (1), 1–12.
35. Subramania, A.; Devi, S.L. Polyaniline Nanofibers by Surfactant-Assisted Dilute Polymerization for Supercapacitor Applications. *Polym. Adv. Technol.* **2008,** *19* (7), 725–727.
36. Gurau, B.; Smotkin, E.S. Methanol Crossover in Direct Methanol Fuel Cells: A Link between Power and Energy Density. *J. Power Sources* **2002,** 112, 339–352.
37. Wei, W.; Cui, X.; Chen, W.; Ivey, D.G. Manganese Oxide-Based Materials as Electrochemical Supercapacitor Electrodes. *Chem. Soc. Rev.* **2011,** *40* (3),1697–1721.
38. Fischer, A.E.; Pettigrew, K.A.; Rolison, D.R.; Stroud, R.M.; Long, J.W. Incorporation of Homogeneous, Nanoscale MnO2 within Ultraporous Carbon Structures via

Self-Limiting Electroless Deposition: Implications for Electrochemical Capacitors. *Nano Lett.* **2007,** *7* (2), 281–286.

39. Chou, J.H.; Kim, Y.K.; Chun, J.Y. Determination of Adsorption Isotherms of Hydrogen and Hydroxide at Pt-Ir Alloy Electrode Interfaces Using the Phase-Shift Method and Correlation Constants. *Int. J. Hydrogen Energy* **2008,** *33,* 762–774.
40. Reddy, R.N.; Reddy, R.G. Porous Structured Vanadium Oxide Electrode Material for Electrochemical Capacitors. *J. Power Sources* **2006,** *156* (2), pp.700–704.
41. Lao, Z. J., Konstantinov, K.; Tournaire, Y.; Ng, S.H.; Wang, G.X.; Liu, H.K. Synthesis of Vanadium Pentoxide Powders with Enhanced Surface-Area for Electrochemical Capacitors. *J. Power Sources* **2006,** *162* (2), 1451–1454.

PART III
Nanomaterials as Bioceramics

CHAPTER 5

SILICATE CERAMICS AND ITS COMPOSITES FOR HARD TISSUE APPLICATIONS

RAJAN CHOUDHARY, LAKSHMI RAVI, and SASIKUMAR SWAMIAPPAN*

Materials Chemistry Division, School of Advanced Sciences, VIT University, Vellore, Tamil Nadu 632014, India.
**E-mail: ssasikumar@vit.ac.in*

CONTENTS

ABSTRACT

In recent times, research focus on silicate biomaterials like bioglass, wollastonite, and Ca–Si–M (M = Mg, Zn, Ti, Zr) have become significant for bone tissue repair applications. It is due to the characteristics of silicate biomaterials as their ability to release Si ions at a concentration that stimulates osteoblast growth and differentiation. Recent reports shows the superiority of porous wollastonite and akermanite ($Ca_2MgSi_2O_7$) ceramic scaffolds in terms of material degradation and in inducing in vivo bone formation, compared to ß-tricalcium phosphate (ß-TCP) ceramic scaffolds, suggesting that silicate ceramics have potential application in bone tissue regeneration. The chemical composition of diopside ($CaMgSi_2O_6$) is similar to that of $CaSiO_3$ and akermanite, but it has a relatively slower degradation rate. Dense diopside ($CaMgSi_2O_6$) bulk ceramics were found to have the ability to induce in vitro apatite formation in simulated body fluids (SBF) and in vivo bone formation. Further in vitro and in vivo studies of dense diopside bulks possess a very good bioactivity and excellent bending strength and fracture toughness. Forsterite (Mg_2SiO_4) is less explored when compared to the calcium silicate and diopside but recent investigations shows it possess good biocompatibility and superior mechanical properties.

In this chapter, different techniques employed in the preparation of these compounds and their in vitro biological activity are discussed in detail. In addition, the bioactivity of different composites and influence of morphology of the particle on its bioactivity are also elaborated.

5.1 BIOMATERIAL

Biomaterial is used to replace the natural body tissues by means of implantation of artificial organs or prostheses and the primary criteria for a material to act as a biomaterial is it should be biocompatible and it should not get rejected by the human body.[1] Biomaterials can be classified into two types as natural biomaterials like collagen and synthetic biomaterial, which can be metals, alloys, ceramics, polymers, composites, and hybrid materials. Materials that can be classified as biomaterial includes hydroxyapatite, other calcium phosphates, alumina, zirconia, silica-based glasses, titanium implants, carbon, stainless steel, porcelain dental, cobalt chrome alloy, polymers, hybrid materials, and composites.[2]

Biomaterials play a major role in the health-care sector and it is highly multidisciplinary in nature as it involves scientists to innovate and prepare the material, engineers to design and manufacture the prosthesis, and physicians to implant it in the human body and to study the response of natural tissues on artificial biomaterials. Global market for biomaterial is huge and it attracts the researchers from various fields hence, it remains evergreen among various disciplines of research. Various applications of biomaterial includes joint replacements, bone plates, bone cement, artificial ligaments, dental implants for tooth fixation, blood vessel prostheses, heart valves, skin repair devices, and contact lenses.[2]

Biomaterials employed within the body can fall into three categories as inert biomaterial, bioresorbable biomaterial, and active biomaterial. Inert biomaterial like titanium implants remains unchanged in physiological conditions and will not undergo any chemical changes as well as it will not initiate any biological response on its surface. Bioresorbable biomaterials like tricalcium phosphate will dissolve in the physiological conditions of the human body and it will be replaced with the natural biomaterials over a period of time. Active biomaterials like calcium silicate bioceramics are the one that actively take part in physiological processes and at the same time it will not be replaced by the natural biomaterials.[3,4]

Everyone may have a simple biomaterial in their body and common tooth fillings represent the first generation of biomaterials, but under some situations we have to rely on more critical implants like joint replacements and cardiovascular implants. Although the current materials are performing successfully, a new generation of biomaterials coming up with improved performance may last longer with better adaptation in the human body and prolonged life.

5.2 BONE—A NATURAL COMPOSITE

Human bone is a natural composite made up of 60% mineral component in the form of hydroxyapatite and 30% matrix contains highly aligned type I collagen and remaining 10% is water. Bone has a complex structure with macro- and micropores mostly interconnected to allow body fluid to carry nutrients and provide a medium where interfacial reactions between hard tissue and soft tissue can occur.[5]

Two different types of bone that are present in human body are cortical and cancellous bones. Cortical bone or compact bone forms the outer thick layer whereas the cancellous bone or spongy bone is formed out of trabeculae, bony struts that provide support. Three types of bone cells present in bone tissues are osteoblasts, osteocytes, and osteoclasts. Osteoblasts harden the protein collagen with minerals to form a new bone. Osteocytes maintain the bone by exchanging nutrients and wastes between the blood and bone tissues. Osteoclasts release minerals back into the blood by destroying the bone.[6]

The skeleton made of bone is the storehouse of calcium and phosphorus ions, which is essential for the functioning of other body systems. The maintenance of a constant level of calcium in the blood and adequate supply of calcium and phosphorus in cells is critical for the function of nerves and muscle. Regulatory hormones maintain the adequate supply of these minerals to bone and other tissues, such as intestine and kidney. When these elements are in short supply, the regulating hormones take them out of the bone to serve vital functions in other systems of the body. In addition, bone responds to the changes in mechanical loading or weight bearing. Thus, skeleton behaves like a calcium reserve where the deposition of calcium from blood and withdrawn them later in times of need. When blood calcium levels get low (hypocalcemia), the bones release calcium to bring it to a normal blood level. When blood calcium level exceeds (hypercalcemia), the extra calcium is stored in the bones or excreted out of the body in urine. Excess removal of Ca and P weakens the bone and leads to the bone disorder and fractures.[7]

The impact of bone diseases and trauma in the developed and developing countries has increased significantly in the last few decades. The surgical procedure that replaces missing bone with material from the patient's own body or by an artificial, synthetic, or natural substitute is called bone grafting. Bioceramics used to prepare bone graft should possess a minimum of the following four properties: (1) it should be chemically compatible; (2) it should provide structural integrity in order to keep the graft in place and intact until the patient's own bone heals around it; (3) it should be soluble to certain extent to permit resorption so that the patient's own bone replace the foreign calcium phosphate; and (4) if it is an *in situ* process with the incorporation of biomolecules such as bone growth proteins that can stimulate bone-forming osteoblast cells into the synthetic bone material, then it is desirable that the process used to form the bone material be carried out at low temperatures to prevent the bimolecular from denaturing.[8]

5.3 VARIOUS PHOSPHATE/SILICATE BIOACTIVE BIOCERAMICS USED IN HARD TISSUE APPLICATIONS

Biologically compatible ceramics are called as bioceramics and synthetically prepared bone mineral is called as biomimetic material as it resembles the natural inorganic substance present in the bone. Both are used in hard tissue application as they can actively interacts with the biological environment and chemically integrate and forms direct chemical bond with the surrounding bone tissues. Biomimetic ceramics are considered as the potential material for bone substitute, because they can form a direct bond with the living bone, without the formation of surrounding fibrous tissue.[9] Hydroxyapatite is considered to be the potential candidate among all biomaterials for the hard tissue repair due to its biomimetic nature and high biocompatibility. The requirement of a bioactive ceramics is that in the presence of human physiological environment, it should produce a biomimetic hydroxyapatite layer on its surface, which provides the bonding interface with the tissues as well as bone.[10]

An artificial implant material with the same morphology and the particle size similar to that of natural bone can be a better bone regenerative material and expected to possess more bone-bonding capability than the other bioceramics with different morphology. There are several calcium phosphate ceramics that are considered biocompatible and used as a bone regenerative material. Of these, most are resorbable and will dissolve when exposed to physiological environments.

Some of these materials include, in order of solubility:

> Tetracalcium phosphate ($Ca_4P_2O_9$) > Amorphous calcium phosphate > α-Tricalcium phosphate ($Ca_3(PO_4)_2$) > β-Tricalcium phosphate ($Ca_3(PO_4)_2$) >> Hydroxyapatite ($Ca_{10}(PO_4)_6(OH)_2$).[11]

Synthetic calcium phosphates are either used independently or in combination to develop a bone substitute material. The in vivo performance of calcium phosphate as a bone substitute depends on its chemical composition as well as other physical characteristics, such as structure, specific surface area, porosity, and particle size. Unlike the other calcium phosphates, hydroxyapatite does not break down under physiological conditions and hydroxyapatite is the major component, and an essential ingredient, of normal bone and teeth. Synthetic hydroxyapatite is used for several applications in medicine either as a bulk ceramic, a ceramic coating, or as one of the components of composite.[12–15] It is a promising material as reinforcing filler for composites, insulating agents, and chromato medium for simple, rapid fraction of proteins and nucleic acids and also it finds importance in many industrial applications, such as catalysis, ion exchange, sensor, and bioceramics. However, poor mechanical properties like low fracture toughness mean that hydroxyapatite cannot be used in bulk form for load bearing applications such as orthopedics.[16]

Recently, extensive attention paid on the development of "bioactive materials," which is an alternate to the biomimetic materials and is the thrust area of research in the field of tissue engineering. The bioactive materials like calcium phosphates, Bioglass, Silicate bioceramics, etc. were used in the hard tissue regeneration applications and these materials are having the ability to induce the formation of the hydroxyapatite layer on their surface in in vitro conditions when immersed in SBF solution,[17] which has ion concentrations similar to human blood plasma.[18] Particularly, much attention paid on silicate biomaterials since Si is directly involved in bone growth mineralization and Si has been widely incorporated into calcium phosphates bioceramics in order to enhance their bioactivity. Silicon-doped calcium phosphates are showing improved bioactivity compared to calcium phosphates.[19,20]

Alternatively, calcium silicate is considered as an ideal candidate for the artificial bone implants in hard tissue regeneration and bone tissue engineering, which is also widely used in various medical applications, such as implants in clinical bone repair and regeneration materials, clinical tissue regeneration and tissue engineering, bioactive coating of metallic implants, and also for drug delivery applications. The formation rate of hydroxyapatite on the surface of the calcium silicate is proved to be faster than the any other biocompatible glasses, glass ceramics, and calcium phosphates, such as hydroxyapatite due to the presence of silicate group, which actively takes part in the metabolic processes during bone formation.[21,22]

Silicon deficiency leads to the abnormal bone formation and during the bone formation osteoblastic proliferation is enhanced due to the presence of silicon ions. Hence, silicon is found to be an essential trace element for the bone cell activity.[23] Pseudowollastonite was reported to possess good osteoblastic formation, which shows that the silicon ion present in the enhance the bone cell proliferation but Si bonded with calcium is proved to have enhanced bioactivity as the ionic dissolution plays a major role in the bond formation between the implant and surrounding bone tissue. Wollastonite is expected to induce osteogenesis by satisfying all this criteria; hence, it can be used as an artificial bone implant material.[24–26]

Calcium silicate is a chain-silicate mineral and its structure is similar to that of ABO_3 pervoskite, consists of a network of covalently bonded silica that is interrupted and modified by Ca^{2+} cations. The weakly bonded network modified Ca^{2+} ion is released into the solution when exposed to the simulated body fluid as it get exchanged for hydrogen ions, resulting in the formation of Si—OH, which is the primary reason for its bioactivity. The advantage of using calcium silicate materials in hard tissue regeneration is the unique characteristics and structure of the surface, including the stability of the stable ordered pore network, high pore volume, surface area, adjustable pore size, and easily functionalized surface modification of specific locations.[27,28]

In the development of silicate biomaterials, larnite (Ca_2SiO_4), hatrurite (Ca_3SiO_5), sodium calcium silicate (Na_2CaSiO_4), akermanite ($Ca_2MgSi_2O_7$), diopside ($CaMgSi_2O_6$), and bredigite ($Ca_7MgSi_4O_{16}$) were found to be bioactive materials. In the recent years, much attention was paid on the calcium silicate ceramic materials particularly wollastonite ($CaSiO_3$), due to its higher in vitro bioactivity and biocompatibility. Further, many composites like titania–wollastonite,[25] hydroxyapatite-wollastonite,[29] and polycaprolactone-wollastonite composites[30] were developed for biomedical applications.

Wollastonite ($CaSiO_3$) is a calcium silicate received considerable attention due to its excellent bone-bonding ability than the hydroxyapatite and tricalcium phosphate. A significant characteristic of wollastonite is its ability to bond with the living bone through the formation of an apatite interface layer. In the recent years, wollastonite has received much attention due to its higher bioactivity than any of the other biocompatible glass and glass ceramics. The bioactivity of wollastonite was analyzed by the rate of hydroxyapatite formation on its surface in presence of simulated body fluid (SBF) and the bioactivity of wollastonite is due the silicon ion present in it as it plays an essential role in the formation of new bone by metabolic process.[21,22,31–34]

5.4 PREPARATION OF CALCIUM SILICATE BIOCERAMICS BY VARIOUS TECHNIQUES

Various methods employed to prepare the bioceramics are wet chemical methods (precipitation), solid-state reactions, combustion method, sol–gel, hydrothermal techniques, mechanochemical synthesis, flash pyrolysis, Dual phase mixing, microwave synthesis, electrochemical synthesis, and sonochemical synthesis. The conventional method of synthesis like solid-state reaction requires high energy and needs longer duration for the formation of products. Generally, calcium silicates are synthesized at very high temperature which is in the range of 1350–1500°C.[35,36]

The preparation of wollastonite was reported for, sol–gel combustion method,[22,31–33,48] hydrothermal method,[37] sol–gel method,[38] solution combustion method,[39] coprecipitation,[35] citrate–nitrate gel combustion method,[40] and microwave synthesis.[41] Moreover, wollastonite is also synthesized by biowaste source such as eggshell as calcium source.[32] The preparation procedure plays a vital role in the bioactivity of the calcium silicates as the sol–gel-synthesized wollastonite is found to have higher bioactivity than the wollastonite prepared by solid state method. However, the agglomerates produced were homogeneous in sol–gel synthesis than the products synthesized by any of the above mentioned methods.[25]

The phase transformation temperature of the solution combustion synthesized wollastonite powder was found to be lower than the powders obtained by solid-state method.[39] Wollastonite synthesized by microwave method by using biowaste and eggshell as a source of calcium is found to possess a good bioactivity.[41] Wollastonite synthesized by the pyrolysis of chicken eggshells followed by sol–gel process and the product was reported to possess dielectric property and it is used in the field of biomedical engineering.[42]

5.5 BIOACTIVITY OF PURE CALCIUM SILICATE BIOCERAMICS

5.5.1 BIOACTIVITY OF BULK WOLLASTONITE

The in vitro bioactivity of the wollastonite scaffold is studied by placing the scaffold in the simulated body fluid medium at 37°C in an incubator with the change of SBF after every 24 h. Periodically, the phase composition of the scaffold surface will be examined by the XRD in order to check the deposition of hydroxyapatite. When $CaSiO_3$ scaffold immersed in SBF, calcium and silicon ions gets released from its structure and reacts with the other constituents present in the simulated body fluid. Calcium ions released from the surface of wollastonite coating is found to increase the ionic activity of the apatite formation in the SBF leading to the formation of hydroxyapatite on the coated surface.[43,44]

When a bioceramic scaffold is placed in the simulated body fluid, one side of the scaffold will rest on the surface of the flask hence, it will have limited exposure to the simulated body fluid, whereas the other side of the scaffold will be fully exposed to the body fluid. Amount of simulated body fluid present on the surface of the wollastonite scaffold plays a major role in the biomineralization process.[33] The surface of the wollastonite scaffold was analyzed by the end of 5th and 10th day after bioactivity studies.

The XRD pattern (Figs. 5.1 and 5.2) of side 1 shows more deposition when compared with side 2, which may be due to the more circulation of SBF on side 1 as it is exposed to SBF. Side 2 is kept over the glass beaker hence, the circulation of SBF is very poor and as a result the deposition of hydroxyapatite is found to be very less. The scaffold was kept for 5 more days by reversing the sides, the analysis on 15th day shows the deposition of apatite layer on both the sides, which indicates that the circulation of SBF on the surface of the scaffold favors the deposition of hydroxyapatite layer.

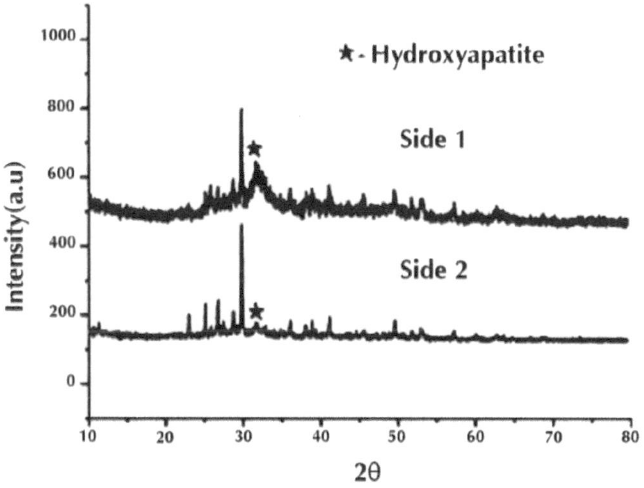

FIGURE 5.1 XRD pattern of wollastonite pellets immersed in SBF for 5 days.[33]

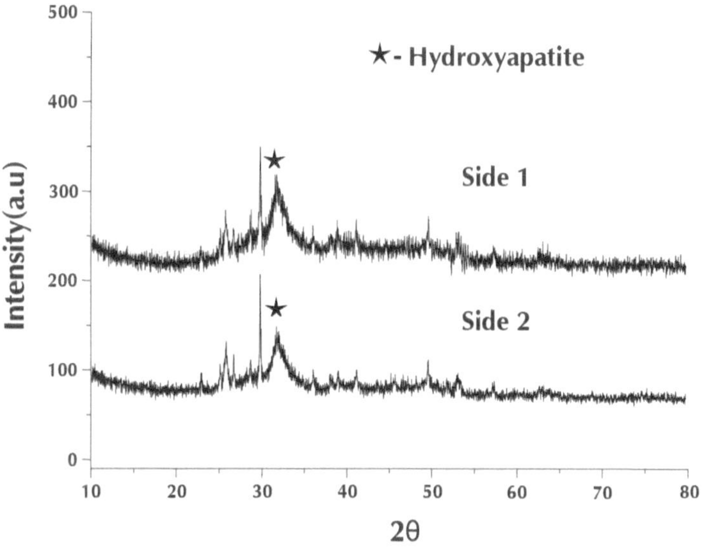

FIGURE 5.2 XRD pattern of wollastonite pellets immersed in SBF for 10 days.[33]

5.5.2 BIOACTIVITY OF POROUS NANOCRYSTALLINE WOLLASTONITE

Surface morphology and particle size of the bioceramic materials greatly affect the in vitro bioactivity characteristics in the SBF solution. Particularly, role of porous materials in the bone tissue engineering is a promising research and these porous materials play major role in bone growth by influencing the migration, proliferation, and differentiation of the cells.[45]

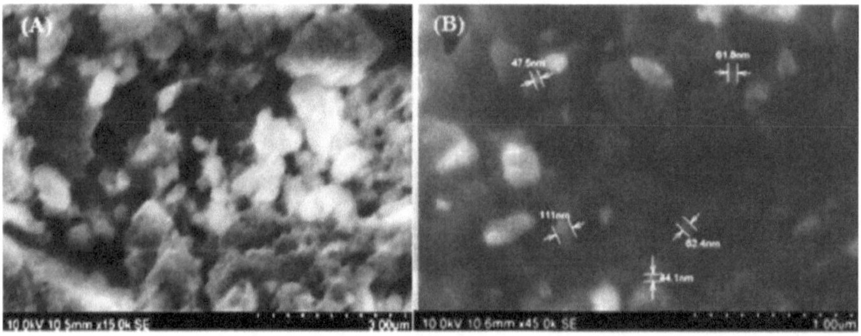

FIGURE 5.3 Scanning electron microscopy (SEM) images of the calcined wollastonite powder.[22]

The SEM images (Fig. 5.3) of the nanocrystalline wollastonite powders used for bioactivity studies shows that the particles are highly agglomerated and porous in nature. Further, SEM images (Fig. 5.3b) reveal that the crystallite size of wollastonite powder is in the range of 44–111 nm and the particles exhibits flakes like morphology. Further, the porous materials have the advantage to maintain the ionic equilibrium, when the ceramic scaffold is immersed in the SBF solution.[22]

Figure 5.4 shows the topography of the scaffold analyzed by AFM before the nucleation of HAP and after the nucleation of HAP. The as-prepared scaffolds average roughness were investigated by using AFM are found to be 183.89 nm. High roughness might be due to uneven surface morphology, coarser particles, and porous nature. Thus, porous nature may influence its biological behavior in SBF by allowing the body fluid to flow inside the pores and enhance its hydroxyapatite induction on the surface.[31]

XRD pattern (Fig. 5.5) of the scaffolds after bioactivity studies reveal the formation of hydroxyapatite on the surface of wollastonite at the end of 5th and 10th day of incubation. All the XRD patterns indicate good deposition of hydroxyapatite layer on both the sides, which conforms good bioactivity of the nanocrystalline wollastonite scaffolds. The XRD pattern (Fig. 5.5) clearly reveals that the bioactivity is found to be uniform on both the sides in terms of hydroxyapatite deposition, which contradict with earlier results as it indicates that the circulation of SBF plays major role in the deposition of hydroxyapatite. This may be due to the porous structure of the wollastonite, which influence on the dissolution and precipitation processes.[22]

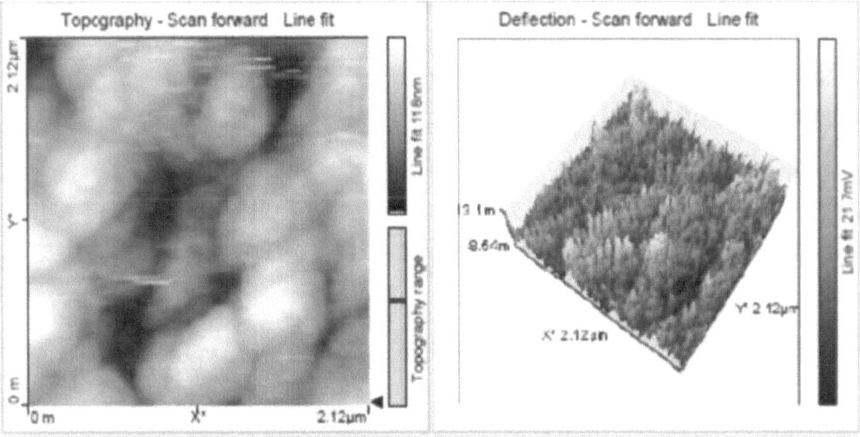

FIGURE 5.4 Topography of the scaffold analyzed by AFM.[31]

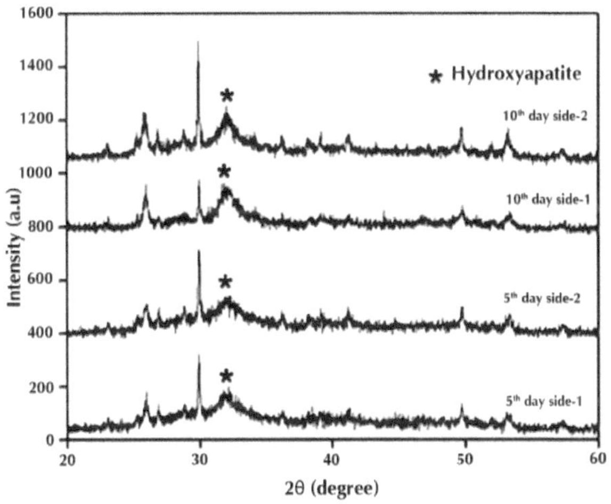

FIGURE 5.5 XRD pattern of the wollastonite pellets immersed in simulated body fluid.[22]

When the scaffold is immersed in the physiological solution, dissolution process will occur. In this dissolution process, the ions released from the scaffold will influence the pH and ionic concentrations at SBF/scaffold interface. Therefore, the scaffold surface exposed to the SBF will have higher ionic concentration than the bottom side. As a result, the higher hydroxyapatite precipitation on the top side of the scaffold is expected to be high when compared with the bottom side. In the present system, the pores present in the wollastonite particles might have induced the homogeneous distribution of the ionic concentration, which results in uniform deposition of hydroxyapatite on both the sides of the scaffold. As a result, almost there is no difference in the XRD pattern of the scaffold recorded on 5th day and 10th day of incubation.[22]

5.5.3 INFLUENCE OF MORPHOLOGY OF THE PARTICLES ON THE BIOACTIVITY OF BULK WOLLASTONITE

Scale factor plays an important role in the rapid deposition of hydroxyl apatite layer. The kinetics of apatite nucleation of the synthesized sample depends on the scale factor of the compound. If the compound possesses nanocrystalline structure then it will have maximum number of nucleation sites for the deposition of apatite crystals. This happens because of the high surface energy of the grain boundary, leads to the rapid precipitation of apatite crystals.[46,47] The needle-like morphology provides high surface area hence, it possesses higher bioactivity.

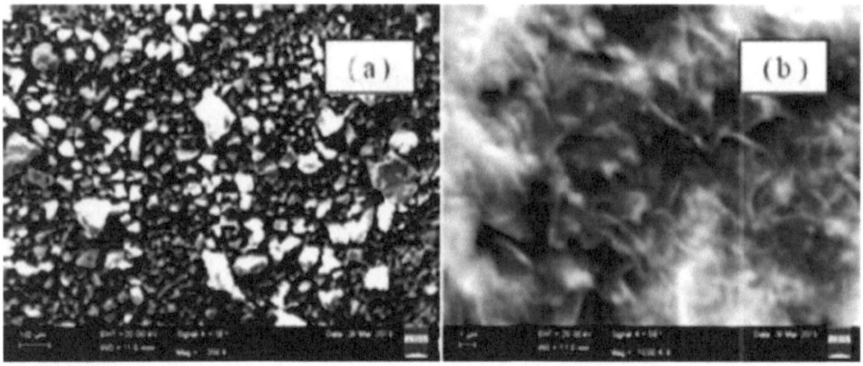

FIGURE 5.6 SEM images of the wollastonite powder prepared using urea as a fuel by sol–gel combustion method.[48]

The SEM images (Fig. 5.6) of the wollastonite powder prepared by using urea as a fuel by sol–gel combustion method shows the particles are heterogeneous in nature and the particle size is found to be in the range of 50–100 μm. At higher magnification (Fig. 5.6b) it is observed that the crystallites are with needle like morphology and highly agglomerated.

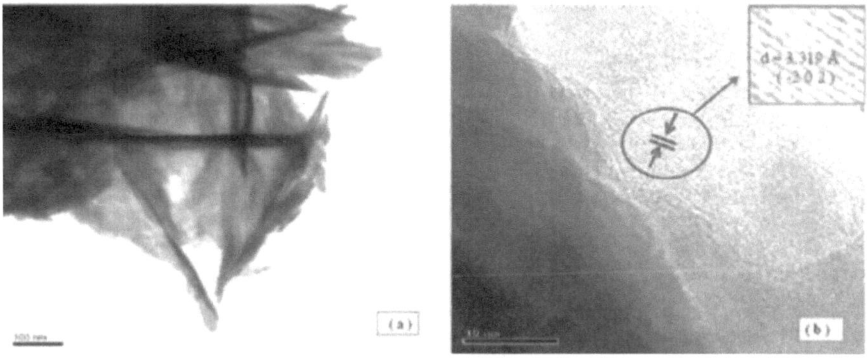

FIGURE 5.7 Transmission electron microscopy images of the wollastonite powder prepared using urea as a fuel by sol–gel combustion method.[48]

TEM micrographs (Fig. 5.7a) and HRTEM (Fig. 5.7b) of the as synthesized wollastonite sample shows needle-like morphology with a mean particle size of 25–28 nm in diameter. The average length of the particle falls in the range of 0.2–0.5 microns. The interplanar distance taken from the zone axis along the direction [−2 0 2] was measured to be 3.319 Å.

The primary factor required for the bioactive surface is when it is immersed in SBF expected to nucleate the hydroxyapatite on its surface should have the chemical composition similar to that of human bone's inorganic constituent. If the ceramic compound possesses good in vitro bioactivity, then it is assumed to have good bone bonding capability during implantation.

The sintered pellet placed in SBF for 21 days was analyzed by powder XRD for bioactivity shows the deposition of hydroxyapatite. The XRD pattern (Fig. 5.8) of the wollastonite pellet after 21 days shows that intensity of the diffraction peaks corresponding to wollastonite becomes less intense. XRD pattern (Fig. 5.8b) of the surface taken after soaking in SBF for 14 days shows the intensity of HAP is more prominent than the wollastonite peaks. After 21 days of immersion period, almost the entire surface of the wollastonite pellet is covered by the crystalline apatite layer and the intensity of the wollastonite peaks appears to be very less and the intensity of HAP peaks (Fig. 5.8c) are very high.[48]

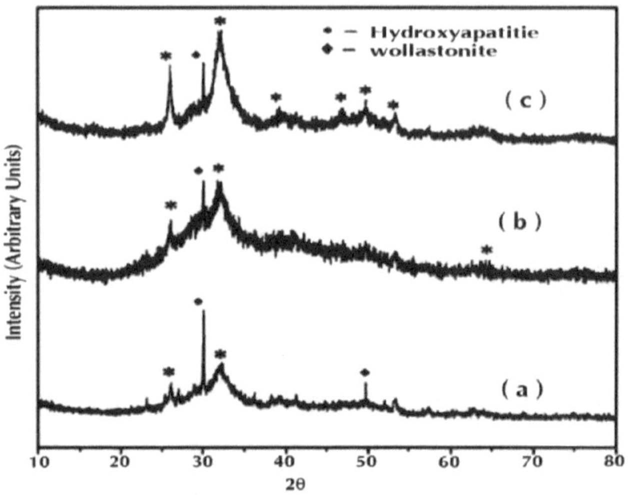

FIGURE 5.8 XRD pattern of the wollastonite sample with needle-like morphology immersed in simulated body fluid.[48]

5.5.4 TOPOGRAPHY OF THE WOLLASTONITE SURFACE AFTER BIOACTIVITY STUDIES

Wollastonite was prepared by sol–gel combustion synthesis by using glycine as a fuel and made as a scaffold with the dimension of 13 mm diameter at 2 mm thickness by using hydraulic pellet press. The scaffold was kept in 30 mL simulated body fluid for the time period of 21 days in the incubator at the temperature of 37°C.

The XRD pattern (Fig. 5.9) of the scaffold immersed in SBF for various period shows major diffraction peaks of 2θ peaks at the angles 26.28°, 32.41°, 50.05°, and 53.59° due to the nucleation hydroxyapatite. The diffraction pattern was exactly matches with the standard hydroxyapatite JCPDS data (Card 01-074-0566). At the end of 21st day, the scaffold immersed in the simulated body fluid shows completely covered layer of well crystalline hydroxyapatite phase on the surface of wollastonite, which is evident from the absence of wollastonite peaks in the XRD pattern (Fig. 5.9e). This reveals that the wollastonite is highly suitable for bone regenerative applications.

SEM images of wollastonite scaffolds before and after soaking in the SBF for 21 days (Fig. 5.10) shows change in morphology of the surface. The SEM image (Fig. 5.10a) of the synthesized wollastonite possesses flakes like morphology and porous in nature. The average width of the flakes ranges from 50 to 60 nm. The change in surface morphology (Fig. 5.10b) of the

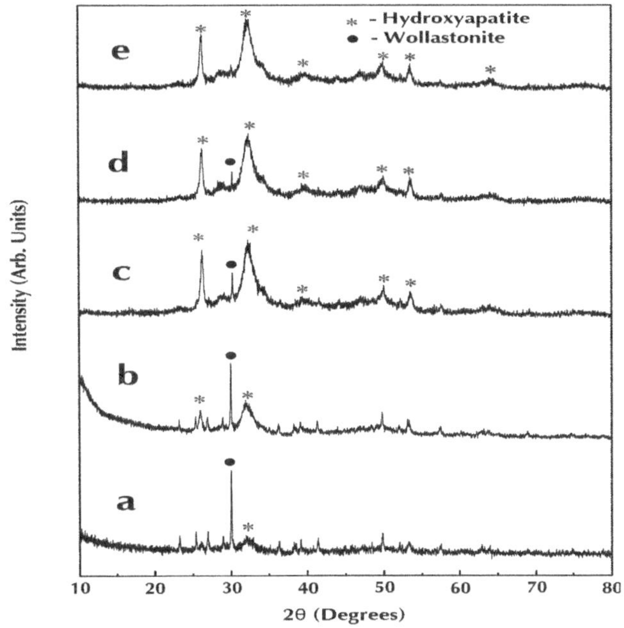

FIGURE 5.9 XRD pattern of the wollastonite scaffold surface after bioactivity studies.[33]

scaffold is observed after 21 days due to the nucleation of hydroxyapatite on the surface of the scaffold. The upper surface of the scaffold is completely covered by the continuous and dense layer of hydroxyapatite. The morphology of the nucleated hydroxyapatite was irregular spheroids.

FIGURE 5.10 SEM images of wollastonite scaffold surface before and after bioactivity studies.[33]

5.6 DIOPSIDE

Diopside ($CaMgSi_2O_6$) belongs to a group of silicate biomaterials was discovered in 1806. The term "diopside" derived from a Greek *dis* and *opse,* which means double appearance[49] implies as reference to two possible orientations of vertical prism.[50] Often deposits naturally in the regions of India, China, Myanmar, South Africa, Italy, and Russia, etc.[49] as a pale to dark green crystals.[51] It is a monoclinic structured pyroxene mineral with high melting point (1391°C). The lattice parameters of diopside are as follows:

$$A = 9.75000\ \text{Å};\ b = 8.92600\ \text{Å};\ c = 5.25100\ \text{Å};\ \alpha = 90.000;\ \beta = 105.900;\ \Upsilon = 90.000.$$[52]

5.6.1 *STRUCTURE OF DIOPSIDE*

The coordination structure of diopside constitutes silicon atom surrounded by four oxygen atoms forming tetrahedron network. The tetrahedron is connected by common oxygen atom in the 3:1 ratio of oxygen atom to silicon atom to form single infinite chain parallel to c crystallographic axis of the crystal, lying side by side and held together by calcium and magnesium atoms. These parallel chains provide high strength to the diopside.[53] Calcium fills M2 site with 8 coordinations, while magnesium octahedrally coordinated at M1 site.[52] Figure 5.1 shows diopside structure of single chain linked with SiO_4 tetrahedron.

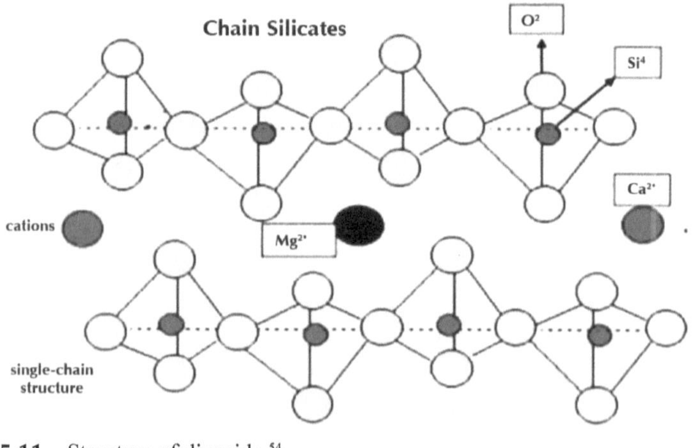

FIGURE 5.11 Structure of diopside.[54]

Hench et al. in 1970 projected his focus on biologically active glass, Bio-glass[55], representing a group of surface reactive materials, which form direct bonding with tissues or bones through specific biological response of tissues toward the interface of the materials and capable of bonding to bone by biological fixation.[56] Important characteristic of biological fixation is that chemical bonding occurs across the interface between an implant and the tissue with strength equal to or greater than that of the bone. Silicate bioceramics are preferred over calcium phosphate bioceramics as they are capable of withstanding more stress[56] and ability to release silicate ions at a definite concentration, which promotes osteoblasts growth and proliferation.[57] It has been already reported that growth and nucleation of hydroxyapatite layer deposition on silicate bioceramic and composites surface is influenced by presence of silicon.[58,59] So, today researchers are concentrated on the silicate bioceramics as candidates for repairing damaged tissues, drug delivery, and biomedical applications. In the search of potential silicate bioceramics wollastonite, merwinite, akermanite, diopside, bredigite, and many other silicate and composites were found to be bioactive ceramics.

The currently existing bone substitutes do not match the required standards for implants particularly lacking adequate mechanical strength and quick degradation rate.[60] Among all above mentioned silicate bioceramics, diopside possesses lower degradation rate, improved mechanical property, and excellent bioactivity.[57] The reason for the improved compressive strength and mechanical stability of diopside over other silicate bioceramics is the significant difference in their composition. The major cause for low degradation of diopside is due to magnesium. The bond energy of Mg—O is comparatively higher than Ca—O bond, which inhibits the release of calcium, magnesium from crystal lattice as a result solubility of diopside decreases.[61] Degradation of diopside can be controlled by varying magnesium content.[57] Concerned to mechanical stability studies of bioceramics implants bending strength (40–45 MPa) is quite lower than that of bone (50–150 MPa) proves to be mechanically not compatible[62] but diopside processes superior mechanical property with fracture toughness 3.5 MPa $m^{1/2}$ and bending strength of 300 MPa double to that of bone.[63]

5.6.2 SYNTHESIS OF DIOPSIDE

Several conventional methods have been employed for the synthesis of pure silicate bioceramics using different starting materials. Some common methods used for the synthesis of pure diopside are discussed in Table 5.1.

TABLE 5.1 Preparation Techniques for Diopside.

Synthesis methods	Starting materials	Synthetic conditions	Comments
Coprecipitation method	$Ca(NO_3)_2.4H_2O$, $Mg(NO_3)_2.6H_2O$, TEOS	Sintered at 1300°C for 2 h	Diopside phase formation achieved[57]
Sintering method	$CaCO_3$, MgO, SiO_2	Mixture burnt at 1100°C for 2 h and sintered at 1300°C 2 h	Highly crystalline diopside peaks were observed[63]
Solid solution method	$CaCO_3$, MgO, SiO_2	Mixture burnt at 1100°C 2 h and sintered at 1300°C	XRD pattern confirms diopside formation[64]
Hydrothermal method	CaO, MgO, Dolomite lime, silica gel (SiO_2 nH_2O)	Calcination at higher temperature (900–1350°C)	The diffractograms of the synthesized diopside are indexed and lattice parameters were calculated using computer program[65]
Sol–gel method	$Ca(NO_3)_2.4H_2O$, $MgCl_2.6H_2O$, Ethanol (solvent), TEOS	Calcined at 700°C for 2 h and sintered at 1100°C for 2 h	Akermanite and Monticellite were observed as primary phase at 700°C Single phasic diopside was obtained at 1100°C. Crystallization temperature of dried gel powder into diopside was observed at 751.4°C[66]
Sol–gel method	$Ca(NO_3)_2.4H_2O$, $MgCl_2.6H_2O$, Ethanol, TEOS	Calcined at 700°C for 2 h and 1100°C for 24 h	No characteristic peaks were observed at 700°C. XRD pattern shows presence of intense diopside peaks as major phase at 1100°C. Nano diopside with particle sizes 35–65 nm. Applicable for bone filler or hybrid composite in bone tissue regeneration[67]

TABLE 5.1 *(Continued)*

Synthesis methods	Starting materials	Synthetic conditions	Comments
Sol–gel Combustion Method	Eggshell as calcium source, $Mg(NO_3)_2.6H_2O$, TEOS, L-alanine, Conc. Nitric acid	Combusted at 400°C 2 h Calcined at 850°C 6 h	XRD data reveals the formation of single phasic diopside at low temperature Suggesting the conversion of biowaste into some useful biomaterials[68]
Sol–gel Combustion Method	$Ca(NO_3)_2.4H_2O$, $Mg(NO_3)_2.6H_2O$, TEOS, Citric acid, Succinic acid, Conc. Nitric acid	Combusted at 400°C for 30 min Calcined at 900°C and 1000°C 6h	At 900°C diopside as major phase and akermanite as a secondary phase. At 1000°C shows reduced content of akermanite peaks[69]

5.6.3 BIOLOGICAL STUDIES

Bioactivity studies of silicate bioceramics testifies by formation of hydroxyapatite (HAp) layer on the surface of sintered body, which arises with the passage of socking time, when investigated by socking in simulated body fluid (SBF).[70–73] Simulated body fluid having inorganic ion concentrations similar to those in human blood plasma thus, serve as our body environment by maintaining all internal conditions required to form apatite layer.[74,75] The bones present in mammalian bodies are natural composite materials with a complex structure and are composed of 70% calcium phosphate (hydroxyapatite) and 30% protein (collagen) by volume. Silicate bioceramics has the ability to mimic hydroxyapatite present in bones and with the passage of time bioceramics replace themselves as Hap, and finally, implant becomes a part of bone. The mechanism of tissue attachment or bond formation is directly proportional to the type of tissue response toward the engraft interface because implanted materials are not always inert in nature toward living tissues, they arouse a response from the living tissue depending on the reactivity of implant with body environment either interfacial bond is formed or sometimes damage surrounding tissues (toxic effect) thus, tissue response is related to bioactivity of a material.[76]

Literature reports reveals that in vitro bioactivity of silicate bioceramics is due to the Si-rich layer, which accelerate the formation of hydroxyapatite on its surface and controls the precipitation kinetics of bone like apatite deposition when it was immersed in simulated body fluid.[77,78] Sintered diopside pellets[79] when trailed to study their in vitro bioactivity, the results showed that after 3 days the surface was covered by uniform leaf-like apatite particles and these particles after 7 days grew in their sizes with the passage of soaking time. The variations in element concentration of SBF was also analyzed to study the influence of soaking time on apatite layer formation and the findings suggested that the dissociation of calcium ions in SBF from the sintered diopside during early stage of soaking induces silica rich layer Si—OH that causes apatite nucleation on the immersed surface. Also, magnesium ions does not participate in apatite formation as it does not get dissociated in SBF and most of its proportions remains as such in sintered diopside pellet.

FIGURE 5.12 SEM images of diopside showing influence of soaking time on surface morphology: (b) 3 days and (c) 7 days.[79]

Yamamoto et.al.[63] studied comparative apatite precipitation ability of diopside, akermanite, pseudowollastonite, protoenastatite, and forsterite stirred in PBS (phosphate buffer solution) separately at 37°C for 1 h. The materials were collected from PBS by decantation method. Influence of calcium and magnesium release in PBS was also investigated to study their influence on apatite precipitation as calcium is a major constituent of hydroxyapatite. The results shows poor calcium ions release from each material (0.02–0.06 mg) while release of magnesium ions in PBS correlates the apatite layer precipitation on the surface soaked materials. Enhanced precipitation of apatite was observed on the surface of materials with lower release of magnesium ions. A previous report also proves that magnesium plays a major role in remineralization of bone but in vitro studies shows poor hydroxyapatite deposition occurs in solutions having higher concentration of magnesium.[80,81] Thus, diopside, pseudowollastonite, and protoenastatite precipitated little HAp on the soaked surface, while akermanite and forsterite no HAp precipitation. It was also concluded that diopside can be more applicable for biomedical applications as it has potential to precipitate Hap deposition as well as low magnesium elution and can make hybrid with titanium.[82]

Porosity plays a vital role in bioactivity of a biomaterial. Previous studies revealed that the features of the substrate, such as the presences of porosity and silanol groups, seem to be crucial in the apatite formation and degradation. It helps to form mesh-like interconnected network with the neighboring bone or tissues providing large surface area, which facilitates improvement in mechanical strength as well as capable of withstanding more complex stress and tissue remain viable and healthy.[76] Still optimum porosity parameters are not yet confirmed.

Wu[83] and his fellow researchers studied the effects of porosity on bioactivity, mechanical strength, cytocompatability, and degradation rate of diopside scaffold. Diopside scaffolds prepared by polymer sponge method reveals highly porous interconnected surface morphology with pore size of 300 μm. The compressive strength (1360 ± 370 kPa) and compressive modulus (68 ± 20 MPa) of diopside scaffold were compared with compressive strength (100 kPa) of hydroxyapatite scaffold. The compressive strength and compressive modulus of diopside were similar till 14 days of immersion in SBF but after 14 days slight decrease in mechanical strength by 30% was noticed. Difference in sintering properties and variation in composition impart improved mechanical strength and mechanical stability to diopside when compared with wollastonite, hydroxyapatite, and bioglass. Diopside scaffolds also induces apatite layer deposition in SBF with sustained silicon release and weight loss of 2% after 28 days of immersion in SBF signifies slow degradation rate of diopside in physiological body environment as compared to other bioceramics, such as akermanite 18%,[84] bioglass 20%,[85] and wollastonite 26%.[86] Cytocompatability studies of diopside scaffold indicate that pore structure with proper interconnectivity supports human osteoblast cell proliferation and growth. The ALP activity of diopside scaffolds increases with increase in culture time. These findings suggest diopside as highly bioactive silicate bioceramic with improved characterstics.

In previous reports, Wu et al. investigated the influence of porosity on different biological properties and later he also studied the effect of compositional difference on in vitro bioactivity, degradation, and cytocompatability of three silicate bioceramics, such as diopside, akermanite, and bregidite.[57] Degradation rate was analyzed by soaking samples separately in 100 mL (Tris-HCl) buffer at 37°C, pH 7.4 for 28 days to measure accurate weight loss before and after soaking. The activation energy of silicon release from bioceramics was determined by Arrhenius equation.

$$\ln [Si] = \ln [Si_0] - E_a/KT$$ [87]

Bregidite shows increased loss in weight, akermanite with intermediate and diopside with lowest while activation energy of silicon release results indicate that lowest release from bregidite, intermediate from akermanite, and highest from diopside. These findings were described on the basis of magnesium content and also prove previous reports that activation energy is inversely related to silicon ion release and weight loss.[88] Elution of silicon ion at different degrees and degradation rate also affects the in vitro bioactivity of three bioceramics. The in vitro bioactivity was carried out for 7 days in SBF to analyze apatite formation. The results shows decrease in apatite deposition from bregidite to diopside. Thus, magnesium concentration indirectly affects degradation and bioactivity of silicate bioceramics.

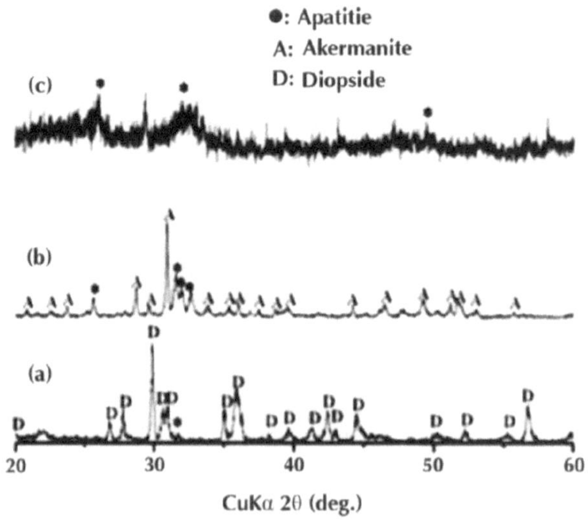

FIGURE 5.13 XRD patterns of in vitro bioactivity after 7 days: (a) diopside, (b) akermanite, and (c) bregidite.[57]

Cytocompatability studies of silicate bioceramics is highly influenced by pH[88] and silicon ion release.[89,90] Ionic concentration of Ca, Mg, and Si in SBF was also analyzed to carry out detailed and comparative study on dissolution of ions at different concentration and their stimulatory effect on osteoblast proliferation. It was suggested that bregidite showed increased silicon ion concentration, faster degradation, and higher pH value in cell culture extracts finally inhibiting cell proliferation. Thus, silicate bioceramics show stimulatory cell proliferation at lower ionic concentrations only. Magnesium and silicon plays key role in degradation, dissolution, apatite formation ability, and cell proliferation of bioceramics. In near future in vivo studies need to investigate to confirm above discussed results.

Toru Nonami and Sadami Tsutsumi[64] conducted in vivo studies of diopside by implanting in rabbit bone cavity and dental implant in monkey bone cavity. Mechanical strength, weight loss in lactic acid, and physiological salt solution was also evaluated in comparison to hydroxyapatite. Bending strength 300 MPa, fracture toughness 3.5 MPa $m^{1/2}$, young's modulus 170 GPa, and poison's ratio 0.35 of single phasic highly crystalline sintered diopside showed improved mechanical properties. These values are 2–3 times higher than that of hydroxyapatite. Weight loss of diopside in lactic and physiological salt solution was 2.8 and 0.05 comparatively lower than hydroxyapatite 16.5 and 0.13. The diopside implant kept in bone cavity of rabbit came in contact with newly developed bone after 12 weeks and the interface between newly grown bone and diopside possess concentration gradient as shown in figure. Deposition of phases of both diopside and apatite was confirmed by XRD patterns. Similarly, the dental implant placed in bone cavity of monkey reveals crystal growth at the intermediate layer between diopside and newly developed bone. These crystals show close resemblance with teeth and dentin. Current findings reports bone like apatite crystallizes on the surface of diopside implants, thereby inducing apatite formation in the presence of body fluids.

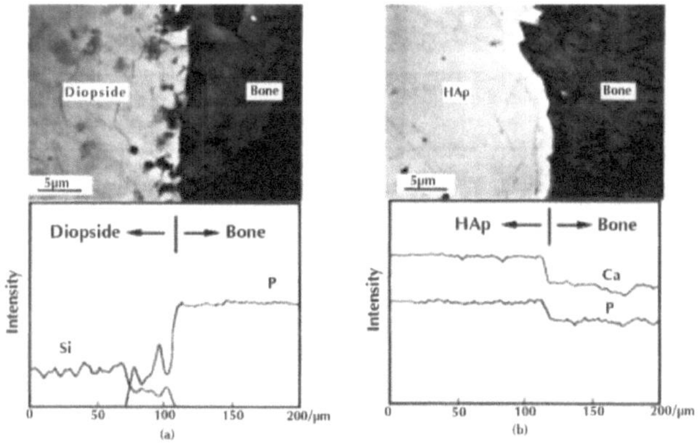

FIGURE 5.14 SEM images and XMA patterns of the interface after 12 weeks implantation: (a) diopside and (b) Hap.[64]

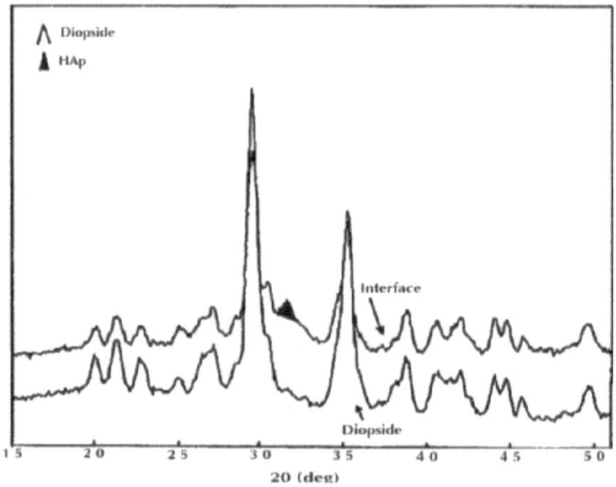

FIGURE 5.15 Microbeam XRD patterns of the interface and inside diopside after 12 weeks implantation.[64]

5.7 HYBRID COMPOSITES

A composite material is one which is composed of at least two or more constituent elements with different physical and chemical properties working together to produce material properties that are different or superior to the properties of those elements on their own.[91] Generally, composites are added to increase the strength and hardness of the matrix. Ceramic composites are first choice of advanced materials due to their light weight, high strength, and toughness.[92] However, as technology has grown over time the ceramic field has elaborated to include a wider range of other compositions, such as scaffold and composite preparation used for variety of applications.

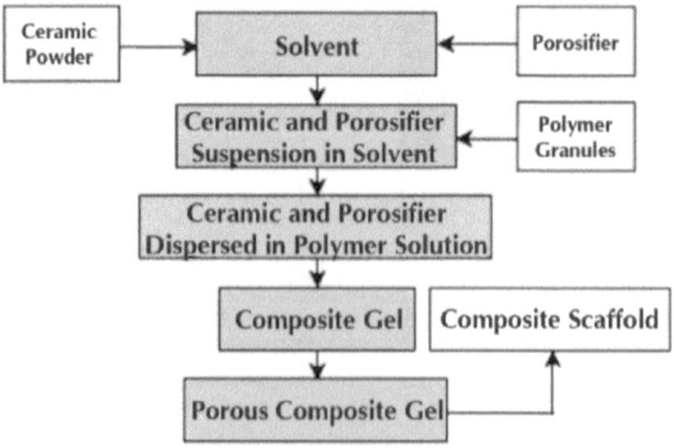

FIGURE 5.16 Preparation of bioactive and biodegradable composites.[93]

With an excellent combination of mechanical stability, strength, and toughness along with bioinert properties and low wear rates, a superior type of oxide called alumina and zirconia are now used in applications such as femoral heads for total hip prostheses.[94] Thus, Mufan Zhang et al.[95] investigated Al_2O_3/diopside ceramic composite prepared by uniaxial hot-pressing method for improved biological activity. Different compositional ratios of alumina to diopside were prepared and studied the influence of composition on in vitro bioactivity of composites by soaking in SBF for 9 days. After 9 days, the composites with 80:20 ratio of alumina/diopside (D_{20}) indicates lath-like layer formed on the composite surface while no evidence of apatite deposition was noticed on composite with 100% alumina (D_1). The excess addition of diopside also resulted in inferior mechanical properties. These results evidencing that the bioactivity of Al_2O_3/diopside ceramic composite increased with increase in diopside content. Thus, utilization of proper amount of diopside should be employed to prepare composite with alumina in order to obtain enhanced biological activity and improved mechanical strength.

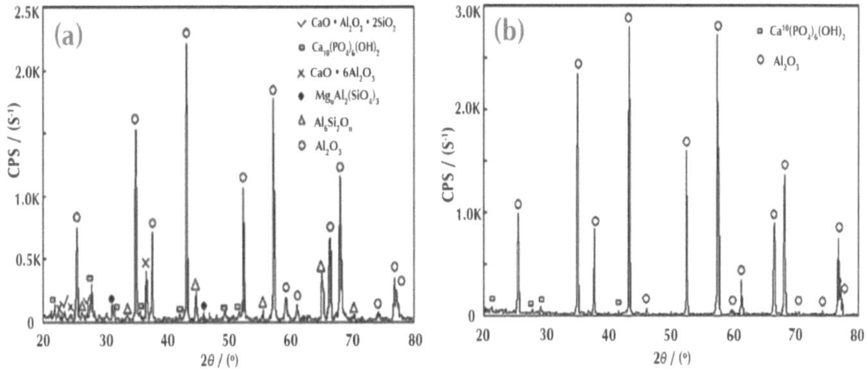

FIGURE 5.17 XRD patterns of alumina/diopside composites: (a) D_{20} and (b) D_1, after immersing in SBF for 9 days.[95]

A major drawback of silicate bioceramics is their brittle character, which can lead to poor implant strength. The brittle nature of glass ceramics can be eliminated by preparing composites with polymers. Ceramic/polymer composites with specific composition show improved strength and mechanical stability with respect to the ceramic phase whereas, their toughness and bioactivity increases with respect to the polymer phase.[96] The effects of polymers on the brittleness, fracture toughness, and bioactivity of ceramics have already been reported.[97] In the current study, diopside/chitin composites[69] were prepared with variation in their composition (M20, M40, and M60) to examine the effect of composition on deposition of hydroxyapatite on the surface of scaffolds soaked in SBF for 30 days. XRD patterns of the three diopside/chitin composites, the ratio which mimics natural bone (M60, 30 days) reveals good bioactivity and remarkable apatite deposition on the surface of the composites. The hydroxyapatite phase on the surface of M60 composite increases with increase in soaking time.

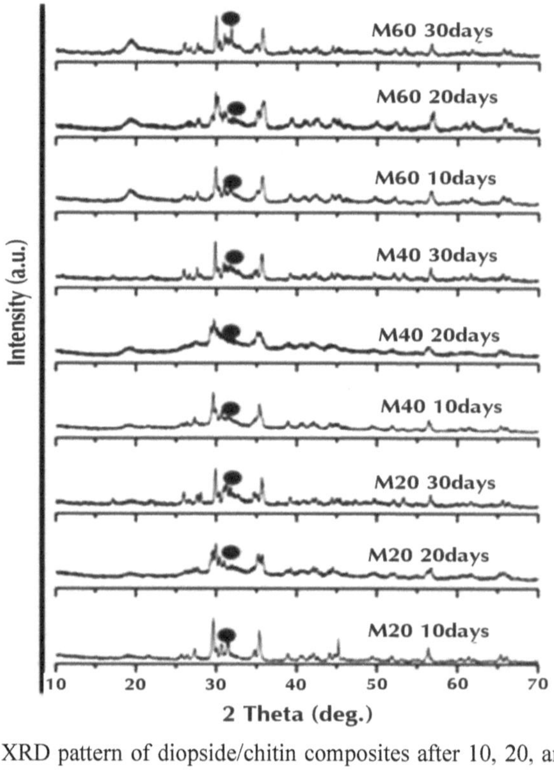

FIGURE 5.18 XRD pattern of diopside/chitin composites after 10, 20, and 30 days of in vitro bioactivity.[69]

Excellent bioactivity can be attained by forming composites with bioactive ceramics. Ceramic particulate reinforcement has resulted to more biomaterials for implant applications that include ceramic/ceramic, ceramic/polymer, and ceramic/metal composites. Ceramic/ceramic composites enjoy superiority on account of resemblance with bone minerals, showing perfect biocompatibility and are capable to get shaped into definite sizes.[98] Researchers now a day's are more focused on improved biocompatibility, bioactivity, and mechanical strength by using ceramic/ceramic composites. In the present context, Zhang et al.[99] prepared hydroxyapatite/diopside composites to assess the influence of ceramic/ceramic composites on mechanical strength and bioactivity by immersing in SBF. After 9 days the surface of composite with 90:10 ratio of hydroxyapatite/diopside composite (D_{10}) was covered with thick uniform bright mineral layers and XRD pattern shows the presence of Hap, SiO_2, and $CaSiO_3$ while the surface hydroxyapatite composite (D_0) comprises of only hydroxyapatite phase.

Elemental analysis describes the presence of NaCl on the surface that might result due to poor washing of composite surface after bioactivity. Hydroxyapatite/diopside ceramics seems to possess higher flexural strength (80 ± 5 MPa), fracture toughness (1.2 ± 0.3 MPa $m^{1/2}$) but reduced vicker's hardness (5.4 ± 1.0 GPa) than hydroxyapatite ceramics. These results indicate that hydroxyapatite/diopside composites have improved mechanical strength and better ability to deposit Hap layer on their surface as compared to hydroxyapatite ceramic alone.

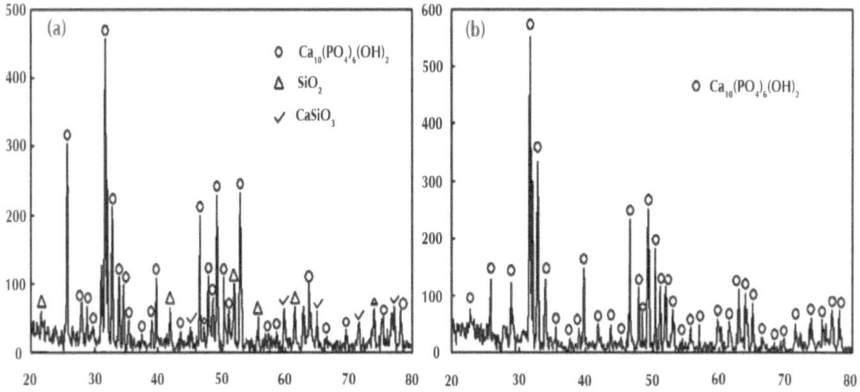

FIGURE 5.19 TF-XRD patterns of hydroxyapatite/diopside composites: (a) D_{10} and (b) D_0, after immersing in SBF for 9 days.[99]

5.8 DRUG DELIVERY STUDIES

A new achievement is made in the field of silicate bioceramics that along with bone substituent they can be used for drug delivery applications. Silicate bioceramics can be produced with uncontrollable porosity so mesoporous and microspheres bioceramics are currently most demanding for controlled drug release through implants. Porous DP (diopside) microsphere[100] was prepared by cross-linking the alginate in the $CaCl_2$ solution and carbon powders was used as porogens to control the porosity and drug loading. Carbon was later removed from diopside microspheres by thermal treatment at 1000°C. Dexamethazone, glucocorticoid class of drug with antiinflammatory and immuneosuppresant effects was used as drug to study drug loading/ release ability of diopside microspheres. SEM images shows that the number of microspheres with interconnected pores (enhance drug loading) on the surface of diopside was higher with the increase in carbon content. The inner pore size of diopside microspheres without carbon was <1 μm and with the increase in carbon content it raised up to 20 μm as shown in figure.

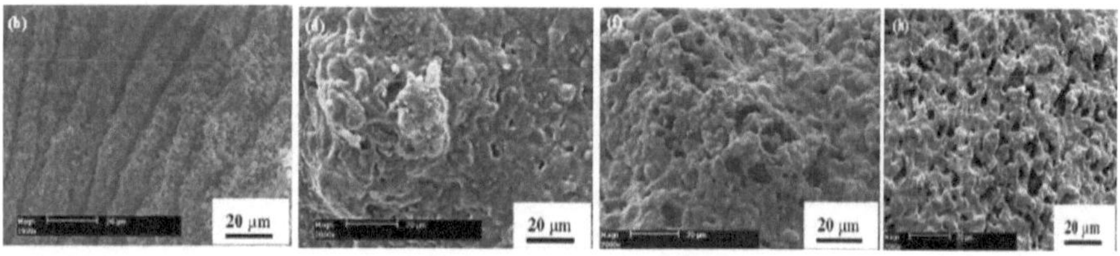

FIGURE 5.20 SEM images of diopside microspheres with different carbon contents: (b) no carbon, (d) 20%, (f) 40%, and (h) 60%.[100]

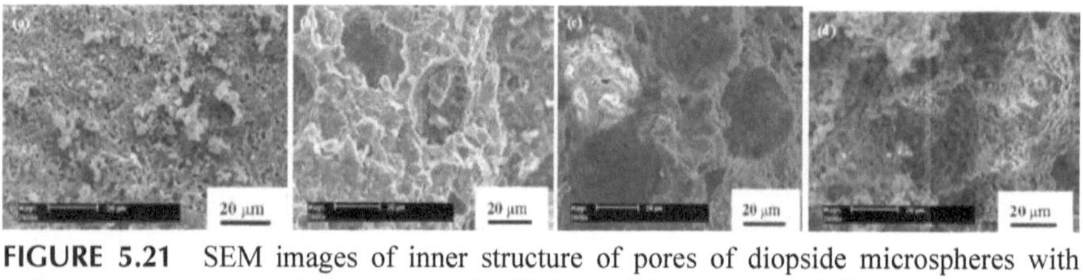

FIGURE 5.21 SEM images of inner structure of pores of diopside microspheres with different carbon content: (a) no carbon, (b) 20%, (c) 40%, and (d) 60%.[100]

The diopside microspheres were soaked in SBF to assess degradation rate and apatite formation ability. Degradation rate of diopside microspheres increases with increase in porosity and soaking time. The diopside microsphere with 63% porosity shows elevated degradation also apatite layer get deposited on the surface all four diopside microspheres and different surface morphologies was noticed. Unlike degradation, porosity does not have much influence on release kinetics of dexamethazone because diopside microspheres with lower size (400 μm) shows considerably accelerated release of drug as compared to DP microspheres with greater size (800 μm). This might have resulted due to high surface area of DP microspheres that will result in tightly bond with the drug molecules.[101]

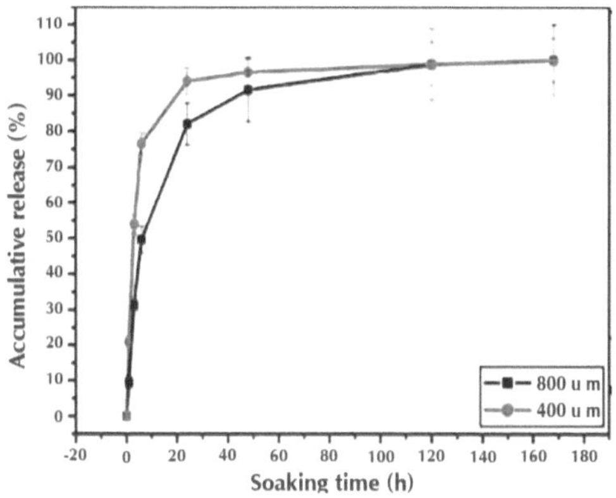

FIGURE 5.22 Influence of microspheres size on dexamethazone release kinetics.[100]

The surface of diopside microspheres was customized by varying the concentration (2.5 and 5%) of Poly(lactide-co-glycolide) (PLGA) polymer to enhance and control the drug loading/drug release ability. PLGA film covers the surface of DP microspheres, aids in blockage of macrospheres created by carbon porogens and few proportion of PLGA gets bind with dexamethazone to promote drug loading ability. PLGA concentration plays a key role in both drug loading and release mechanism from DP microspheres. With the increase in concentration of PLGA coatings on DP microspheres, thick layers gets deposited that result into enhanced drug loading and slow release. DP microspheres with 12.7 and 63.3% porosity showed increased drug release while 27.7% porosity shows negative drug release.

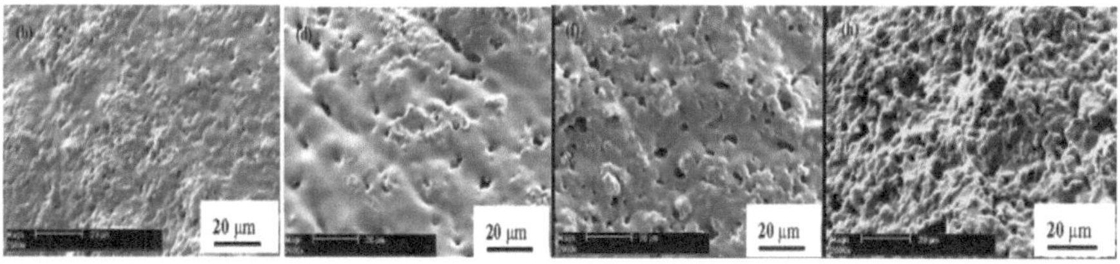

FIGURE 5.23 High magnification SEM images of DP microspheres with different porosities after modification by 5% PLGA: (b) 12.7%, (d) 27.7%, (f) 40.3%, and (h) 63.3%.[100]

Therefore, current findings shows that the factors responsible for drug loading and release ability from DP microspheres are variations in PLGA concentration, porosity, and complex inner structures of pores. The dissolution Ca and Si ions from the ceramic surface help in neutralizing the acidic by-products degraded from PLGA coatings with the passage of soaking time and maintains pH to an adequate level favorable for biological activities.[102,103] As a result DP microspheres possess good bioactivity, optimum degradability, and a controllable drug release.

5.9 CONCLUSION

Silicate bioceramics have evolved as a potential biomaterial for artificial bone and dental root implants due to its excellent bioactivity and improved mechanical strength compared to calcium phosphates. Different methods have been employed to prepare single phasic highly crystalline diopside. The main disadvantage of discussed synthetic methods is high thermal temperature required for sintering, which could have influence on microstructure, bioactivity, and mechanical strength. Researchers are focusing on modifying some conventional methods to obtain pure diopside at a comparatively low temperature, which could promote their bioactivity as well as mechanical properties. Surface characteristics and crystallite size play a major role in determining biological studies of diopside. Rough, irregular, agglomerated surface morphologies with nanoscale particle size have high surface area thereby inducing remarkable hydroxyapatite deposition, osteoblast cell nucleation, and proliferation on the surface of silicate bioceramics. Few studies show that external parameters, such as ionic concentration of SBF, soaking time, pH, and temperature have adverse on bioactivity.

The presence of magnesium in diopside composition prevents rapid degradation and makes a stable structure as compared to other silicate and calcium phosphate bioceramics. Incorporation of biopolymers, bio-inert metals, and bioceramics in diopside matrix shows superior bioactivity and improved fracture toughness, hardness, bending strength compared to bone, and hydroxyapatite. The composites with ratio that mimics natural bone show excellent bioactivity but the accurate volume fraction or compositional variation above and below, which biological response is influenced or composite may not possess desired bioactivity is not yet reported. Only few reports describes about in vivo behavior of diopside implant and in vitro investigation of diopside as potential material for controlled as well as targeted delivery of drugs. Hence, the biological activity of silicate bioceramics has to be understood through various in vitro and in vivo studies to play major role as first choice of advanced bioceramic implants.

5.10 FORSTERITE

Forsterite is a magnesium rich silicate bioceramic belonging to the group of olivine with chemical formula Mg_2SiO_4 and named after the English naturalist, Jacob Forster. Forsterite is an abundant mineral occur at a depth of 400 km in the earth's mantle and also results from metamorphism of magnesium limestone and dolostones. Few metamorphosed serpentinities contain even pure forsterite[104]. It is naturally distributed throughout the world particularly in the regions of Pakistan, Italy, Egypt, USA, Germany, etc.[105] Some physical properties of forsterite are[106]:

- Glassy lustrous with transparent to translucent appearance.
- Green or pale yellow to white colored solid mass.
- Higher melting point: 1890°C.
- Crystal structure: orthorhombic.
- Cell dimensions: a = 4.754 Å; b = 10.1971 Å; c = 5.9806 Å
- $\alpha = 90.000$; $\beta = 90.000$; $\beta = 90.000$.
- X-Ray powder diffraction pattern:

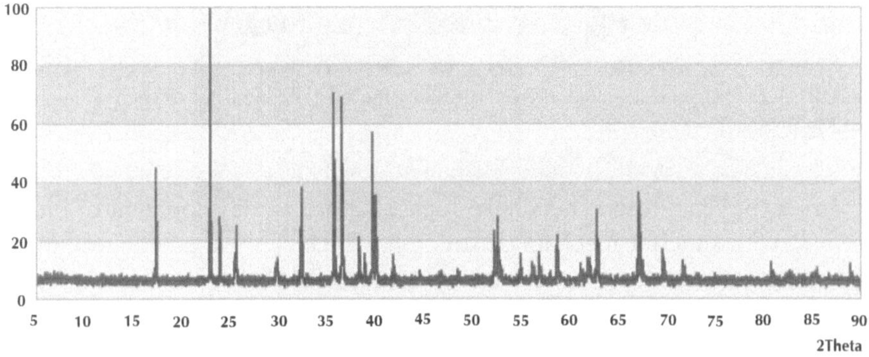

FIGURE 5.24 XRD pattern of pure forsterite.[106]

5.10.1 STRUCTURE

Structure of forsterite consists of SiO_4^{4-} (anion) and Mg^{2+} (cation) in molar 1:2 molar ratio. The central position in the structure is occupied by silicon atom and each oxygen atoms are covalently bonded to it by single bond. Due to this covalent bonding slight negative charge develops on all four oxygen atoms and maintains distance between each other to reduce repulsion force by forming tetrahedral geometry. The cations reside in M1 and M2 octahedral sites. This arrangement of structure packing provides forsterite a dense nature as shown in figure. This structure possesses pbnm space group and 2m/2m/2m dipyramidal space point, which correspond to orthorhombic crystal structure. It is also reported that in olivine structure magnesium ion is totally replaceable by iron (Fe^{2+}). As per forsterite structure cations fills two octahedral sites, iron is also capable of forming two different cations, such as Fe^{2+} and Fe^{3+}, presence of same charge and ionic radii as magnesium provides information that iron can be easily substituted with magnesium in forsterite structure.[104]

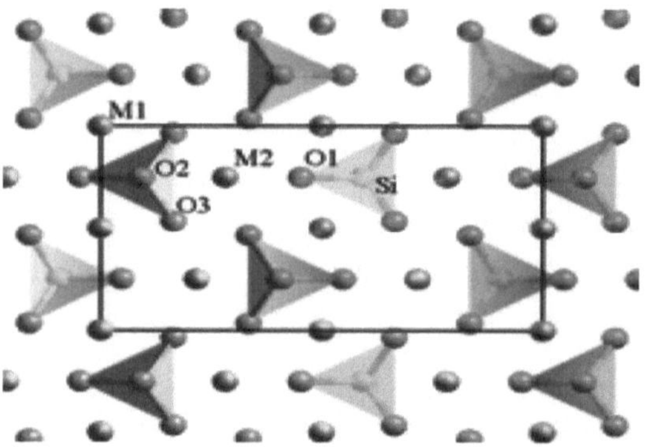

FIGURE 5.25 Crystal structure of forsterite along the axis: red indicates oxygen, Si in pink, and Mg in blue.[103]

In recent years researchers have focused more on development of bioceramics with superior mechanical strength, which can withstand more load and stress and excellent bioactivity, cytocompatible behavior. Thus, a wide range of silicate bioceramics as biomaterials possess high demand in biological applications ranging from bone implant to repairing of damaged tissues and biomedical pumps. Dentistry has also advanced with bioceramics as dental fillers or implants matches to the patient's natural teeth both in appearance as well as functions. In near future bioceramics might find applications in gene therapy and potential vehicle for drug delivery purposes for implantable medical devices. Innovations have been made in biomedical field to combine material research, manufacturing, and medicinal societies in order to search technological advancements to discover their new application fields.

Owing to resistance against corrosion, ability to tolerate high temperature, high compressive strength, and wear resistance forsterite has emerged as biomaterials for biomedical applications for repairing and reconstructing damaged musculo-skeletal system or defected bone structures.[107] Single phasic forsterite powders have been successfully synthesized by different methods. Some common methods utilized for synthesis are tabulated below along with complete procedure, observations, and graphically illustrated by XRD patterns (Table 5.2).

TABLE 5.2 Different Methods Used to Synthesize Forsterite

Synthesis methods	Starting materials	Synthetic conditions	Comments
Polymer matrix method	$Mg(NO_3)_2.6H_2O$, PVA, colloidal silica, sucrose, dil. nitric acid	Stirred at 80°C 2h, heated at 200°C 4 h, and calcined at 500–1000°C 3 h	Forsterite crystallizes at 800°C, pure forsterite noticed at 800°C, crystallite size: 10–30 nm, particle size> 20 nm[108].
Microwave sintering method	$Mg(OH)_2$, silica gel	Homogeneous mixture of raw materials was ball milled at RT for 0.25, 10, 20, 30, and 40 h. Calcined by microwave heating at 500–1200°C	Nanosized single phasic forsterite formed at 900°C, particle size: 45–64.5 nm[109]
Sol–gel method	$Mg(NO_3)_2.6H_2O$, PVA, colloidal silica, sucrose, nitric acid	2 h stirred, heated at 80°C 2 h, aged 24 h, dried at 100°C, and calcined at 800°C 2 h	Pure forsterite formed at 800°C and crystallite size: 17–20 nm, particle size: 25–45 nm[110]
Sol–gel method	$Mg(NO_3)_2.6H_2O$, PVA, TEOS, sucrose, nitric acid	Stirred at 80°C 2h, heated at 100°C, and calcined at 800–1000°C	Crystalline forsterite obtained at 1000°C with periclase (MgO) as secondary phase, crystallite size: 45–60 nm[111]
Sol–gel method	$Mg(NO_3)_2.6H_2O$, colloidal SiO_2	Stirred 3 h, aged 24 h, dried at 120°C 48 h, and calcined between 1100–1200°C 3 h	Highly crystalline forsterite phase observed at 1200°C with particle size: 5–50 μm but below this temp enstatite minor peak was noticed[112]
Solid-state method	MgO, $Mg_3Si_4(OH)_2$	Sonicated, ball milled, and sintered at 1200–1500°C 1 h	Appearance of MgO, enstatite peaks upto 1200°C, intense peaks of forsterite at 1400°C with less intense MgO peaks[113]

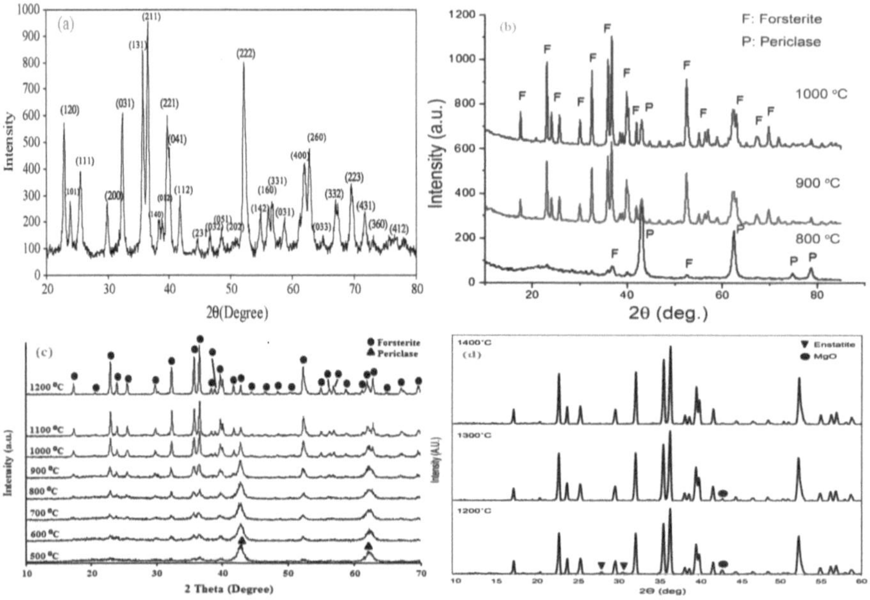

FIGURE 5.26 XRD patterns of forsterite calcined at different temperatures. (a) 800°C[108], (b) 1000°C[111], (c) 1200°C[112], and (d) 1400°C.[113]

5.10.2 MECHANICAL PROPERTIES

Bioactive glass ceramics are widely used for biomedical applications, such as bone substitute, orthopaedic, or dental implants. Those materials, which form direct bonding with tissues or bones through specific biological response of tissues toward the interface of the materials are termed as "bioactive materials or biomaterials."[107] Biomaterials intended for biomedical applications must satisfy three major requirements as biocompatible, biodegradable, and mechanically stable.[114] However, numerous biomaterials are investigated for their biological response in physiological environment to induce apatite deposition among all hydroxyapatite is the most widely investigated biomaterial. Due to its inferior mechanical strength more focused research has been carried out to design a biomaterial with superior mechanical properties as compared to bone. Therefore, silicate bioceramics evolved as potential biomaterials with mechanical properties as per implant requirements. Silicate bioceramics containing magnesium in their crystal structure are reported to have good mechanical properties like diopside, akermanite, bredigite, merwinite, forsterite, enstatite, etc. Among all these, forsterite is reported to possess capability to withstand more load and stress due to presence of magnesium that do not allow the rapid degradation of ions and maintains stable structure. Recent reports regarding mechanical stability studies of different biomaterials in comparison to bone are tabulated below:

Mechanical properties	Trabecular bone[115]	Cortical bone[115]	Hydroxyapatite[116]	Bioglass 45S5[116]
Compressive strength (MPa)	0.1–16	130–200	500–1000	500
Tensile strength (MPa)	n.a.	50–151	40–300	42
Compressive modulus (GPa)	0.12–1.1	11.5–17	120–150[117]	n.a.
Young's modulus (GPa)	0.05–0.5	7–30	80–110	35
Fracture's toughness (MPa $m^{1/2}$)	n.a.	2–12	0.6–1	0.7–1.1

FIGURE 5.27 Different mechanical properties of trabecular bone, cortical bone, hydroxyapatite, and Bioglass 45S5.

Ramesh et al.[113] investigated the effect on thermally and nonthermally treated samples before sintering on mechanical properties and sinterability. Variations in Vicker's hardness (2.3 GPa at 1200°C and 7.7 GPa at 1500°C) for thermally treated samples after sintering while nonthermally treated samples after sintering show < 1.7 GPa. This decline in hardness might be due to preparatory routes used for forsterite, which removes secondary phase and promotes the bulk density of sintered powders. Similarly, poor fracture toughness for thermally treated samples and 5.16 MPa.$m^{1/2}$ for nonthermally treated samples calculated by Niihara's equation. This value crosses the lower limit of cortical bone 2 MPa.$m^{1/2}$. Elastic property of forsterite was found to be dependent on grain size that changes with the increase in sintering temperature as a result steep rise in Young's modulus 89.7 GPa was observed between 1400–1500°C in thermally treated samples while in nonthermally treated samples with 77.7 GPa up to 1300°C. As sintering temperature increased to 1300–1500°C growth in grain size caused deterioration of Young modulus to 11.5 GPa. Thus, above results conclude that Vicker's hardness and fracture toughness are dependent on density and elastic property is grain size dependent. These parameters are highly influenced by synthesis methods.

Faithi and Kharaziha[118] reported comparative mechanical stability of nanoforsterite prepared by different methods. The sintered nanoforsterite shows fracture toughness (3.61 MPa $m^{1/2}$) and hardness (940 Hv) while two-step sintered sol–gel derived nanoforsterite show fracture toughness (4.3 MPa $m^{1/2}$) and hardness (1102 Hv) was higher than that of hydroxyapatite, diopside, akermanite, and bregidite. These superior stress properties suggest nanoforsterite as potential replacement to hydroxyapatite for hard tissue engineering applications. Similarly, they[119] have also investigated the biocompatibility of two-step sintered sol–gel derived forsterite dissolution on osteoblast cell. After 7 days, results reveal slow cell proliferation in samples with high concentration of forsterite (50–200 mg/mL) while increased proliferation at low concentration (6.25–50 mg/mL).

Recently, Mirhadi[120] also prepared nanoforsterite scaffolds by two-step sintering method and investigated the effects of porosity on mechanical property. The porosity of forsterite scaffolds at 1300–1500°C was 80–88%, while decreased to 58–79% at 1600°C. This might be due to the formation of SiO_2 liquid phase at 1557°C, which closes the open pores on the surface.[121] Decrease in porosity improves better connectivity between forsterite particles thereby superior compressive strength in the range of 0.03–24.16 MPa was observed at 1600°C while at 1300–1500°C temperature compressive strength was equivalent to human spongy bone (0.2–4 MPa). This study concludes that sintering temperature between 1300–1400°C is inappropriate for tight bonding between particles that might cause improvement in mechanical properties. Later, Ghomi et al.[122] fabricated forsterite scaffolds by gel casting method to study the parameters, which influences mechanical strength. The crystallite size, porosity, apparent density, compressive strength, and elastic modulus at different temperatures are shown in table. All above discussed findings reveal that different processing methods, thermal treatment at different temperatures and porosity causes variations in mechanical properties.

Temperature	Crystallite size (nm) (S.D.[a])	Apparent density (gr cm^{-3}) (S.D.)	Total porosity (%) (S.D.)	Compressive strength (MPa) (S.D.)	Elastic modulus (MPa) (S.D.)
Forsterite powder	23 (±1)	3.14 ((±0.1)	–	–	–
900°C	26 (±2)	0.45 (±0.04)	86 (±1)	2.06 (±0.09)	145 (±9)
1000°C	28 (±1)	0.48 (±0.08)	85 (±1)	2.19 (±0.06)	165 (±12)
1100°C	31 (±1)	0.55 (±0.05)	83 (±2)	2.31 (±0.07)	171 (±21)
1200°C	35 (±2)	0.61 (±0.06)	81 (±1)	2.43(±0.11)	182 (±19)

FIGURE 5.28 Influence of sintering temperature on different properties of forsterite.[122]

5.10.3 BIOLOGICAL RESPONSE OF FORSTERITE BIOCERAMICS

Several in vitro bioactivities have been trailed to study the deposition of hydroxyapatite layer on the surfaces soaked in SBF for different period of time intervals and characterized by different techniques to confirm it. Some reported biological reviews of forsterite are discussed in the current section. M. Kharaziha and M. H. Fathi[110] synthesized forsterite nanopowders at a low temperature to investigate the mechanism of apatite deposition on nanoforsterite and influence of nanoparticles on bioactivity and degradation rate. After 28 days of bioactivity, results shows that with the increase in soaking time the surface of forsterite is covered with hydroxyapatite as major phase. At early stage of bioactivity, Mg ions are exchanged with H^+ ions present in SBF resulting in the formation of negatively charged surface of Si—OH^- layer. Presence of positively charged Ca and P ions in SBF are attracted toward silica rich layer leading to the formation of apatite on the surface. Therefore, steep rise in Mg concentration and pH level in SBF was noticed with decreased in concentration of Ca from 2.5 to 1.78 mM and P 1 to 0.55 mM. It is obvious from this dissolution behavior that more Ca and P ions were consumed at the interface for the apatite layer formation. From above findings authors also suggests the need of in vivo activity to explore the applicability of forsterite for biomedical implants.

Besides improved bending strength, fracture toughness, and low degradation makes forsterite as eligible bone substitute to commercially available hydroxyapatite ceramics still few reports show poor apatite formation on the surface of forsterite.[123] In the present report,[124] forsterite was investigated to study bone like apatite deposition on forsterite after soaking in SBF and degradation rate in Ringer's solution, respectively. As per previous reports, it was noticed that increase of pH and Mg ion concentration with decrease in Ca ion in the SBF. These variations in the concentration of ions and pH and degradation reveal the release of these ions in SBF accelerates apatite layer deposition in the surface of forsterite. In vitro bioactivity (28 days) shows nanostructured and low crystallinity of forsterite possess good bioactivity and degradable in physiological body environment.

Nanostructure nature of bioceramics has drawn more attention of biomaterial scientists due to their higher surface area, solubility, and decreased surface stability. These advanced properties of nanobioceramics could be helpful in forming bone grafts for biomedical applications. Therefore, recently[111] nanoforsterite was synthesized at 900°C with periclase as secondary phase and assayed for cytotoxicity tests at different concentrations (6.25, 12.5, and 25%) and biological reaction between forsterite and SBF. MTT test [3-(4,5-Dimethylthiazole-2-yl)-2,5-Diphenyltetrazolium Bromide] confirms positive but insignificant response of different concentration on osteoblast growth without cytotoxic effects on cultured forsterite samples. In vitro bioactivity result shows existence of forsterite as primary phase and brucite as secondary phase and poor deposition of hydroxyapatite on the surface soaked for 28 days.

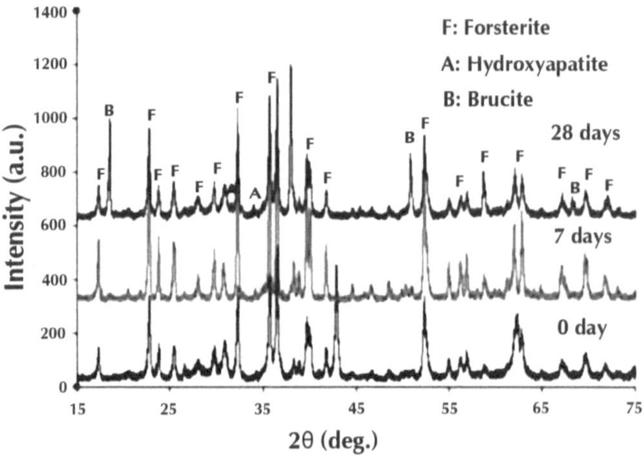

FIGURE 5.29 XRD pattern of nanoforsterite after 28 days in vitro bioactivity.[111]

Siyu ni et al.[112] cultured forsterite ceramics to evaluate cellular biocompatibility studies at different time intervals of 4 and 24 h. Initially, after 4 h, small rounded cells spread over the surface with different morphologies, while after 24 h cells are found to be tightly and closely attached at the sample surface with elongated morphologies. Moreover, MTT test on calvaria of neonatal proves stimulation of forsterite ceramics toward proliferation of osteoblast cells after 7 days without cytotoxicity. Proliferation rate of cultured samples are found to be time and surface morphology dependent. Later, Siyu Ni et al.[125] compared bioactivity, degradation, and cytocompatability of three different bioceramics, such as calcium silicate, tricalcium phosphate, and forsterite. The degaradation rate was analyzed by activation energy of released Si, P ions from the bioceramics and weight loss by Tris-HCl buffer solution. Calcium silicate degraded more rapidly compared to tricalcium phosphate and forsterite. In vitro bioactivity of immered surface in SBF shows poor bone-like apatite deposition on forsterite and tricalcium phosphate, while excellent deposition on calcium silicate surface. Further, cellular studies reveal the capability of calcium silicate and forsterite ions to induce proliferation of osteoblast cells at a certain concentration limit. These findings might be due to difference in their composition, morphology structure, and particle size.

5.11 COMPOSITES OF FORSTERITE

Bone tissue engineering combines cells and a biodegradable three-dimensional scaffold to repair diseased or damaged bone tissue. Challenges are set by the design and fabrication of the synthetic tissue scaffold and the engineering of tissue constructs in vitro and in vivo. In bone tissue engineering, bioactive glasses and related bioactive composite materials represent promising scaffolding materials.

Bone tissue engineering seeks to restore and maintain the function of human bone tissues using the combination of cell biology, materials science, and engineering principles. The three main ingredients for tissue engineering are therefore, harvested cells, recombinant signaling molecules, and three-dimensional matrices. Cells and signaling molecules, such as growth factors are seeded into highly porous biodegradable scaffolds, cultured in vitro, and subsequently the scaffolds are implanted into bone defects to induce and direct the growth of new bone. Signaling molecules can be coated onto the scaffolds or directly incorporated into them. Hence, the first and foremost function of a scaffold is its role as the substratum that allows cells to attach, proliferate, differentiate (i.e., transform from a nonspecific or primitive state into cells exhibiting the bone specific functions), and organize into normal, healthy bone as the scaffold degrades. A major hurdle in the design of tissue engineering scaffolds is that most materials are not simultaneously mechanically competent and bioresorbable, that is, mechanically strong materials are usually bioinert, while degradable materials tend to be mechanically weak. Hence, the fabrication of composites comprising biodegradable polymers and bioactive glass becomes a suitable option to fulfill the requirements of bioactivity, degradability, and mechanical competence.

Bioglass/Forsterite hybrid composites[126] were prepared with different composition (NC0-100% BG; NC1-90/10; NC2-80/20; NC3-70/30) of forsterite by cold press molding method to improve their mechanical and biological properties. Mechanical properties result shows with the increase in forsterite content highest fracture toughness of NC3-0.22 MPa $m^{1/2}$ while reduced values of elastic modulus (NC0-490 MPa to NC3-110 MPa) and yield stress (NC3-4.3 MPa). This investigation of mechanical properties of hybrid composites reveals that increase in forsterite content in composite play a key role in elastic modulus, fracture toughness, and yield stress. In vitro apatite deposition on hybrid composites surface soaked in SBF for 14 days was confirmed by Ca/P ratio with respect to stoichiometric apatite ratio. The Ca/P ratio of different composition hybrid composites show increase in values with the increase in forsterite content NC0-1.69 near to while other show nonstoichiometric apatite ratio NC1-1.76, NC2-2.52, NC3-3.21. These findings were supported by FTIR spectra indicating presence of carbonated apatite and pure apatite peaks.

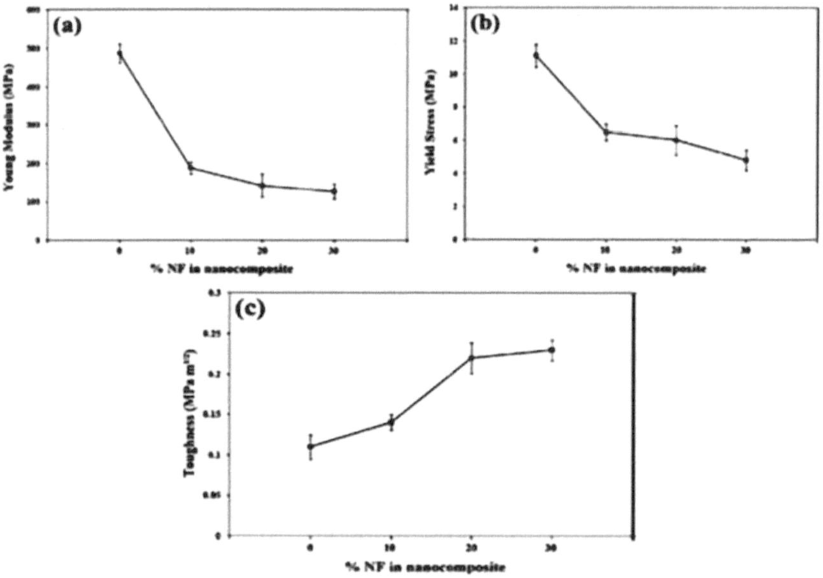

FIGURE 5.30 Mechanical properties of bioglass/forsterite hybrid composites: (a) elastic modulus, (b) yield stress, and (c) fracture toughness.[126]

A major drawback of silicate bioceramics is their brittle character, which can lead to poor implant strength. The brittle nature of glass ceramics can be eliminated by preparing composites with polymers. Ceramic/polymer composites with specific composition show improved strength and mechanical stability with respect to the ceramic phase whereas, their toughness and bioactivity increases with respect to the polymer phase[127]. In this research, polycaprolactone (polymer) and nanoforsterite (ceramic) biocomposite scaffold[128] was fabricated by solvent casting and particle leaching method to study the influence of forsterite content on mechanical property, bioactivity, biodegradability, and cytotoxicity. Mechanical properties, such as elastic modulus (3.1–6.9 MPa) and compressive strength (0.0024–0.3 MPa) of scaffolds was promoted with the increase in amount of forsterite to 30 wt.%. These superior properties might have resulted due to incorporation of forsterite as stiff filler in polymer matrix, thereby enhancing mechanical stability and reduced porosity. Moreover, the critical limit to increase forsterite content was 30 wt.% beyond which scaffold resulted in inferior mechanical properties. The degradation rate of nanocomposite and pure PCL scaffolds was analyzed by soaking in PBS for different time intervals. At initial stage of degradation study slight weight increase is noticed in pure PCL scaffolds that might be due to absorption of water by polymer and later slow weight loss was observed but rapid weight loss in nanocomposite which continued throughout the incubation period. This higher percentage weight loss in nanocomposite scaffold was dependant on degradation of nanoforsterite, release of ions from nanocomposite and increase in pH of PBS.

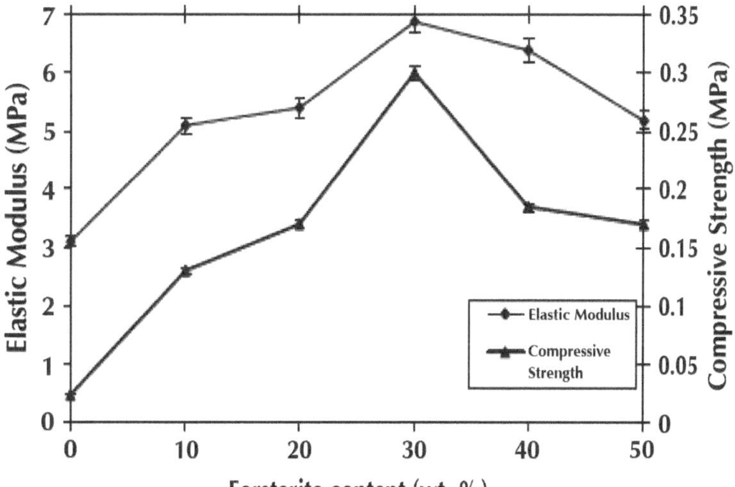

FIGURE 5.31 Elastic modulus and compressive strength of nanocomposites with different forsterite content.[128]

Biological response of nanocomposite (PCL-10, 30, and 50 wt.%) and pure PCL scaffolds to deposit apatite layer on soaked surface in SBF for 3 weeks was analyzed. The presence of Ca:P ratio on soaked surface was compared with Ca:P ratio (1.63) of carbonated hydroxyapatite. Presence of Ca and P on the surface of pure PCL scaffold was not detected thus confirming no apatite layer was deposited. The surface of nanocomposites indicate different Ca:P ratio (PCL10 wt.%—1.59, 30 wt.%—1.54 and 50 wt.%—1.26). This finding proves that increase in forsterite content decreases the values of Ca:P ratio hence, reduced bioactivity. Cellular toxicity test was investigated among PCL and nanocomposite scaffolds to study the influence of forsterite content on cell growth and development after 3 days of incubation period. Pure PCL scaffold show 50% cell growth in comparison to polystyrene as control and it was reported that the hydrophobic nature of PCL causes limited cell proliferation. MTT test for nanocomposite scaffolds states superior cell growth than PCL scaffolds and increase in nanoforsterite content to 50 wt.% does not inhibit proliferation and cellular growth. Similarly, study on sarcoma osteogenic cell lines (SaOs-2) for cell attachment and proliferation proves the better activity of nanocomposite scaffolds over pure PCL scaffolds and as the content of nanoforsterite in composite is increased no cellular toxic response was observed. The Si ions in ceramic matrix are negatively charged due to reduced isoelectric point, this peculiar behavior of silica establishes binding with different functional groups when comes in contact with physiological environment containing growth factors, proteins, and enzymes. This binding promotes favorable environment for cell growth and development.[129] Therefore, dissolution of Mg and Si ions play a key role in determining the cellular response in vitro and addition of polymer in nanoforsterite proves to be excellent scaffold with superior mechanical and biological properties.

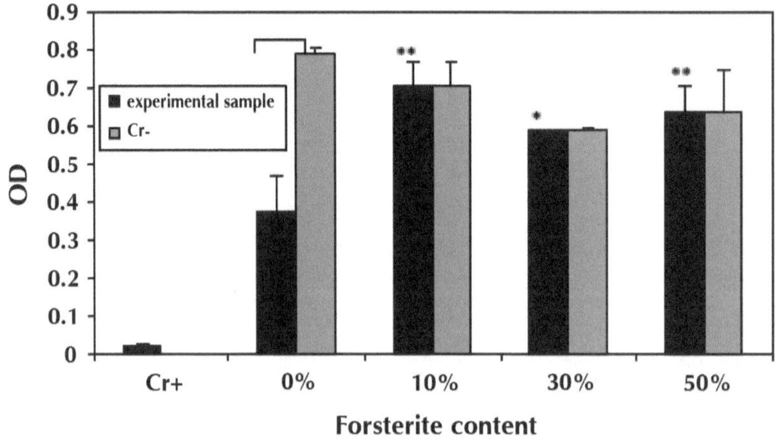

FIGURE 5.32 MTT assay of pure PCL and nanocomposite scaffolds after 3 days of cell culture (Cr^+: positive control; Cr^-: negative control).[128]

5.12 FORSTERITE-COATED BIOCOMPOSITES

Scientists are studying high tech materials to replace diseased or damaged tissues, cells, and even entire injured part using novel biomaterials. Total replacement of damaged tissues and bone are possible by metallic implants, which have capability to withstand load and stress. Highly porous surface morphology imparts low mechanical behavior to the implants. Thus, attempts have been made to develop porous, mechanical stable and bioactive implants without toxic responses in the physiological body environment. Besides, synthetic hydroxyapatite has drawn more attention in biomedical field as it is believed to have similar composition and resorbable nature as natural apatite but its inferior capability to bear load has restricted their applications to bone dental filler or substitute for nonload bearing parts. In this regard, studies are more focused toward biomaterials with porous nature to enhance their fracture toughness, hardness and compressive strength applicable for stress bearing implants.

Emadi et al.[130] reported nanoforsterite coating on porous hydroxyapatite struts to achieve improved properties. The coated porous scaffold showed 80% porosity with average pore size 740 μm, grain size ranging between 35–80 nm. Mechanical studies showed huge variation in compressive strength 1.61 MPa compared to conventional hydroxyapatite as 0.12 MPa and excellent bioactivity in simulated body fluid. These findings prove forsterite has evolved as promising biomaterial with improved mechanical and biological properties.

Although metallic bioceramics provide high mechanical stability to the implant but inert, corrosive nature, and poor bioactivity in body fluid reduces their demand for implantable materials. Thus, researches are now focused on development of bioceramic coating on metal implants to enhance their biological response in the body. Several reports reveal the study of bioceramic coating on titanium alloy[131], Yttria-stabilized zirconia[132], stainless steel,[133,134] etc. to study their improved biological properties.

In this context, current report Mazrooei Sebdani[135] developed hydroxyapatite-forsterite-bioactive glass nanocomposites by sol–gel method to apply coating on 316L stainless steel by dip coating technique. The nanocomposite prepared for coating contains different weight percentage content of nanoforsterite as (0, 10, 20, and 30%) while bioactive glass nanopowder (10%) into hydroxyapatite sol to investigate the influence of forsterite content on bioactivity. Finally, the coated 316L stainless steel was thermally treated at different temperatures to optimize the suitable temperature required to achieve phase purity of different nanocomposite coated samples hydroxyapatite-forsterite-bioactive glass. The optimum sintering temperature to obtain single phasic coated nanocomposites with different forsterite content was 600°C. It was also noticed that with the increase in sintering temperature to 800°C cause highly crystallinity and hydroxyapatite gets transformed into its secondary phases (β-TCP and CaO). The crystallite size was found to be lower than 100 nm and it increases with increase in forsterite content.

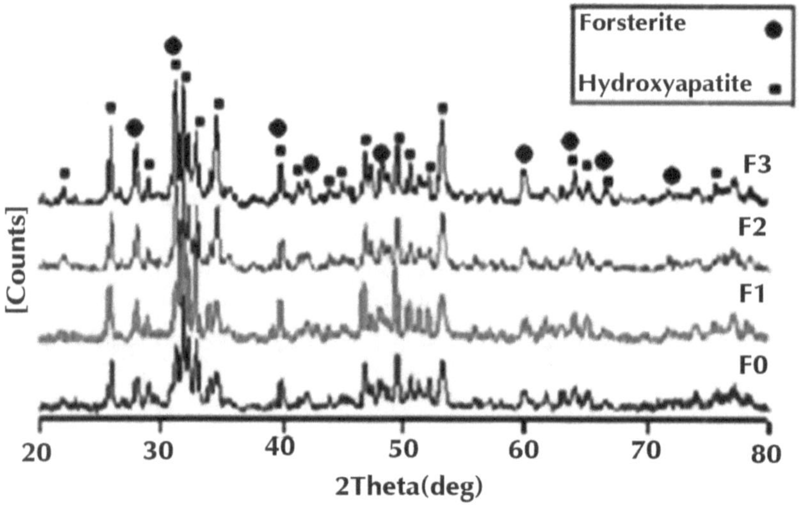

FIGURE 5.33 XRD pattern of nanocomposite coated 316L stainless steel sintered at 600°C containing different forsterite content (F0-0wt.%, F1-10wt.%, F2-20wt.%, and F3-wt.30%).[135]

The single phasic nanocomposite coated 316L stainless steel samples were soaked in SBF for 4 weeks to testify their ability to induce apatite layer deposition on their surfaces. In vitro bioactivity report reveals that after 14 days, the soaked surface was covered by apatite and it was also observed that incubation period and forsterite content in coatings highly affect precipitation of apatite producing different surface morphologies.

FIGURE 5.34 SEM images of nanocomposite coated 316L stainless steel containing different forsterite content: (a) F0-0 wt.%, (b) F1-10 wt.%, (c) F2-20 wt.%, and (d) F3-30 wt.%.[135]

KEYWORDS

- **bioceramics**
- **wollastonite**
- **diopside**
- **forsterite**
- **bioactivity**
- **hydroxyapatite**

BIBLIOGRAPHY

1. Bhat, S.V. *Biomaterials*, 2nd ed.; Narosa Publishing House: India, 2002.
2. Davis, J.R. (Ed). Overview of Biomaterials and Their Use in Medical Devices. Handbook of Materials for Medical Devices, ASM International, 2003 (DOI: 10.1361/hmmd2003p001).

3. Hench, L.L. Bioceramics: From Concept to Clinic. *J. Am. Ceram. Soc.* **1991**, *74* (7), 1487–1510.
4. Hench, L.L. Bioceramics. *J. Am. Ceram. Soc.* **1998**, *81*, 1705–1728.
5. Fratzl, P.; Weinkamer, R. Nature's Hierarchical Materials. *Prog. Mater. Sci.* **2007**, *52* (8), 1263–1334.
6. Bone, from Wikipedia, the Free Encyclopedia, http://en.wikipedia.org/wiki/Bone (accessed Dec 23, 2014).
7, http://www.vivo.colostate.edu/hbooks/pathphys/endocrine/thyroid/calcium.html (accessed on Dec 23, 2014).
8. Hench, L.L. Biomaterials: A Forecast for the Future. *Biomaterials* **1998**, *19*, 1419–1423.
9. Best, S.M.; Porter, A.E.; Thian, E.S.; Huang, J. Bioceramics: Past, Present and for the Future. *J. Eur. Ceram. Soc.* **2008**, *28*, 1319–1327.
10. Vallet-Regi, M.; Gonzalez-Calbet, J.M. Calcium Phosphates as Substitution of Bone Tissues. *Prog. Solid State Chem.* **2004**, *32*, 1–31.
11. http://www.azom.com/article.aspx?ArticleID=1743# (accessed on Dec23, 2014).
12. Gu, Y.W.; Khor, K.A.; Cheang, P. Bone-Like Apatite Layer Formation on Hydroxyapatite Prepared by Spark Plasma Sintering (SPS). *Biomaterials* **2004**, *25*, 4127–4134.
13. Rapacz-Kmita, A.; Slosarczyk, A.; Paszkiewicz, Z. HAp–ZrO_2 Composite Coatings Prepared by Plasma Spraying for Biomedical Applications. *Ceram. Int.* **2005**, *31*, 567–571.
14. Sasikumar, S.; Vijayaraghavan, R. Solution Combustion Synthesis of Bioceramic Calcium Phosphates by Single and Mixed Fuels—A Comparative Study. *Ceram. Int.* **2008**, *4*, 1373–1379.
15. Sasikumar, S.; Vijayaraghavan, R. Low Temperature Synthesis of Nanocrystalline Hydroxyapatite from Egg Shells by Combustion Method. *Trends Biomater. Artif. Organs*. **2006**, *19* (2), 70–73.
16. Wu, C.; Chang, J. Degradation, Bioactivity, and Cytocompatibility of Diopside, Akermanite, and Bredigite Ceramics. *J. Biomed. Mater. Res. B Appl Biomater.* **2007**, *83*, 153–160.
17. Hench, L.L. Wilson, J. *An Introduction to Bioceramics*. World Scientific Publishing, NJ, USA, 1993.
18. Kokubo, T.; Takadama, H. How Useful is SBF in Predicting In Vivo Bone Bioactivity? *Biomaterials* **2006**, *27*, 2907–2915.
19. Vallet-Regi, M.; Arcos, D. Silicon Substituted Hydroxyapatites: A Method to Upgrade Calcium Phosphate Based Implants. *J. Mater. Chem.* **2005**, *15*, 1509–1516.
20. Patel, N.; Best, S.M.; Bonfield, W.; Gibson, I.R. Hing, K.A.; Damien, E.; Revell, P.A. A Comparative Study on the In Vivo Behavior of Hydroxyapatite and Silicon Substituted Hydroxyapatite Granules. *J. Mater. Sci. Mater. Med.* **2002**, *13*, 1199–1206.
21. Liu, X.; Ding, C.; Chu, P.K. Mechanism of Apatite Formation on Wollastonite Coatings in Simulated Body Fluid. *Biomaterials* **2004**, *25*, 1755–1761.
22. Udduttula, A.; Koppala, S.; Sasikumar S. Sol–Gel Combustion Synthesis of Nanocrystalline Wollastonite by Using Glycine as a Fuel and Its In Vitro Bioactivity Studies. *Trans. Indian Ceram. Soc.* **2013**, *72* (4), 257–260.
23. Hench, L.L. The Story of Bioglass. *J. Mater. Sci. Mater. Med.* **2006**, *17*, 967–978.
24. De Aza, P.N.; Luklinska, Z.B.; Anseau, M.; Guitian F.; De Aza, S. Morphological Studies of Pseudowollastonite for Biomedical Application. *J. Microsc.* **1996**, *182* (1), 24–31.

25. Ortega-Lara, W.; Cortes-Hernandez, D.A.; Best, S.; Brooks, R.; Hernandez-Ramirez, A. Antibacterial Properties, In Vitro Bioactivity and Cell Proliferation of Titania–Wollastonite Composites. *Ceram. Int.* **2010**, *36*, 513–519.
26. Magallanes-Perdomo, M.; De Aza, A.H.; Mateus, A.Y.; Teixeira, S.; Monteiro, F.J.; De Aza, S.; Pena, P. In Vitro Study of the Proliferation and Growth of Human Bone Marrow Cells on Apatite–Wollastonite-2M Glass Ceramics. *Acta Biomater.* **2010**, *6*, 2254–2263.
27. Mohammadi, H.; Hafezi, M.; Nezafati, N.; Heasarki, S.; Nadernezhad, A.; Ghazanfari, S.M.H.; Sepantafar. M. Bioinorganics in Bioactive Calcium Silicate Ceramics for Bone Tissue Repair: Bioactivity and Biological Properties. *J. Ceram. Sci. Technol.* **2014**, *5* (1), 1–12.
28. Wu, C.-T.; Chang, J. Silicate Bioceramics for Bone Tissue Regeneration. *J. Inorg. Mater.* 2013, *28* (1), 29–39.
29. Sprio, S.; Tampieri, A.; Celotti, G.; Landi, E. Development of Hydroxyapatite/Calcium Silicate Composites Addressed to the Design of Load-Bearing Bone Scaffolds. *J. Mech. Behav. Biomed. Mater.* **2009**, *2*, 147–155.
30. Wei, J.; Chen, F.; Shin, J.-W.; Hong, H.; Dai, C.; Su, J.; Liu, C. Preparation and Characterization of Bioactive Mesoporous Wollastonite—Polycaprolactone Composite Scaffold. *Biomaterials* **2009**, *30*, 1080–1088.
31. Lakshmi, R.; Velmurugan, V.; Sasikumar, S. Preparation and Phase Evolution of Wollastonite by Sol–Gel Combustion Method Using Sucrose as the Fuel. *Combust. Sci. Technol.* **2013**, *185* (12), 1777–1785.
32. Udduttula, A.; Sasikumar, S. Bioactive Nanocrystalline Wollastonite Synthesized by Sol–Gel Combustion Method by Using Eggshell Waste as Calcium Source. *Bull. Mater. Sci.* **2014**, *37* (2), 207–212.
33. Lakshmi, R.; Sasikumar, S. Synthesis, Characterization and Bioactivity Studies of Calcium Silicate Bioceramics. *Adv. Mat. Res.* **2012**, *584*, 479–483.
34. Xue, W.; Liu, X.; Zheng, X.; Ding, C. In Vivo Evaluation of Plasma-Sprayed Wollastonite Coating. *Biomaterials* **2005**, *26*, 3455–3460.
35. Yoruc, A. B.; Hazar. Preparation and In Vitro Bioactivity of $CaSiO_3$ Powders. *Ceram. Int.* **2007**, *33*, 687–692.
36. Zhang, N,; Molenda, J.A.; Fournelle, J.H.; Murphy, W.L.; Sahai, N. Effects of Pseudowollastonite ($Casio_3$) Bioceramic on In Vitro Activity of Human Mesenchymal Stem Cells. *Biomaterials* **2010**, *31*, 7653–7665.
37. Lin, K.; Chang, J.; Lu, J. Synthesis of wollastonite nanowires via hydrothermal micro-emulsion methods. *Mater. Lett.* 2006, *60 (24)*, 3007–3010.
38. Meiszterics, A.; Sinko, K. Sol–Gel Derived Calcium Silicate Ceramics. *Colloids Surf. A* **2008**, *319* (1–3), 143–148.
39. Sreekanth Chakradhar, R.P.; Nagabhushana, B.M.; Chandrappa, G.T.; Ramesh, K.P.; Rao, J.L. Solution Combustion Derived Nanocrystalline Macroporous Wollastonite Ceramics: Review. *Mater. Chem. Phys.* **2006**, *95*, 169–175.
40. Huang, X.-H.; Chang, J. Synthesis of Nanocrystalline Wollastonite Powders by Citrate–Nitrate Gel Combustion Method. *Mater. Chem. Phys.* **2009**, *115*, 1–4.
41. Vichaphund, S.; Kitiwan, M.; Atong, D.; Thavorniti, P. Microwave synthesis of wollastonite powder from eggshells. *J. Eur. Ceram. Soc.* **2011**, *31*, 2435–2440.
42. Tangboriboon, N.; Khongnakhon, T.; Kittikul, S.; Kunanuruksapong, R.; Sirivat, A. An Innovative $Casio_3$ Dielectric Material from Eggshells by Sol–Gel Process. *J. Sol–Gel Sci. Technol.* **2011**, *58* (1), 33–41.

43. Liu, X.; Ding, C.; Chu, P.K. Mechanism of Apatite Formation on Wollastonite Coatings in Simulated Body Fluids. *Biomaterials* **2004**, *25*, 1755–1761.
44. Paluszkiewicz, C.; Blazewicz, M.; Podporska, J.; Gumuła, T. Nucleation of Hydroxyapatite Layer on Wollastonite Material Surface: FTIR Studies. *Vib. Spectrosc.* **2008**, *48*, 263–268.
45. Paital, S.R.; Dahotre, N.B. Calcium Phosphate Coatings for Bio-Implant Applications: Materials, Performance Factors and Methodologies. *Mater. Sci. Eng. R.* **2009**, *66*, 1–70.
46. Saiz, E.; Goldman, M.; Gomez-Vega, J.M.; Tomsia, A.P.; Marshall, G.W.; Marshall, S.J. In Vitro Behavior of Silicate Glass Coatings on Ti6Al4V. *Biomaterials* **2002**, *23*, 3749–3756.
47. Montenero, A.; Gnappi, G.; Ferrari, F.; Cesari, M.; Salvioli, E.; Mattogno, L.; Kaciulis, S.; Fini, M. Sol–Gel Derived Hydroxyapatite Coatings on Titanium Substrate. *J. Mater. Sci.* **2000**, *35*, 2791–2797.
48. Lakshmi, R.; Sasikumar, S. Influence of Needle Like Morphology on the Bioactivity of Nano Crystalline Wollastonite: An In-Vitro Study. *Int. J. Nanomedicine.* **2015** 10 (Suppl 1), pp. 129–136.
49. Alfred's Gemstone Library, Sparkles of Nature. http://www.alfredneo.com/?page_id=241 (accessed Sep 18, 2014).
50. Diopside Mineral Data and Information. http://www.mindat.org/min-1294.html (accessed Sep 16, 2014).
51. Diopside, from Wikipedia, the Free Encyclopedia. http://en.wikipedia.org/wiki/Diopside (accessed Sep 16, 2014).
52. Mineral Group Classification, Diagnostic Properties and Structural Information, Diopside ($CaMgSi_2O_6$) http://www.geo.arizona.edu/~mcnamara/diopside.html (assessed Aug 28, 2014).
53. Warren, B.; Bragg, W.L. The Structure of Diopside $CaMg(SiO_3)_2$. *J. Crystallogr. Cryst. Mater.* **1929**, *69* (1), 168.
54. Nagar, R. Preparation of Diopside by Novel Sol–Gel Method Using Rice Husk Ash as Silica Source. B.Tech Thesis. Department of Ceramic Engineering, National Institute of Technology, Rourkela, India. http://ethesis.nitrkl.ac.in/2165/1/Rahul_Nagar.pdf (assessed Sep 29, 2014).
55. Hench, L.L.; Splinter, R.J.; Allen, W.C.; Greenlee, T.K. Bonding Mechanism at the Interface of Bioceramics Prosthetic Materials. *J. Biomed. Mater. Res. Symp.* **1971**, *2*, 117–141.
56. Hench, L.L. Bioceramics: From Concept to Clinic. *J. Am. Ceram. Soc.* **1991**, *74* (7), 1487–1510.
57. Wu, C.; Chang, J. Degradation, Bioactivity, and Cytocompatibility of Diopside, Akermanite, and Bredigite Ceramics. *J. Biomed. Mater. Res. B Appl. Biomater*. **2007**, *83*, 153–160.
58. Kokubo, T. Novel Bioactive Materials. *An. Quim. Int.* **1997**, *93*, 49–55.
59. Padilla, S.; Roman, J.; Sanchez-Salcedo, S.; Vallet-Regi, M. Hydroxyapatite/SiO_2–CaO–P_2O_5 Glass Materials: In Vitro Bioactivity and Biocompatibility. *Acta Biomater.* **2006**, *2* (3), 331–342.
60. Saravanan, C.; Sasikumar, S. Bioactive Diopside (CaMgSi2O6) as a Drug Delivery Carrier—A Review. *Curr. Drug Deliv.* **2012**, *9*, 583–587.
61. Vallet-Regi, V.; Salinas, A.J.; Roman, J.; Gil, M. Effect of Magnesium Content on the *In Vitro* Bioactivity of CaO–MgO–SiO_2–P_2O_5 Sol–Gel Glasses. *J. Mater. Chem.* 1999, *9*, 515–518.

62. Sainz, M.A.; Pena, P.; Serena, S.; Caballero, A. Influence of Design on Bioactivity of Novel $Casio_3$–$Camg(Sio_3)_2$ Bioceramics: *In Vitro* Simulated Body Fluid Test and Thermodynamic Simulation *Acta Biomater.* **2010**, *6*, 2797–2807.
63. Nonami,T.; Tsutsumi, S. Study of Diopside Ceramics for Biomaterials. *J. Mater. Sci.–Mater. Med.* **1999**, *10*, 475–479.
64. Yamamoto, S.; Nonami, T.; Hase, H.; Kawamura, N. Fundamental Study on Apatite Precipitate Ability of CaO-MgO-SiO2 Compounders Employed Pseudo Body Solution of Application for Biomaterials. *J. Aus. Ceram. Soc.* **2012**, *48* (2) 180–184.
65. Bozadjiev, L.; Doncheva, L. Methods for Diopside Synthesis. *J. Univ. Chem. Technol. Metallurgy*. **2006**, *41* (2), 125–128.
66. I Wata, N.Y.; Lee, G.-H.; Tsunakawa, S.; Tokuoka, Y.; Kawashima, N. Preparation of Diopside with Apatite-Forming Ability by Sol–Gel Process Using Metal Alkoxide and Metal Salts. *Colloids Surf. B Biointerfaces* **2004**, 33, 1–6.
67. Ghorbanian, L.; Emadi, R.; Razavi, M.; Shin, H.; Teimouri, A. Synthesis and Characterization of Novel Nanodiopside Bioceramic Powder. *JNS.* **2012**, *2*, 357–361.
68. Choudhary, R.; Raj, R.; Swamiappan, S. Synthesis and Characterization of Diopside by Sol–Gel Combustion Method by Using L-Alanine as Fuel for Biomedical Applications. *J. Indian Chem. Soc.* (accepted manuscript).
69. Mohan, A.; Gandhi, M.; Swamiappan, S. In-Vitro Bioactivity Studies of Chitin/Diopside Composites. *Int. J. Chem. Tech. Res.* (accepted manuscript).
70. Kitsugi, T.; Yamamuro, T.; Nakamura, T.; Higashi, S.; Kakutani, Y.; Hyakuna, K.; Ito, S.; Kokubo, T.; Takagi, M.; Shibuya, T. Bone Bonding Behavior of Three Kinds of Apatite Containing Glass-Ceramics. *J. Biomed. Mater. Res.* **1986**, *20*, 1295–1307.
71. Kitsugi, T.; Nakamura, T.; Yamamuro, T.; Kokubo, T.; Shibuya, T.; Takagi, M. SEM-EPMA of Three Types of apatite-containing glass-ceramics implanted in bone: the variance of a Ca-P rich layer. *J. Biomed. Mater. Res.* **1987**, *21*, 1255–1271.
72. Kitsugi, T.; Yamamuro, T.; Nakamura, T.; Kokubo, T. Bone Bonding Behavior of MgO-CaO-SiO_2-P_2O_5-CaF_2 Glass (Mother Glass of A-W Glass-Ceramics). *J. Biomed. Mater. Res.* **1989**, *23*, 631–648.
73. Ohtsuki, C.; Kushitani, H.; Kokubo, T.; Kotani, S.; Yamamuro, T. Apatite Formation on the Surface of Ceravital-Type Glass-Ceramic in the Body. *J. Biomed. Mater. Res.* **1991**, *25*, 1363–1370.
74. Kokubo, T. Surface chemistry of bioactive glass-ceramics. *J. Non Cryst. Solids.* **1990**, *120*, 138 –151.
75. Cho, S.-B.; Nakanishi, K.; Kokubo, T; Soga, N. Dependence of Apatite Formation on Silica Gel on Its Structure: Effect of Heat Treatment. *J. Am. Ceram. Soc.* **1995**, *78* (7), 1769–1774.
76. Hench, L.L. Bioceramics: From Concept to Clinic. *J.Am.Ceram.Soc.***1991**, *74* (7), 1487–1510.
77. Wu, C.; Chang, J.; Ni, S.; Wang, J. *In-Vitro* Bioactivity of Akermanite Ceramics. *J. Biomed. Mater. Res.***2006**, *76* (1), 73–80.
78. Hou, X.; Yin, G.; Chen, X.; Liao, X.; Yao, Y.; Huang, Z. Effect of Akermanite Morphology on Precipitation of Bone-Like Apatite. *Appl. Surf. Sci.* **2011**, *257*, 3417–3422.
79. Iwata, N.Y.; Lee, G.-H.; Tsunakawa, S.; Tokuoka, Y.; Kawashima, N. Sintering Behavior and Apatite Formation of Diopside Prepared by Coprecipitation Process. *Colloids Surf. B Biointerfaces*. **2004**, *34*, 239–245.
80. Okazaki, M.; Takahashi, J.; Kimura, H. Specific Physicochemical Properties of Apatites: The Effect of Mg^{2+} Ions. *Den. Mater. J.* **1986**, *5* (5), 571–577.

81. Thiradilok, S.; Feagrin, F. Effects of Magnesium and Fluoride on Acid Resistance of Remineralized Enamel. *Ala. J. Med. Sci.* **1978**, *15* (2), 144–148.
82. Nonami, T.; Takahashi, C.; Yamazaki, J. Synthesis of Diopside by Alkoxide Method and Coating on Titanium. *J. Ceram. Soc. Jpn.* **1995**, *103* (7), 703–708.
83. Wu, C.; Ramaswamy, Y.; Zreiqat, H. Porous Diopside ($CaMgSi_2O_6$) Scaffold: A Promising Bioactive Material for Bone Tissue Engineering. *Acta Biomater*. **2010**, *6*, 2237–2245.
84. Wu. C.; Chang, J.; Zhai, W.; Ni, S.; Wang, J. Porous Akermanite Scaffolds for Bone Tissue Engineering: Preparation, Characterization and In Vitro Studies. *J. Biomed. Mater. Res. B Appl. Biomater*. **2006**, *78* (1), 47–55.
85. Du, R. Preparation and Characterization of Bioactive Glass for Bone Repair. Ph.D. Thesis. Chinese Academy of Sciences, Beijing, China, 2006.
86. Wu, C.; Ramaswamy, Y.; Boughton, P.; Zreiqat, H. Improvement of Mechanical and Biological Properties of Porous $Casio_3$ Scaffolds by Poly(D,L-Lactic Acid) Modification. *Acta Biomater*. **2008**, *4* (2), 343–53.
87. Holland, L. *The Properties of Glass Surface*. Wiley: New York, 1964; p 142.
88. Acros, D.; Greenspan, D.C.; Vallet-Regi, V. A New Quantitative Method to Evaluate the In Vitro Bioactivity of Melt and Sol–Gel Derived Silicate Glasses. *J. Biomed. Mater. Res*. **2003**, *65*, 344–351.
89. El-Ghannam, A.; Ducheyne, P.; Shapiro, I.M. Formation of Surface Reaction Products on Bioactive Glass and Their Effects on the Expression of the Osteoblastic Phenotype and the Deposition of Mineralized Extracellular Matrix. *Biomaterials* **1997**, *18*, 295–303.
90. Valerio, P.; Pereira, M.M.; Goes, A.M.; Leite, M.F. The Effect of Ionic Products from Bioactive Glass Dissolution on Osteoblast Prolferation and Collagen Production. *Biomaterials* **2004**, *25*, 2941–2948.
91. Composite material, from Wikipedia, the free encyclopedia, http://en.wikipedia.org/wiki/Composite material (assessed Sep 29, 2014).
92. Bunsell, A. R. *Handbook of Ceramic Composites.* Springer: USA, 2005.
93. Wang, M. Developing Bioactive Composite Materials for Tissue Replacement. *Biomaterials* **2003**, *24*, 2133–2151.
94. Learmonth, I.D. *Interface in Total Hip Arthroplasty*. Springer-Verlag: London, 2000.
95. Zhang, M.; Liu, C.; Zhang, X.; Pan, S.; Xu, Y. Al_2O_3/Diopside Ceramic Composites and Their Behavior in Simulated Body Fluid. *Ceram. Int*. **2010**, *36*, 2505–2509.
96. Yarlagadda, P.K.; Chandrasekharan, M.; Shyan, J.Y.M. Recent Advances and Current Developments in Tissue Scaffolding. *Bio-Med. Mater. Eng.* **2005**, *15* (3), 159–177.
97. Yunos, D.M.; Bretcanu, O.; Boccaccini, A.R. Polymer-Bioceramic Composites for Tissue Engineering Scaffolds. *J. Mater. Sci.* **2008**, *43*, 4433–4442.
98. Thamaraiselvi, T. V.; Rajeswari, S. Biological Evaluation of Bioceramic Materials—A Review. *Trends Biomater. Artif. Organs*. **2004**, *18*, 9–17.
99. Zhang, M.; Liu, C.; Sun, J.; Zhang, X. Hydroxyapatite/Diopside Ceramic Composites and Their Behaviour in Simulated Body Fluid. *Ceram. Int*. **2011**, *37*, 2025–2029.
101. Wu, C.; Zreiqat, H. Porous Bioactive Diopside ($CaMgSi_2O_6$) Ceramic Microspheres for Drug Delivery. *Acta Biomater.* **2010**, *6*, 820–829.
101. Xia, W.; Chang, J. Well-Ordered Mesoporous Bioactive Glasses (MBG): A Promising Bioactive Drug Delivery System. *J. Control Release.* **2006**, *110* (3), 522–30.
102. Wu, C.; Ramaswamy, Y.; Zhu, Y.; Zheng, R.; Appleyard, R.; Howard, A.; Zreiqat, H. The Effect of Mesoporous Bioactive Glass on the Physiochemical, Biological and

Drug-Release Properties of Poly(DL-Lactide-Co-Glycolide) Films. *Biomaterials* **2009**, *30* (12), 2199–208.
103. Li, H.; Chang, J. PH-Compensation Effect of Bioactive Inorganic Fillers on the Degradation of PLGA. *Comp. Sci. Tech.* **2005**, *65*, 2226–2232.
104. Forsterite, from Wikipedia, the Free Encyclopaedia. http://en.wikipedia.org/wiki/Forsterite (accessed Sep 16, 2014).
105. Forsterite Mineral Data, http://webmineral.com/data/Forsterite.shtml#.VC4jBPmSyjk (accessed Sep 16, 2014).
106. Forsterite Mineral Data and Information, http://www.mindat.org/min-1584.html (accessed Sep 16, 2014).
107. Cao, W.; Hench L.L. Bioactive Materials. *Ceram. Int.* **1996**, *22*, 493–507.
108. Saberi, A.; Alinejad, B.; Negahdari, Z.; Kazemi, F.; Almasi, A. A Novel Method to Low Temperature Synthesis of Nanocrystalline Forsterite. *Mater. Res. Bull.* **2007**, *42*, 666–673.
109. Bafrooei, H.B.; Ebadzadeh, T.; Majidian, H. Microwave Synthesis and Sintering of Forsterite Nanopowder Produced by High Energy Ball Milling. *Ceram. Int.* **2014**, *40*, 2869–2876.
110. Kharaziha, M.; Fathi, M.H. Synthesis and Characterization of Bioactive Forsterite Nanopowder. *Ceram. Int.* **2009**, *35*, 2449–2454.
111. Naghiu, M.A.; Gorea, M.; Mutch, E.; Kristaly, F.; Tomoaia-Cotise, M. Forsterite Nanopowder: Structural Characterization and Biocompatibility Evaluation. *J. Mater. Sci. Technol.* **2013**, *29* (7), 628–632.
112. Ni, S.; Chou, L.; Chang, J. Preparation and Characterization of Forsterite (Mg_2SiO_4) Bioceramics. *Ceram. Int.* **2007**, *33*, 83–88.
113. Ramesh, S.; Yaghoubi, A.; Sara Lee, K.Y.; Christopher Chin, K.M.; Purbolaksono, J.; Hamdi, M.; Hassan, M.A. Nanocrystalline Forsterite for Biomedical Applications: Synthesis, Microstructure and Mechanical Properties. *J. Mech. Behav. Biomed. Mater.* **2013**, *25*, 63–69.
114. Biomaterials: Mechanical Properties, from Wikipedia, the Free Encyclopaedia. http://en.wikipedia.org/wiki/Biomaterials:_Mechanical_Properties (accessed Sep 16, 2014).
115. Gerhardt, L.-C.; Boccaccini, A.R. Bioactive Glass and Glass-Ceramic Scaffolds for Bone Tissue Engineering. *Materials* **2010**, *3*, 3867–3910.
116. Kokubo, T.; Kim, H.-M.; Kawashita, M. Novel Bioactive Materials with Different Mechanical Properties. *Biomaterials*. **2003**, *24*, 2161–2175.
117. Orlovskii, V.P.; Komlev, V.S.; Barinov, S.M. Hydroxyapatite and Hydroxyapatite-Based Ceramics. *Inorg. Mater.* **2002**, *38* (10), 973–984.
118. Fathi, M.H.; Kharaziha, M. Two-Step Sintering of Dense, Nanostructural Forsterite. *Mater. Lett.* **2009**, *63*, 1455–1458.
119. Kharaziha, M.; Fathi, M.H. Improvement of Mechanical Properties and Biocompatibility of Forsterite Bioceramic Addressed to Bone Tissue Engineering Materials. *J. Mech. Behav. Biomed. Mater*. **2010**, *3*, 530–537.
120. Mirhadi, S.M. Synthesis and Characterization of Nanostructured Forsterite Scaffolds Using Two Step Sintering Method. 2014, (doi: http://dx.doi.org/10.1016/j.jallcom.2014.05.032).
121. Douy, A. Aqueous Synthesis of Forsterite (Mg_2SiO_4) and Enstatite ($MgSiO_3$). *J. Sol–Gel Sci. Technol.* **2002**, *24*, 221–228.
122. Ghomi, H.; Jaberzadeh, M.; Fathi, M.H. Novel Fabrication of Forsterite Scaffold with Improved Mechanical Properties. *J. Alloys Compd.* **2011**, *509*, L63–L68.

123. Ni, S.; Chou, L.; Chang, J. Preparation and Characterization of Forsterite (Mg_2SiO_4) Bioceramics. *Ceram. Int.* **2007**, *33*, 83–88.
124. Tavangarian, F.; Emadi, R. Improving Degradation Rate and Apatite Formation Ability of Nanostructure Forsterite. *Ceram. Int.* **2011**, *37*, 2275–2280.
125. Ni, S.; Chang, J. In Vitro Degradation, Bioactivity, and Cytocompatibility of Calcium Silicate, Dimagnesium Silicate, and Tricalcium Phosphate Bioceramics. *J. Biomater. Appl.* **2009**, *24* (2), 139–158.
126. Yazdanpanah, A.; Kamalian, R.; Moztarzadeh, F.; Mozafari, M.; Ravarian, R.; Tayebi, L. Enhancement of Fracture Toughness in Bioactive Glass-Based Nanocomposites with Nanocrystalline Forsterite as Advanced Biomaterials for Bone Tissue Engineering Applications. *Ceram. Int.* **2012**, *38*, 5007–5014.
127. Yarlagadda, P.K.; Chandrasekharan, M.; Shyan, J.Y.M. Recent Advances and Current Developments in Tissue Scaffolding. *Bio-Med. Mater. Eng.* **2005**, *15* (3), 159–177.
128. Diba, M.; Kharaziha, M.; Fathi, M.H.; Gholipourmalekabadi, M.; Samadikuchaksaraei, A. Preparation and Characterization of Polycaprolactone/Forsterite Nanocomposite Porous Scaffolds Designed for Bone Tissue Regeneration. *Compos. Sci. Technol.* **2012**, *72*, 716–723.
129. Zhu, H.; Shen, J.; Feng, X.; Zhang, H.; Guo, Y.; Chen, J. Fabrication and Characterization of Bioactive Silk Fibroin/Wollastonite Composite Scaffolds. *Mater. Sci. Eng. C.* **2010**, *30*, 132–40.
130. Emadi, R.; Tavangarian, F.; Esfahani, S.I.R.; Sheikhhosseini, A.; Kharaziha, M. Nanostructured Forsterite Coating Strengthens Porous Hydroxyapatite for Bone Tissue Engineering. *J. Am. Ceram. Soc.* **2010**, *93* (9), 2679–2683.
131. Ding, S.J.; Ju, C.P.; Chern, J.H. Surface Reaction of Stoichiometric and Calcium-Deficient Hydroxyapatite in Simulated Body Fluid. *J. Mater. Sci. Mater. Med.* **2000**, *11*, 183–190.
132. Fu, L.; Khor, K.A.; Lim, J.P. Processing, Microstructure and Mechanical Properties of Yttria Stabilized Zirconia Reinforced Hydroxyapatite Coatings. *Mater. Sci. Eng. A* **2001**, *316*, 46–51.
133. Liu, D.-M.; Yang, Q.; Troczynski, T. Sol–gel Hydroxyapatite Coatings on Stainless Steel Substrates. *Biomaterials* **2002**, *23*, 691–698.
134. Balamurugan, A.; Kannan, S.; Rajeswari, S. Bioactive Sol-Gel Hydroxyapatite Surface for Biomedical Applications—In Vitro Study. *Trends Biomater. Artif. Organs*. **2002**, *16* (1), 18–20.
135. Sebdani, M.M.; Fathi, M.H. Preparation and Characterization of Hydroxyapatite–Forsterite–Bioactive Glass Nanocomposite Coatings for Biomedical Applications. *Ceram. Int.* **2012**, *38*, 1325–1330.

CHAPTER 6

NANOBIOCERAMIC HYBRID MATERIALS FOR BIOMEDICAL APPLICATIONS

T.M. SRIDHAR[1*] and R. PRAVEEN[2]

[1]*Department of Analytical Chemistry, University of Madras, Guindy Campus, Chennai 600025, India*

[2]*The School of Metallurgy and Materials, University of Birmingham, Birmingham, UK. E-mail: tmsridhar23@gmail.com; sridhar@unom.ac.in*

CONTENTS

ABSTRACT

Bone is a natural functionally graded material, which exhibits two types of structures. One is a dense and stiff structure and the other is a porous, soft load-bearing structure known as cortical and cancellous bone, respectively. This property has inspired the development of functionally graded bioceramic resources as implant materials with the original biotissue. The most important considerations for their selection in the human body are their biocompatibility, corrosion resistance, tissue reactions, surface conditions, and osseointegration (a bone bed formed through direct attachment to bone). Surgical grade stainless steel and titanium and its alloys are widely used in orthopedic and dental restorations. The human body is a very hostile environment and metals and alloy implants are unique that they are exposed to this dynamic environment containing living cells, tissues, and biological fluids. Clinical experience has shown that metallic implants are susceptible to localized corrosion in the human body, releasing metal ions into the surrounding tissues. Common failures of metallic implants have led to the application of biocompatible and corrosion resistant coatings, as well as to surface modification of the alloys.

A bioactive material is one that elicits a specific biological response at the interface, which results in the formation of a bond between the tissues and the material. Hydroxyapiatite ceramics offer attractive properties, such as lack of toxicity, absence of intervening fibrous tissue, the possibility of forming a direct contact with bone, and the possibility to stimulate bone growth. Deposition of layers of nanobioceramic coatings on 316L SS, titanium, and magnesium alloys are a viable solution. Surface engineering, nanobioceramics, and functionally graded coatings are the promising techniques to battle corrosion of biomaterials. Nanobioceramics-based orthopedic implants as scaffolds and coatings combined with tissue engineering would lead to the development of new hybrid biomedical devices with the better understanding of the structure–property relationship. Modification of biomaterial surface properties through control of the characteristic length scale is one of the promising approaches to modulate select cell functions.

6.1 INTRODUCTION

Medical devices are devices made from synthetic or natural materials, which are used to restore the functioning of disabled organs or their parts. These materials are referred to as biomaterials and the devices are called as implants, which work in conjunction with medical intervention. They remain in incessantly contact with blood and tissues. Biomaterials are intended to interact with biological systems. Williams defined biomaterials as nonviable materials used in medical devices, intended to interact with the biological systems.[1] The fundamental requirement of a biomaterial is that the material should coexist in the human body without either having an undesirable or inappropriate effect on the other. Another essential requirement of any biomaterial is described as biocompatibility, which implies the ability of the material to perform with an appropriate host response in a specific application. These medical devices include from the most common hip replacements, prosthetic heart valves to the less common neurological prosthesis, and implant drug delivery systems. These devices when placed inside the body are termed as implants and are intended to remain there for a substantial period of time while prosthesis is a device used for long-term application and is expected to remain in the body for lifetime.

Biomaterials have enhanced the value of life for millions of people during the past 30 years. A few commonly used devices include intraocular lenses, heart valves, pacemakers, dental appliances, total hip replacements, joints, knees, and elbows, etc. Based on the site and area of application in the human body, the size, shape, technological inputs, and materials used vary from device to device. The four major classifications of biomaterials are metals, polymers, ceramics, and composites. The history of metallic biomaterials dates back to 18th century where gold, silver, and copper were used exploiting their antimicrobial properties in poor surgical environment. The emergence of modern era metals and alloys like 316L stainless steel, Co–Cr alloy and Ti–6Al–4V for orthopedic applications has changed the facade of orthopedic implants.

6.2 ORTHOPEDIC IMPLANTS

Though prosthetic materials have been implanted for centuries, the professional discipline of Biomaterial Science and Engineering has only begun to evolve only recently. The performance and biocompatibility of implants are continuously being improved by means of more traditional approaches, such as tribology studies, surface topographic designs, and newly evolving surface electrochemical modifications. These include the development of novel hybrid nanobiomaterials by design of the surface chemistry on a molecular level (allowing chemical reactions with the contacting tissues and cells), purposeful incorporation of material-triggered biological responses (such as preventing blood coagulation in catheters and vascular grafts or providing adhesion molecules for attaching tissues and cells) or use of prosthetic biomaterials with living biological substrates (e.g., meniscus replacement grafts contacting cartilage cells or implants containing insulin producing living cells).

The forms of biomaterial depend on its intended function in the body and are roughly divided into two categories: structural and functional materials. Structural materials are used for withstanding cyclic loading, such as mastication or walking motions. Functional biomaterials on the other hand are applied to release a specific biofunctional material, such as scaffolds, extra cellular matrix, drug delivery system in which the implant materials are inserted in hard and soft tissues. Load bearing implants are usually made from bulk, nonporous materials, but coatings or composite structures may also be used to achieve improved mechanical and interfacial chemical properties. Implants that serve to fill the space or augment existing bone tissues are used in the form of powders, particulates, or porous materials, which may be bioinert of bioactive.

6.3 METALS AND ALLOYS

The use of surgical metal implants in humans was first recorded in 1562 when a gold prosthesis was used to close a defect in a cleft palate.[2] These were primarily developed to function as prosthetic or fixation devices. Their shortcomings were notably inadequate strength and inertness, which resulted in the adoption of type 316L stainless steel, cobalt–chromium, and titanium and its alloys. These shortcomings resulted from intrinsic factors related to the interatomic bonding in these materials as well as from the microstructures that were obtained during processing.

Metals and alloys have a wide range of applications as implants, which include devices for fracture fixation, partial and total joint replacement, external splints, braces and traction apparatus, as well as dental amalgams. The high modulus and yield point coupled with the ductility of metals make them suitable for bearing high loads without leading to large deformations and permanent dimensional changes. The composition most commonly used for orthopedic implant fabrication consists of stainless steels (primarily for temporary fixation devices), cobalt-based alloys (cast and wrought), and increasingly—thermomechanically-processed titanium alloys. There has been an increasing interest in some β-titanium alloys due to their low Young's modulus ($E \approx 70$ GPa). In addition, the TiNi shape memory alloy have attracted much attention due to its ability to reproduce original shape upon exposure to body temperature and its pseudo-elastic properties ($E_{eff} \approx 40$ GPa), which allow the processing of low stiffness, high spring back, and orthodontic wires. Current dental implants are often made of commercially pure (CP) titanium or titanium alloys, while noble metals (e.g., Pt and Pt–Ir) also find use as electrodes in cardiac pacing and other neuromuscular stimulatory devices.

The most widely used stainless steel for implant applications is type 316L SS. This is an austenitic stainless steel that can be hardened by cold working. The austenitic phase is stabilized by nickel. This type of steel has low carbon content (max 0.03 wt.%) compared to other stainless steels. The lower carbon content provides an improved corrosion resistance in saline and chloride environments, which simulate the physiological environment. The presence of 2–4 wt.% molybdenum increases the resistance to pitting corrosion. The most important alloying constituent in stainless steels is chromium, which should have a concentration of at least 12 wt.% to allow the development of a passive chromium oxide layer that provides the corrosion resistance.

6.3.1 BREAKDOWN OF IMPLANT APPARATUS

Metallic orthopedic implants are often mounted on to the skeletal system of the human body as constituents of reconstructive devices (e.g., hip or knee joint replacement) or fracture fixation products (e.g., plates, screws, and nails). The design of these implants are dictated by the anatomy and restricted by the physiology of the skeletal structure of the human body. The mechanical and chemical stabilities and biocompatibility of the implant materials in the environment of tissues and body fluids are of fundamental importance for the successful treatment of bone fractures and bone replacements. An orthopedic implant is considered to have failed if it has to be prematurely removed from the body.

Unfortunately, many failures of stainless steel implants have been reported and they are related to corrosion. Clinical experience has shown that such implants are susceptible to crevice and pitting corrosion in the human body, causing release of metallic ions into the surrounding tissue. This causes local irritations or systemic effects and in some cases the removal of implants is indispensable. Moreover, allergenic reactions sometimes occur as well. Pure Ti and Ti–6Al–4V alloy are known to be more susceptible to wear than stainless steels, thus, generating greater amounts of metallic particles for a loose functional implant. Although cobalt–chromium alloys are generally known to have excellent wear and corrosion resistance in vivo, they do occasionally generate particles and release ions to their surroundings by mechanisms of corrosion and wear. Cobalt can be carcinogenic and may cause inheritable damage in exposed humans. Hence, it is difficult to demonstrate an advantage of titanium or cobalt–chromium alloys over stainless steels for in vivo applications. These shortcomings limit the use of titanium and cobalt–chromium alloys for implant applications.[3]

Corrosion is the major problem affecting the service life of orthopedic implants.[4–6] and it has two effects. First, the implant might weaken, resulting in a premature failure. Secondly, tissue reactions might occur due to the release of corrosion products from the implant. No metallic material is totally resistant to corrosion or ionization within living tissues. In vivo studies have shown that the implantation of devices from most alloys significantly increases the concentrations of various ions adjacent to the tissues. Moreover, once a foreign material is implanted, there are several ways in which the body might react unfavorably. The presence of the implant might inhibit the defense mechanisms of the body, leading to infection and requiring the removal of the implant. If infection does not occur or is overcome, the tissue response may range from mild edema to chronic inflammation, and alteration in bone, and tissue structures. These issues necessitate the requirement that materials used in making implants be inert or well tolerated by the body. The response of the body to an inert implant will be the development of a fibrous collagen sheath of low cellularity, which encapsulates the implant and separates it from the normal tissue. In some cases, the capsule has a well-defined boundary but in other cases, it extends irregularly and gets diffused into the surrounding muscles. The thickness of the fibrous sheath depends on the corrosion resistance of the material. The materials producing the thinnest sheaths are regarded as best tolerated by the body. In industry, common practices to control this phenomenon include the addition of inhibitors, changing the pH of the electrolyte, and its composition, reducing temperature, applying protective coatings, applying electrical current, etc. Unfortunately, most of these approaches cannot be used for orthopedic implants since the body environment is fixed and cannot be altered without destructive biological effects.

The failures with metallic implants is accompanied by the hostile cell environment leading us to explore the alternative class of biomaterials namely polymers, bioceramics, and composites.

6.4 POLYMERS

Polymers are extensively used in fabricating implants for various applications in the human body, especially in hip and knee ligaments, articulating components, orthopedics, ophthalmology, plastic, and reconstructive surgery, neuro surgery and as artificial organs. Polymethylmethacrylate (PMMA) has a long history of use in orthopedics as bone cement. There are three grades of polyethylene: low-density polyethylene (LDPE); high-density polyethylene (HDPE); and ultra-high molecular weight polyethylene (UHMWPE), which are also widely used in surgery. UHMWPE is widely used for joint replacements as it processes superior load-bearing surfaces. However, these materials have disadvantages of friction and wear characteristics. In case of polyethylene total knee components, a reduction in molecular weight and surface degradation of the components adjacent to load bearing surfaces is observed. In silicone implants, the silicone-gel filling causes damage to patient's immune system and other disorders. The use of Teflon (PTFE) in temporo mandibular joints causes fragmentation and damage to surrounding tissues.

Generally speaking, polymers have poorer mechanical properties than bone. But the possibility to be mechanically strengthened and to be biodegradable makes polymers very promising as candidates for bone replacement. The improvement of the mechanical properties of polymer can be achieved by either the modification of the structure of the polymer, or the strengthening of the polymer with fiber and/or filler. The use of polymer matrix composites for bone replacement may offer the advantages of avoiding the problem of stress shielding, eliminating the need for a second surgical procedure to remove the implants if the implants can be made biodegradable, and also the elimination of the ion release problem of metal implants. The possibility to make the composites as strong as cortical bone and to improve the material's bioactivity or bone-bonding activity by adding of a secondary reinforcing phase makes the composites very attractive. Fibers and mineral filler particles can be used to reinforce the polymer materials as well as to improve the bone-bonding properties of the composites.

The use of particulate fillers to reinforce polymeric biomaterials is quite important and quite successful in clinical applications, like dental restorative resins and bone cement. The use of a bioactive filler, such as hydroxyapatite (HAP) or Bioglass particles to reinforce a polymer may improve both the mechanical properties and the bone-bonding properties. When implanted in vivo, such composites will induce bone formation or bone ingrowth and as the biodegradable polymer matrix degrades the implant will finally be replaced by bone tissue. In this way, the load can be gradually transferred to the newly formed bone. Based upon this idea, several hydroxyapatite reinforced biodegradable polymer composites have been developed such as HAP/polyhydroxybutyrate.[7–10]

6.5 BIOCERAMICS

Ceramics play a major role in biomaterials due to their excellent biocompatibility and they are known as bioceramics, which are extensively calcium phosphate based, resembling the composition of the skeletal structure.[11] Bioceramics can be divided into three major groups based on the nature of reactivity with the physiological environment. The three main groups of bioceramics are inert, surface reactive or bioactive, and resorbable.[12,13] Inert bioceramics can remain in a physiological environment for longer periods without leaching out any toxic chemicals nor by reacting or activating with the neighboring cells and tissues.[14,15] Surface reactive ceramics or bioactive ceramics can react with the tissue in a physiological environment forming a chemical bond.[12,14,15] The skeletal system has the capacity to repair itself, although the capabilities diminish with age and disease. The ideal solution to the problem of interfacial safety is the use of biomaterials to augment the body's own reparative process. These bioactive materials stimulate bone growth and form a firm bond with the bone and this process is designated as osteointergration. On the other hand, resorbable ceramics degrade along with the heeling of the diseased site without causing any harm to the environment around the tissue.[14,15] Resorbable implants, such as tricalcium phosphate and other calcium phosphate phases and some bioglasses are based on this concept. Chemical and mechanical problems exist in the development and use of resorbable biomaterials. First, the products of resorption must be compatible with the cellular metabolic processes. This compatibility is a highly restrictive requirement. The capacities of the cellular environment, cardiovascular, and lymphatic systems to process and transport of resorption products must match the rate of resorption.[16] The second problem is that the strength of the resorbable implant decreases as chemical breakdown occurs. Eventually, strength is restored by the growth of the replacement tissues, which is the subject of implantation. However, during the interim period the mechanical properties required for the function are difficult to maintain.

6.5.1 BIOACTIVE CERAMICS

Bioactive materials offer an alternative to the establishment of a stable implant-tissue interface. A bioactive material is one that elicits a specific biological response at the interface, which results in the formation of a bond between the tissues and the material. This concept is based on the control of surface chemistry of the material. A bioactive implant reacts chemically with the body fluids in a manner that is compatible with the repair process of the tissues. Formation of the fibrous capsule is prevented by the adhesion of repairing tissues. Because the chemical reactions are restricted to the surface, the material does not degrade in strength. The important point is that failure does not occur at the interface as HAP bonds to soft tissues as well as to bone, with adherence strength greater than the cohesive strength of the collagen bundle fibers of the soft connective tissues.

6.5.2 MAJOR BONE CONSTITUENT: HYDROXYAPATITE

The interest in calcium phosphates (CaP)-based bioceramics stems from the fact that bones and teeth contain a high percentage of mineralized calcium phosphate known as hydroxyapatite (HAP). The similarity between the in vivo mineral and basic calcium phosphate has been proved long ago by powder X-ray diffraction (XRD) studies. Basic calcium phosphates precipitated from aqueous solutions form a nonstoichiometric series partly because of substitutions by other ions and surface effects due to extremely small crystallite sizes formed by this low-solubility mineral. These properties coupled with the fact that the kinetics due to large, complex unit cell of apatite favor dissolution, and the possible existence of one or more distinct precursor phases contributes to its structural complexity and its size.

The major deviations from stoichiometry in apatitic calcium phosphates as observed by XRD patterns are calcium and hydroxyl deficiency, which are respectively substituted by magnesium, lanthanum, sodium, and strontium for calcium ions, carbonate ion for hydroxyl and/or phosphate ions, fluoride or chloride substitution for hydroxyl ions, acid phosphates being substituted for orthophosphate ions and substituting for calcium ions. Apatitic calcium phosphates can have crystallites ranging from a few tens to thousands of Angstrom.

Hydroxyapatite has a hexagonal structure of the space group $P6_3/m$, with basal plane edge a = 9.432 Å and height c = 6.81 Å. The cell contents are given by the formula $Ca_{10}(PO_4)_6(OH)_2$. The hydroxyl ions lie in projection at the corners of the rhombic base of the unit cell forming columns of hydroxyls with a spacing of half the unit-cell height. Six of the calcium ions are associated with these hydroxyls, forming equilateral triangles centered on and perpendicular to the hydroxyl columns. Four calcium ions lie along two separate columns at heights half way between the calcium triangles. Oxygen from the orthophosphate tetrahedra coordinates these calcium ions. The hydroxyl calcium ions are displaced by 0.3 Å from the plane of the calcium triangles. The hydroxyls are oriented such that the oxygen–hydrogen bond is along the column axis but does not cross the plane of calcium triangle. The solubility constants of hydroxyapatite in aqueous solutions vary within the range 10^{-49}–10^{-58}. The chemical and crystallographic similarity of HAP to bone mineral allows it to form very strong biological bonds.

The chemical and crystallographic similarity of HAP to bone mineral allows it to form very strong biological bonds and functions as a smart bioactive material. The material has been shown to have a simulating effect on bone formation known as osseoinduction. It enhances osseointergration and there are indications that chemical bonding may occur between HAP and bone. The long-term stability of HAP is guaranteed in particular by the phase purity, because a chemical breakdown of the layer is induced by higher solubility of foreign phases. The biological performance of HAP layers is based on a rapid osteointegration with a pronounced increase in the load carrying capacity of the implant.

6.5.3 THERMAL BEHAVIOR

The manufacturing of calcium phosphates for biomedical application includes heat treatment of devices for various purposes including sintering for the formation of nanostructures. At high temperatures, the thermal behavior of the HAP structure may be modified. Results from thermogravimetric analysis, X-ray diffraction, and FTIR absorption analysis show the presence of two types of water namely absorbed water and lattice water. Stoichiometric HAP contains constitutional water in the form of OH^- ions, this water can be driven off at high temperatures (~ 1200 °C), first producing a partially dehydrated HAP, which presumably contains one O^{2-} ion of each water molecule that has been lost. At high temperatures, HAP can be totally or partially dehydrated. Above 900°C a small weight loss is recorded and from 1050°C onwards the hydroxyapatite may decompose to give beta tricalclium phosphate (TCP) $\beta Ca_3(PO_4)_2$ and $Ca_4P_2O_9$. A further increase in temperatures above 1350 °C, β-$Ca_3(PO_4)_2$ irreversibly transforms to α-$Ca_3(PO_4)_2$. Bioresorbable properties are exhibited by α and β-TCP and these ceramics degrade to varying degrees in the order α-TCP>β-TCP>HAP.[17]

HAP has been widely used not only as a biomedical implant material but also as a biological chromatography support material for protein purification. It is also currently used for fractionation and purification of a wide range of biological molecules, such as subclass of enzymes, antibody fragments, and nucleic acids. A crystalline HAP column is used commonly in high-performance liquid chromatography (HPLC) with typical ceramic beads of the size 20–80 μm.

Calcium phosphate ceramics are probably not as good bone inducers as autogenous material but they do enhance bone growth. The exact chemical composition of the reaction layer between the implant and bone has not been completely understood. The possible explanations are furnished by crystal growth mechanisms, that is, either precipitation from a supersaturated solution or epitaxial growth. A slow resorption of a degradable ceramic provides excess calcium and phosphate, which further crystallizes toward HAP. Another possibility more likely applicable in almost all nondegradable ceramics is the epitaxial growth of new crystals upon the implant surface.[18–20]

Among the various classes of bioceramics involving calcium phosphate groups, hydroxyapatite on implantation in the body forms a chemical bond between hydroxyapatite and bone.[21–23] Materials like Bioglass®, Apatite–Wollastonite glass-ceramics, and calcium phosphates are few other examples of developed osteoconductive materials, which integrate with bone. Recently, researchers showed a vast interest in bioceramics for creating new materials with osteoconductive properties for biomedical applications.

Hydroxyapatite ceramics obviously show attractive properties, such as lack of toxicity, absence of intervening fibrous tissue, the possibility of forming a direct contact with bone, and the possibility to stimulate bone growth. However, its brittle nature and poor mechanical properties curtail its use directly as an implantation material. This has paved the way for the development of HAP coatings on commonly used type 316L SS, Co–Cr alloys, and Ti and its alloys for use as various types of bone replacements and fractures with the mechanical strength provided by the metals. This would assist toward direct mineralization of bone which creates a strong bonding between bone and implanted materials.

Bioactive resorbable materials offer an alternative to the achievement of a stable implant–tissue interface. A bioactive material is one that elicits a specific biological response at the interface of the material, which results in the formation of a bond between the tissues and the material. This concept is based on the control of surface chemistry of the material. A bioactive implant reacts chemically with the body fluids in a manner that is compatible with the repair process of the tissues. Formation of the fibrous capsule is prevented by the adhesion of repairing tissues. Because the chemical reactions are restricted to the surface, the material does not degrade in strength, unlike resorbable or porous implants. The important finding is that failure does not occur through the interface. Hydroxyapatite bonds to soft tissues as well as bone, with adherence strength greater than the cohesive strength of the collagen bundle fibers of the soft connective tissues.

6.5.4 BIOACTIVE GLASS AND GLASS-CERAMICS AS HYBRID BIOACTIVE BIOCERAMICS

Table 6.1 shows the list of bioactive glasses and glass-ceramic compositions that are currently in practice for medical treatments. Initially, two materials were developed for implantation: Bioglass®, which is an amorphous glass based on the general composition shown in Table 6.1 and Ceravital® a ceramic-based one which is also shown in Table 6.1.[24,25] These materials were developed to enable direct bonding with the tissue that would ensure the stability of the implanted material. They bond directly to the bone and while dissolving over time, they can stimulate new bone growth. Through this process they have the ability to restore diseased or damaged bone to its original state and function Their composition and structure are of paramount importance as this controls the rate of release of products from the surface. A very low release rate would not allow bonding to occur, while a higher rate might lead to cell necrosis.

Bioglass® is brittle and not appropriate for load bearing applications.[26] Furthermore, a new machinable mica–apatite glass-ceramic was developed based on the general composition of SiO_2–Al_2O_3–MgO–Na_2O–K_2O–F–CaO–P_2O_5, which is osteo-conductive like Bioglass®.[27,28] Apatite–wollastonite (A–W) glass-ceramics consists of a congregation of minute apatite specks that are successfully toughened by wollastonite for bone tissue applications. A–W is another hybrid glass-ceramic that consists of two main crystalline phases; apatite [$Ca_{10}(PO_4)_6(OH,F)_2$] and wollastonite [$CaO{\cdot}SiO_2$].[24,26,27] The dense and consistent apatite–wollastonite glass-ceramics is based on 34.0 SiO_2-44.7 CaO-16.2 P_2O_5-4.6 MgO (Cerabone®) glass composition have a tensile strength that varies between 100 and 200 MPa and modulus of elasticity in the range of 60–100 GPa. The compression and tensile strength of human bone ranges from 140 to 200 MPa and the elastic modulus ranges from 19 to 20 GPa. The compression strength of the latter is very similar to the compression strength of apatite–wollastonite glass-ceramics as mentioned above.[28,29]

Another hybrid A–W glass-ceramic was developed by bringing together micrometric apatite particles together with wollastonite for load bearing applications. It is an impenetrable and uniform combination, which is a mixture of 28 wt% Wollastonite, 34 wt%Oxyfluoroapatite, and 28 wt% glass.[30] These A-–W glass-ceramics possess good mechanical properties required for orthopedic implants like the bending strength, fracture toughness, and Young's modulus, which are the highest among bioactive glass and glass-ceramics. These properties facilitate their applications in major density based load bearing applications.

TABLE 6.1 Types of Bioactive Glass and Glass-Ceramics by Composition (Dubok 2000).

Composition by Mass %	Bioactive Glass		Bioactive Glass-Ceramics			
	45S5	S45PZ	Ceravital®	Cerabone® A/W	Ilmaplant®	Bioverit®
Na_2O	24.5	24	5–10	0	4.6	3–8
K_2O	0	–	0.5–3.0	0	0.2	3–8
MgO	0	–	2.5–5.0	4.6	2.8	2–21
CaO	24.5	22	30–35	44.7	31.9	10–34
Al_2O_3	0	–	0	0	0	8–15
SiO_2	45.0	45	40–50	34.0	44.3	19–54
P_2O_5	6.0	7	10–50	16.2	11.2	2–10
CaF_2	0	–	–	0.5	5.0	3–23
B_2O_3	0	2	–	–	–	–
Phase	Glass	Glass	Apatite, glass	Apatite, β-wollastonite, glass	Apatite, β-wollastonite, glass	Apatite, Fluoro-phlogopite, glass

Bioglasses are susceptible to both general corrosion and pitting attack in aqueous solution as they contain a silicate glassy phase. The corrosion process involves a proton from the water penetrating the glassy network and replacing an alkali ion, which is then released into solution. An OH^- ion is produced, which destroys the Si—O—Si bonds, forming Si—O bonds, which then interacts with water and goes into solution. The release of Si—O bond is believed to be important in the initiation of bone mineralization and in producing a bond between the bone and the bioglass. The use of bioglass with ceramics is at a relatively early stage and much work is required to develop and explain its degradation in the body and to study in detail the effects on the adjacent tissue.[31–33]

6.5.5 BIORESORBABLE CERAMICS

The skeletal system has the capacity to repair itself, although the capabilities diminish with age and disease. The ideal solution to the problem of interfacial safety is the use of biomaterials to augment the body's own reparative process.[31] Resorbable implants, such as tricalcium phosphate and other calcium phosphate phases and some bioglasses are based on this concept. Bioresorbable ceramics are also known as biodegradable ceramics; the advantages of these ceramics are limitless because they can degrade in the body avoiding the need for a second operation or any passive internal damage occurred during an operation. After implanting bioresorbable ceramics, they react with the surrounding tissues and help the tissues to grow along as the materials degrade away.[20,30] This phenomenon is called osteo-induction. In orthopedics, these newly formed tissues are the new bones formed at the site of implant. These kind of ceramics release the chemicals, which form the new bone is formed at a much accelerated rate when compared with other bioactive materials, but it is very difficult to achieve due to the rate of resorption along with the stability and strength of the materials.[31,34] Resorbable ceramics do not produce any toxic substances while degrading; only very few ceramics have been successful as resorbable materials and one of them is tri-calcium phosphate (TCP) that has been in applications for regeneration and restoration of bone defects. Two chemical and mechanical problems exist in the development and use of resorbable biomaterials.[35] First, the products of resorption must be compatible with the cellular metabolic processes. This compatibility is a highly restrictive requirement. The capacities of the cellular environment, cardiovascular, and lymphatic systems to process and transport of resorption products must match the rate of resorption. The second problem is that the strength of the resorbable implant decreases as chemical breakdown occurs. Eventually strength is restored by the growth of the replacement tissues, which is the subject of implantation. However, during the interim period the mechanical properties required for the function are difficult to maintain.

6.5.6 BIOINERT CERAMICS

Bioinert ceramics do not react with tissues in a biological environment. These types of ceramics stay in the biological environment for longer period. Alumina and zirconia are good examples of inert materials and are used as implants in both dental and orthopedic applications.[30,36] These inert ceramics have very low surface reactivity and they have superior mechanical properties when compared with other materials. These materials are used as coatings of hip implants due to minimal friction after implantation. If these materials are implanted and undisturbed, the new bone will grow around the implant.[37–39]

Alumina, also known as aluminum oxide (Al_2O_3), is widely used oxide ceramics and its applications are limitless in both engineering and medical fields. Alumina is the most chemically inactive among the inert ceramics, used in reconstructive surgery. It has mechanical properties include good hardness compared to most ceramics, chemical durability and stability even at high temperatures where no phase transformation occurs. Due to its rigidity and hardness, alumina has found its main uses in hard tissue, structural applications in orthopedic surgery, and dentistry. Its bioinertness property has improved its use in nearly load free applications in ear, nose, and throat surgery and in some soft tissue replacements. The strength of alumina is sufficient for use under compressive loads, but even high-purity alumina exhibits slow crack growth, static and cyclic fatigue, and low toughness under tensile loads in physiological environments. Alumina offers no advantage over metals with regard to stress shielding of bone and interface stability because of its high modulus of elasticity and the formation of a thin, fibrin capsule weakens the interfacial stability.[40]

High purity alumina ceramics with purity greater than 99.99% have been developed for orthopedic applications as alternatives for metal alloys for hip replacement applications. For this purpose, the material must be free of porosity and it must have a fine and homogeneous microstructure, as this will allow the material to achieve a long-term stability. The application of alumina with purity of 99.99% is used in total hip replacement. This alumina possesses a mixed combination of excellent corrosion resistance, good biocompatibility and high wear resistance, and high strength. The reasons for the excellent wear and friction behavior of (Al_2O_3) are associated with the surface energy and surface smoothness of this ceramic.[41] However, the main problem with alumina is its brittleness, which makes parts made with it to be prone to fracture. This hard but brittle combination of material properties means that certain design restrictions apply.

Biocompatibility studies of alumina have been tested and its efficacy and safety has been reported by many researchers. Histopathological studies carried out by implanting it in the eye sockets of albino rabbits for 8 weeks indicate that the results showed no signs of implant rejection or prolapse of the implanted piece Noiri et al.[42] After a period of 4 weeks of implantation, fibroblast proliferation and vascular invasion were noted and by eighth week, tissue growth was noted in the pores of the implant. Single crystal alumina screws and pins were implanted in the femoral bone of mature rabbits. Changes in the implant-bone interface were observed. Alumina was never in direct contact with the bone and cytotoxicity was not observed in the interface.[40,43]

Zirconia: It is a well-known polymorph that occurs in three forms: monoclinic (M), cubic (C), and tetragonal (T). Pure zirconia is monoclinic at room temperature. This phase is stable up to 1170 °C. Above this temperature it transforms into tetragonal and then into cubic phase at 2370 °C. Cracks are generally observed in pure zirconia on sintering in the temperature range of 1500–1700°C due to the stresses generated during the expansion of the material and further fractures in to pieces at room temperature. To overcome these issues stabilizing oxides, which include Y_2O_3, La_2O_3, CeO_2, CaO, MgO, etc. are added to pure zirconia. The addition of oxides generate multiphase materials known as partially stabilized zirconia (PSZ) whose microstructure at room temperature generally consists of cubic zirconia as the major phase, with monoclinic and tetragonal zirconia precipitates as the minor phase.[50] The precipitates are fully coherent with the cubic lattice, forming on a nanometer scale with lenticular morphology (approximately 200 nm diameter and 75 nm thick) parallel to the three cubic axes. Following sintering or solution annealing in the cubic solid solution single phase field (approximately >1850°C) these precipitates are nucleated and grown at lower temperatures (approximately 1100°C) within the two-phase tetragonal solid solution plus cubic solid solution phase field; a process termed "aging."[44–47] These ceramics processes very interesting mechanical properties due to the transformation toughening mechanisms operating in their microstructure that can give to components made out of them when compared with other ceramic materials.

Zirconia is one of the most promising restorative biomaterial, because it has very favorable mechanical and chemical properties suitable for biomedical applications. Zirconia ceramics (ZrO_2) are becoming a widespread biomaterial in dentistry and dental implantology. The use of zirconia ceramics as biomaterials has been investigated for the past three decades but developments are in progress for application in other medical devices. Current developments have concentrated on the chemistry of precursors used in forming and sintering processes and on surface finish of components. It is commonly used as total hip replacement ball heads. In vitro experimentation showed absence of mutations and good viability of cells cultured on this material.[45] Zirconia cores for fixed partial dentures (FPD) on anterior and posterior teeth and on implants are now available. Mechanical properties of zirconium oxide FPDs have proved superior to those of other metal-free restorations. Clinical evaluations, which have been ongoing, indicate a good success rate for zirconia FPDs. Zirconia implant abutments can also be used to improve the aesthetic outcome of implant-supported rehabilitations.

Zirconia is also isolated from tissues by a thin fibrous capsule and therefore, it does not improve interfacial stability over metal implants. The ageing process of the ceramic strongly influences the variability of zirconia to in vivo degradation. Zirconia is prone to ageing in the presence of water due to its metastability.[45] Currently, research is in progress to develop alumina–zirconia composites as an alternative to monolithic alumina and zirconia. It is based on the concept that moderate toughness of alumina and ageing in zirconia would combined to develop bioinert composites. These composites would facilitate to take advantage from zirconia's transformation toughening without the major drawback associated with its transformation under stress or body fluid condition. In the recent literature concerning alumina–zirconia composites for biomedical applications, different compositions have been tested, from the zirconia rich to the alumina rich side.[30,47] A composite material processed with 80% tetragonal zirconia polycrystals (ZrO_2–TZP) and 20% alumina were reported to have outstanding mechanical and tribological properties. The alumina-toughened zirconia (ATZ) Bio-Hips, developed by Metoxit AG (Thayngen, Switzerland), has a bending strength of up to 2000 MPa, indicating that it can withstand loads that are four times greater than conventional alumina implants.[48]

Carbon: is a versatile element and exists in a variety of forms. Good compatibility of carbonaceous materials with bone and other tissue and the similarity of the mechanical properties of carbon to those of bone indicate that carbon is an exciting candidate for orthopedic implants.[49] Unlike metals, polymers and other ceramics, these carbonaceous materials do not suffer from fatigue. However, their intrinsic brittleness and low tensile strength limits their use in major load bearing applications. The mechanical bonding between the carbon fiber reinforced carbon and host tissue was investigated. The bonding is developed 3 months after intrabone implantation and is accompanied by a decrease of the implant strength.

Carbon in its various forms namely diamond, graphite, pyrolitic low-temperature isotropic (LTI), and ultra-low temperature isotropic (ULTI) carbon and vitreous carbon are applied as implant materials. These carbon-based materials are highly useful for many tissue interfacing applications (e.g., disc occluders in heart valves). They do not normally invoke undesirable host responses. The commonly used heart valves are made of LTI carbon containing the orifice and the occluder. This is because of their excellent resistance to blood clot formation and long fatigue life. Pyrolytic and glassy carbons are used in clinical application as cardiovascular, orthopedic, and dental areas for long-term percutaneous applications. Intrinsic brittleness and low tensile strength is a key concern that must be addressed while using them for major load bearing applications.

6.6 COMPOSITES

Composites are the result of the mixture of two different or more materials or phases taking advantage of the salient features of each constituent employed leading to a material with improved properties.[50] It is essential that each component of the composite be biocompatible to avoid degradation between interfaces of the constituents. Glass ionomer cements are an example of such a material. Fiber-reinforced polymers (FRP) are the most widely investigated composites for medical device applications. These have included the development of low stiffness femoral components for hip joint arthroplasty, radiolucent, and biodegradable fracture fixation devices, fracture-resistant bone cements, wear, creep, and fracture-resistant articulation components. The uncertain lifetime under complex stress state, degradation, and low mechanical strengths curtail their use.

Natural composites display a network consisting of arrangements, which are particulate, porous, and fibrous structural features ranging from several orders of micrometers. Composite materials offer a range of fine properties in comparison with homogeneous materials. These properties offer scope to material technologists to engineer the material properties with precise control over them. There is the potential for stiff, strong, lightweight materials as well as for highly resilient and compliant materials. The major requirement of composites for biomedical applications is that each constituent of the composite must be biocompatible. The interface formed connecting the components should not be tarnished by the body environment. A few commonly used biomedical applications of composites are dental filling composites, dental molds, methyl methacrylate bone cement and ultrahigh molecular weight polyethylene, and orthopedic implants with porous surfaces. Dental composite resins have virtually replaced the controversial amalgams, which were commonly a couple of decades ago and most commonly in developing countries as dental fillings.[41,51] The dental composite resins consist of a polymer matrix and inflexible inorganic additive substances, which form the polymer composite matrix.

In case of bone composites are used as bone cements by incorporating a certain percentage of constituent materials. The property of PMMA cement can be increased in terms of its stiffness and fatigue life by addition of bone particles.[51] The property of bioresorption of bone and bone particles occurs at the interface with the site of implantation leading to the affected area being replaced by ingrown new bone tissue. But analogous implant materials with similar mechanical properties have to be prepared where the bone is to be replaced. Consequently, another approach to attainment of properties analogous to bone is to stiffen a biocompatible synthetic polymer such as PE (Polyethylene) with higher modulus ceramic (second phase) such as hydroxyapatite. Machining the surface of the composite exposes the bioactive phase. Thus, the mechanical properties of PE-HA are close to or superior to that of bone.

Natural bone tissue resembles a nanocomposite structure in several senses that is, being a porous polymer ceramic material, a lamellar material, and a fiber-matrix material, which provides appropriate physical and biological properties. To develop an ideal bone scaffold nanocomposites are the best choice as they can be engineered to possess the composition, structure, and properties of native bone.

6.7 BIOCERAMICS–TISSUE INTERFACES

The mechanism of tissue attachment is directly related to the type of tissue response at the implant interface. The relative levels of reaction of the implant influence the thickness of the interfacial zone or layer between the material and tissues. The in growth of tissues into the pores takes place on the surface of microporous bioceramics. The increased interfacial area between the implant and the tissues results in increased inertial resistance. It is capable of withstanding more complex stress states than implants that reach only biological fixation. All of these bioactive materials form an interfacial bond with the adjacent tissue. However, the type of bonding, its strength, and the thickness of the bonding zone differ for different materials.

6.7.1 THE STIMULATION OF BONE GROWTH

Although calcium phosphate ceramics are probably not as good bone inducers as autogenous materials, they do enhance bone growth. The way in which calcium phosphate ceramics enhance bone growth is not completely understood. Possible explanations are furnished by crystal growth mechanisms, that is, either precipitation from a supersaturated solution or epitaxial growth. A slow resorption of a degradable ceramic provides excess calcium and phosphate, which further crystallizes to HAP.[4] Another possibility, which is more applicable to almost all nondegradable ceramics is the epitaxial growth of new crystals on the implant surface. The bonding begins when a narrow electron dense band with no distinct surface details covers the implant. Loosely arranged collagen cells position themselves between the ceramic surface and cells. The adsorption of proteins to the biomaterial surfaces is favorable for the interactions between cells and biomaterials. Calcium phosphate ceramics have the excellent advantage of facilitating this adsorption. Among the calcium phosphates, hydroxyapatite has the largest capacity of adsorbing proteins to its surface.[52] As the implant site matures, the bone-bonding zone gets integrated with the natural bone as given in Figure 6.1. The bone-bonding substance on the surface of the implants is very similar in character to natural bone and may be characterized as amorphous in structure, heavily mineralized and rich in mucopolysaccharides.[34,53] Thus, researchers all over the world have been trying to develop surfaces which can result into faster osseointegration. Implant design (macro scale)

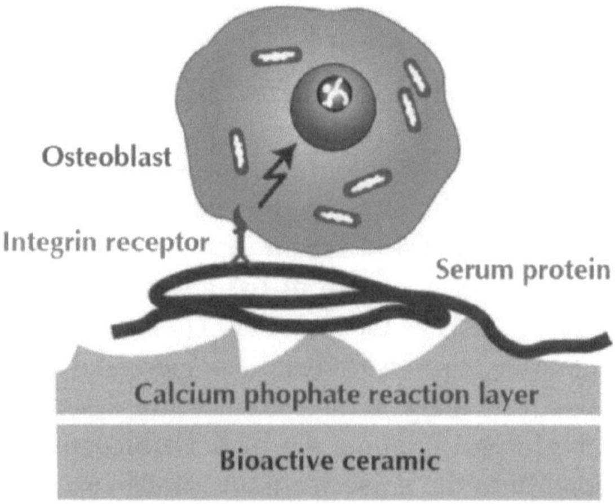

FIGURE 6.1 Illustration of bioactive behavior of bioceramics [5,6] (From Mueller, W.J., Poult. Sci., 40, 1562, 1961. with permission).

and aspects related to the surface properties in micro- and nanometric scale can affect osseointegration. However, those properties (surface composition and energy, topography and roughness) are interrelated.

But the human hard tissue bone is an interesting and intriguing tissue, which is highly adaptable and shows specific response to the placement of biomaterials in its vicinity. Replacement and reconstruction of bone has been attempted over the years. Damage to the bones as fractures due to accidents, bone cancer, and diseases has accelerated the pace of research to discover new materials and technologies to develop implants. In recent years, the number of materials used for bone reconstruction has been on the rise. Especially, long-term replacement and augmentation of bone using rigid materials are gaining focus. At every point of time, there occurs a necessity to review the trends in such rapidly changing areas of research. When bioactive coatings are made on materials, resorption or damage to such coatings during or after placement of implant exposes the true nature of material, sowing the seed to implant failure in the future. This is one typical reason toward development of biointegrative materials. Osseointegration is simply the mechanical retention in bone with excellent biocompatibility and it is quantitatively evaluated from the strength of bone material interface. This strength can be enhanced by the mechanical retention features deliberately incorporated on the implants.

Human teeth are natural composites, which comprise nano-HAP rods (typically smaller than 100 nm) arranged in lamellae and bound to collagen. Due to their nanosize, HAP crystals have a very large surface area, which enables homogenous resorption by osteoclasts.[54,55] Nanosized HAP particles are desirable when biocompatibility is considered. In recent years, with the rapid development of nanometric HAP materials and an understanding of its characteristic has led to many new ways of using the nanometer HAP. The shape of HAP crystal can affect many characteristics of HAP, such as surface characteristics, bioactivity, and so on. In addition, nanometric biomaterials are in favor of cellular adhesion and proliferation, synthesis of alkaline phosphatase, and the deposition of calcium-containing minerals on the surface of these materials. Therefore, nano-HAP exhibits superiority in the field of orthopedic implants for its improved biological and biomechanical properties. Moreover, new developments on the production of nanosized HAP particles have led to many new applications. Nanosized HAP particles have been used as a resourceful drug delivery agent to the retard multiplication of cancer cells. It is reported that nano-HAP particles have suppressive effect on the proliferation of tumor cells. Apparently different characteristics of HAP particles, such as morphology, size, and crystallinity, have shown different biological consequences.

The drawbacks and salient features of bioceramics and metallic implants have led to the development of a combination of both along with nanotechnological modification of HAP, metallic surfaces, or both. This has led to the advancement of a new generation of materials, such as surface coatings, scaffolds, tissue engineering, biomimetics, and graded structures.

6.8 NANO BIOCERAMICS HYBRID SURFACE COATINGS ON IMPLANTS USING NANOTECHNOLOGY

Metallic implants have played a vital role in orthopedics by providing superior mechanical properties and strength to the implants but corrosion is one of the serious issues. An alternative option is to coat the metallic implants with bioceramics and can be used for both dental implants and hip joint prosthesis. But, there is long way to go as these implants are not available on a commercial scale in all markets across several countries. Bioceramics coatings are commercially available in certain countries and several researchers are working on the nanosurface technologies to develop hybrid surface coatings to develop stable coated implants. The bioceramics coating process on a metallic substrate is quite complicated, and several methods are available in this sense. The biocompatibility restrictions have curtailed the development of coatings by choosing several successful coating techniques used in aerospace, nuclear, and space applications to develop the coated products. These coatings or surface modification techniques should not cause any cytotoxic and allergic affects to the tissues and these coatings are developed for use in human beings and not industries thus, signifying the care that is needed. A great deal of the clinical success depends on the selected coating technique, since the quality and durability of the interface attachment greatly depends on the purity, particle size, chemical composition of the coating, layer thickness, and surface morphology of the substrate. An additional advantage of bioceramics coating is the reduction of release of ions from the metal alloy. The bioceramics coating represents a truly effective barrier that hinders the metallic ion kinetics of release toward the living body.

The bioceramic coating thickness can be varied obtained by using various coating techniques, which varies from less than 1 μm for ion implantation and sol–gel coatings, to 1–40 μm for PVD and CVD coatings, to 1–100 μm for electrophoretic deposition, to greater than 1000 μm for coatings produced by thermal spraying, dipping, etc. Coating thickness is an important parameter in bioapplications, since thick coatings tend to crack and fail prematurely.[34] A coating thickness greater than 100 μm can potentially introduce fatigue failure under tensile loading. The optimum thickness of bioceramic coatings needed for orthopedic applications are under active under investigations.

The ability to modify the chemistry and surface morphology of the coating by fine control of bath composition, pH, and temperature makes electrochemical deposition a versatile process for deposition of coatings on implants, with a tailored body response.[56–58] A number of novel methods for coating HAP have been proposed offering the potential for better control of film structure, such as dip coating, hot isostatic pressing, flame spraying, ion beam deposition, laser ablation, and electrochemical deposition along with plasma spraying, which has been widely studied over the past few decades Plasma spraying is a high-temperature and line-of-sight process, which has been used commercially to produce bioceramics coatings. But, some potential problems include exposure of substrates to intense heat, residual thermal stresses in coatings, poor adherence to the substrate, chemical inhomogeneity, high porosity, and the inability to coat complex shapes with internal cavities. In plasma spraying process due to the high temperatures encountered phase transformation of HAP occurs and impure phases ($Ca_3(PO_4)_2$, $Ca_4P_2O_9$, CaO, oxyhydroxyapatite, etc.) and amorphous calcium phosphate are formed and well characterized these sprayed coatings. HAP will transform into metastable phases with a higher solubility than crystalline HAP, such as tricalcium phosphate (TCP), tetra-calcium phosphate (TTCP), and calcium oxide (CaO). These impurity and amorphous calcium phosphate have higher dissolution rate than crystalline HAP in aqueous solutions, and it has a problem with decreasing the structural homogeneity, and the mechanical properties for long-term clinical applications of the coatings.

Response of body to implant placement and to the implant itself is well orchestrated in every respect and follows a highly specific sequence of events, tracking of which lucidly gives the future status of the implant. The sequence is briefly described here. Initial placement is almost always traumatic, inviting inflammatory response. This phase is followed by normal bone-healing sequence and the fate of the implant is decided in this stage. Bioactive coating endure support at this stage, by promoting adhesion of osteoblasts to the material and enhancing the bone growth.

Although different deposition methods have been applied in the last years, sol–gel method offers a good alternative since the synthesis temperatures are low and it can be applied to a great number of substrates, including those which would oxidize at higher temperatures.[34] Sol–gel technology offers a chemically homogenous and pure product and has been used for HAP production since 1988. The precursors play a vital role in preparation of HAP coatings via sol–gel technique. Hijon et al.[59] have deposited single-phase HAP coatings on Ti6–Al–4V by the sol–gel dipping technique form aqueous solutions containing triethyl phosphite and calcium nitrate. Balamurugan[57] et al. have reported that the coating thickness alters the shear strength and corrosion resistance of sol-–gel derived apatite films of 316L SS. Rod-like HAP nanostructures prepared by a simple sol–gel precipitation method that show similar morphology, size, and crystallinity to HAP crystals of human teeth have been reported.[72] Further work is focused on a novel way to control the sintering of these HAP nanorods and the possibility of simulating a human enamel structure.

6.8.1 ELECTROPHORETIC DEPOSITION (EPD)

EPD is evolving as an important method in ceramic processing because it enables the formation of thin ceramic films and nanostructured layers. It is also an important tool for preparing thick ceramic films and body shaping. The interest in EPD for biomedical applications stems from a variety of reasons, mainly the possibility of depositing stoichiometric, high-purity material to a degree not easily obtainable by other processing techniques. The deposition of submicron powders offers advantages in fabrication of ceramic coatings with dense packing, good sinterability, and homogenous microstructure. The development and application of EPD for deposition of HAP have been reviewed.[58,60–62] Sridhar et al. have developed new EPD protocols for the application of thin layers of HAP on the surface of type 316L SS and studied their electrochemical properties in vitro.[63,64] Both the coating weight and thickness increased with increasing voltage and coating duration. During deposition, the electric field drives the particles toward the electrode and exerts pressure. The use of high voltage has the advantage of shorter deposition times and higher deposition thicknesses. However, higher voltages resulted in significant hydrogen evolution at the cathode, which in turn increased the porosity of the deposit. Loss of thicker coatings from the metal surface was observed due to decohesion that occurred between the surface and the coating. Particle congestion at the cathode resulted in weakly bonded coatings. During the deposition process a coating voltage ranging from 30–90 V was applied for 1–5 min. The coatings obtained were uniformly adhered and no significant hydrogen evolution was observed. Secondary ion mass spectrometry (SIMS) studies of the coatings showed higher intensity levels of calcium signal after sputtering for 3000 s (Fig. 6.2), indicating the diffusion of the coating into the base alloy. The optimum coating parameters for EPD of HAP on type 316L SS were identified at 60 V for 3 min, followed by vacuum (10^{-5} torr) sintering at 800C for 1 h.[65,66]

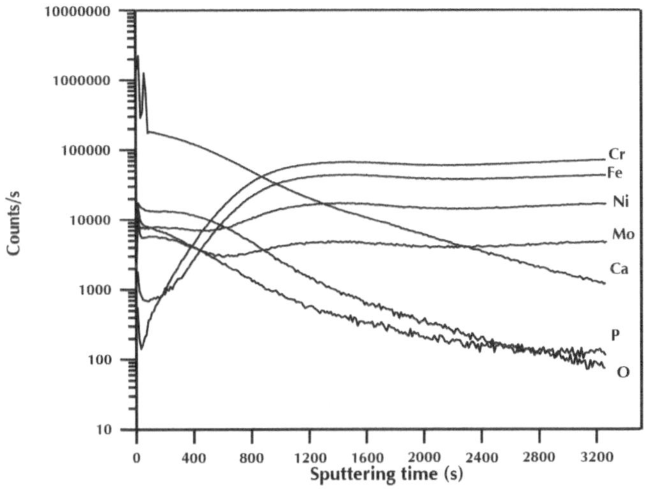

FIGURE 6.2 SIMS elemental depth profile of HAP-coated type 316L SS.

In clinical practice delamination of coatings has been observed at the interface between coating and substrate. This is due to a mismatch in the thermal expansion coefficients of HAP and titanium/titanium alloy. In order to improve the adhesion, a proper coating is required. Thin layers of nano-HAP were coated on the surface of pure Ti by EPD from 2% suspension in isopropyl alcohol. Coatings were carried out at a constant voltage from 70 to 90 V at different time intervals followed by sintering in air at 800 °C for 1 h. The XRD patterns of the nano-HAP coatings show diffraction peaks given in Figure 6.3 with minimal line broadening and high intensities, which indicates highly crystalline, stoichiometric hydroxyapatite. No other calcium phosphate phases were observed. It is revealed from diffraction pattern of HAP that first peak was found at 26° as reported by others for HAP coatings. The surface morphologies of the coated samples obtained by electrophoretic deposition are given in Figure 6.3 shows uniform distribution of the particles, indicating that coatings are dense and microporous. No cracking of coating was observed. This result confirms that EPD is a viable technique for developing nano-HAP coatings on implant metals and alloys.

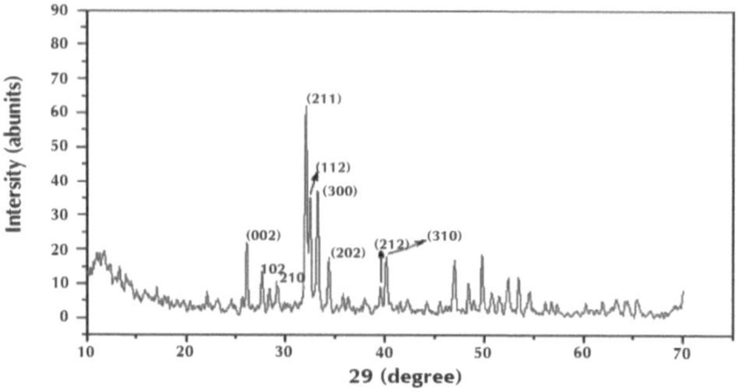

FIGURE 6.3 XRD Pattern and SEM micrographs of nanoHAP-coated pure Ti by EPD after sintering at 800°C.

Recently, it has been reported that nanostructured coatings fabricated by EDP have high chemical homogeneity, reduced flaw size and microstructural uniformity, and require a lower sintering temperature for densification. The nanometer-sized grains and the high volume fraction of grain boundaries in nanostructured HAP can increase osteoblasts adhesion, proliferation, and mineralization. Studies on EPD of HAP on different implant alloys have been reported by several investigators.[67,68] HAP-coated 316L stainless steel fabricated by EPD exhibited higher corrosion resistance than the substrate in simulated body fluids, with nobler open-circuit potential and higher pitting and protection potentials.[63,65] The corrosion behavior of EPD HAP coatings on the Ti alloys has not been fully studied with respect to the morphology and size of the starting HAP particles.

Furthermore, HAP coatings reinforced with carbon nanotubes (CNTs) was reported to be a viable route for forming surface composite with higher hardness, elastic modulus, and interlaminar shear strength than that of the monolithic HAP layers. Addition of CNTs also helps to prevent peeling off of the coating layers by acting as reinforcement network with the deposit. Cathodic EPD has been used for fabricating bioactive HAP coatings with and without CNTs on Ti6–Al–4V followed by vacuum sintering at 800 °C.[69] Moreover, carbon nanotubes (CNTs) were also used to reinforce the HAP coating for enhancing its hardness. The improved properties could be attributed to the use of submicron-sized HAP particles in the low temperature EDP process. Compared with monolithic HAP coating, the CNT-reinforced HAP coating markedly increased the coating hardness without compromising the corrosion resistance or adhesion strength.

6.8.2 ELECTROCHEMICAL DEPOSITION

Electrochemical processes can be an excellent tool to produce one-dimensional structures due to the provision of a two-dimensional reaction interface and a precise control of critical reaction parameters.[70,71]. The occurrence of lateral self-organized phenomena on the reaction interface can therefore, be used to grow three-dimensional nanostructures by electrochemical methods. Especially, the electrochemical formation of self-organized porous nanostructures has attracted much interest, since they possess wide applications in biological nanopatterning, high-density recording media, and templates for nanomaterials.

Electrochemical techniques have been used for two purposes on the surfaces of orthopedic implants, that is, for modifying the oxide layers of the surface and deposition of calcium phosphates. Much interest in electrodeposition has evolved due to (1) the low temperatures involved, which enable formation of highly crystalline deposits with low solubility in body fluids and low residual stresses, (2) the ability to coat porous, geometrically complex, or non-line-of-sight surfaces, (3) the ability to control the thickness, composition, and microstructure of the deposit, (4) the possible improvement of the substrate/coating bond strength, and (5) the availability and low cost of equipment.[71,72]

Electrocrystallization of nanohydroxyapatite on titanium was achieved by cathodic polarization in solution containing calcium nitrate and ammonium dihydrogen phosphate. The composition and pH of the bath were found to significantly affect the nature and surface morphology of the deposit. The formation of well-crystallized HAP at $pH_0 = 6.0$ occurs at any temperature between 70 and 95 °C, whereas, at $pH_0 = 4.2$, less-crystallized, thicker, and more porous coatings that contained traces of octacalcium phosphate were observed.[72] The influence of potassium chloride and sodium nitrite on the composition and surface morphology of the deposit was also evaluated. A speciation-precipitation model was applied to better understand the effect of bath conditions. The topography of different coatings was also evaluated by means of ex situ AFM imaging. Figure 6.4 shows some typical AFM images. The deflection image is more sensitive than topography images to delegate spatial information such as sharp edges. This suggests that nano-HAP coatings formed by electrocrystallization are more biomimetic with respect to their structure and morphology.

FIGURE 6.4 AFM images of the top surface of coatings produced at: (a) pH = 6.0, T = 90 °C, t = 5 min; (b) pH = 6.0, T = 85 °C, t = 2 h; and (c) pH = 6.0, T = 90 °C, t = 2 h, in the presence of 0.10 M KCl. A 3D view of the latter sample is presented in (d).[83]

Electrochemical modification of the surface morphology of titanium surfaces are processed by spark anodization, which typically leads to formation of rough porous TiO_2 layers.[73,74] The sparking occurs because of a high applied voltage, leading to dielectric breakdown of the oxide layer and discharge events. The nanotube layers were fabricated by electrochemical anodization of titanium in fluoride-containing electrolytes. Various nanotube lengths, layers with an individual tube diameter of 100 nm were grown to a thickness of approximately 2 m or 500 nm. The presence of nanotubes on titanium surface enhances the apatite formation and that the 2 μm thick nanotube layer triggers deposition faster than the thinner layers. Tubes annealed to anatase, or a mixture of anatase and rutile are clearly more efficient in promoting apatite formation than the tubes in their "as-formed" amorphous state. Electrochemically grown and annealed TiO_2 nanotube arrays having anatase structure are expected to be a good precursor system for the formation of HAP.[75] The initial and later stages of apatite formation from simulated body fluid on titania with different surface morphologies (compact or nanotubular) and different crystal structures (anatase or amorphous) have been reported. In the initial stages of apatite growth, more nuclei are formed on the nanotubular surface than on flat compact TiO_2. While the crystallographic structure of the substrate plays a less important role than the morphology in the initial nucleation stages, it is of great importance in the later stages of apatite crystal growth. The nanotubular morphology combined with an anatase structure leads to the formation of apatite layers with a thickness of >6 nm in less than 2 days.

Electrochemically deposited nanograined calcium phosphate coatings were produced on titanium alloy substrates using aqueous electrolyte maintained at acidic pH by Narayanan et al.[88] Ultrasonated bath produced coatings containing dicalcium phosphate dehydrate and the grain sizes were in the range of 50–100 nm. An electrochemical method of producing nanocrystalline hydroxyapatite coatings on titanium surface is reported by Narayanan et al.[76,77] The bath contained $Ca(NO_3)_2$ and $NH_4H_2PO_4$ in the molar ratio 1.67:1. The electrolyte was maintained at physiological pH and was ultrasonically agitated throughout the time of electrolysis. Coatings contained mono hydroxyapatite phase whose crystal sizes were lower than 30 nm. These sizes are comparable to the size of the bone hydroxyapatite crystals. Small globules of hydroxyapatite covered the coating surface completely. Ultrasonic agitation promoted the formation of nanocrystalline structure, which will help in better attachment of bone tissues to the implant surface. They contained very small crystals of the size of few nanometers, which were confirmed by the bright field transmission electron micrographs and electron diffraction patterns.

Fluorine, which exists in human bone and enamel, can be incorporated into HAP crystal structure by substituting OH^- groups with F^- ions to form fluoridated hydroxyapatite (FHA). FHA possesses lower solubility than pure HAP, while maintaining the comparable bioactivity and biocompatibility. Many researchers focused on the application of FHA as bioactive coatings to provide both early stability and long-term performance.[78] In comparison with HAP coating, FHA coatings could provide lower dissolution, better apatite-like layer deposition, better protein adsorption, comparable, or better cell attachment and improved alkaline phosphatase activity in cell culture.[79] Hydroxyapatite (HA) and fluoridated hydroxyapatite (FHA) coatings were deposited on titanium substrates using an electrochemical technique. Typical apatite structures were obtained for all the coatings after electrodeposition and subsequent posttreatment, including alkaline immersion and vacuum calcination. The coatings were uniform and dense, with a thickness of ~5 μm. Compared with pure Ti, FHA, and HA coatings exhibited higher biological affinity like cell proliferation and alkaline phosphatase activity. Clinical applications suggest that a moderate content of F, such as $Ca_5(PO_4)_3(OH)_{0.375-0.5}F_{0.5-0.625}$, be most suitable as a compromise among cell attachment, cell proliferation, apatite deposition, and dissolution resistance.

The concept of graded coating or multilayered coating has been under development more than a decade for biomedical applications.[80] The concept here is to develop layers of coatings one above the other with good mechanical properties and each layer would its own functional properties. In case of implants the topmost layer would interact with blood hence, it is desired to be microporous and favorable for cell attachment.[81,82] Coatings containing bioglasses or bioceramics are not ideal for orthopedic functions as they possess high brittleness and low wear resistance. Bioinert ceramics like zirconia are used to toughen bioglass in order to obtain favorable mechanical and tissue compatible properties of the coatings. The reinforcing phase in the form of ZrO_2 nanoparticles are selected due to its excellent biocompatibility and also because of its exceptional mechanical properties. The addition of stabilizers to zirconia further enhances its mechanical properties and in yttria-stabilized zirconia, yttria supports a tetragonal structure known as tetragonal zirconia polycrystals (TZP). Its enhanced strength is attributable to one of its properties termed as transformation toughening. Pure zirconia and yttria stabilized zirconia have wider applications in hip and knee prostheses, hip joint heads, temporary supports, tibial plates, and dental crowns. Zirconia exhibits three well-established polymorphs, the monoclinic, tetragonal, and cubic phases. Zirconia does not exert mutagenic and carcinogenic effects toward osteoblasts growing on it. The physical, chemical, and biological properties of the bioglass-reinforced yttria-stabilized composite layer on Ti6–Al–4V titanium substrates have been developed.[80] The Ti6–Al–4V substrate was deposited with yttria-stabilized zirconia—YSZ as the base layer of thickness ≈4–5 μm, to inhibit metal ion leach out from the substrate and bioglass–zirconia-reinforced composite as the second layer of thickness ≈15 μm, which would react with surrounding bone tissue to enhance bone formation and implant fixation. The deposition of these two layers on the substrate was carried out using the most viable electrophoretic deposition (EPD) technique. Biocompatible yttria-stabilized zirconia (YSZ) in the form of nanoparticles and sol–gel-derived bioglass in the form of microparticles were chosen as precursors for coating. Studies indicate that the topmost layer of zirconia reinforced bioglass bilayer system promoted significant bioactivity on the other hand it demonstrates an enhanced corrosion resistance property along with an increase in the mechanical strength under load bearing conditions in comparison with the monolayer YSZ coating on Ti–6Al–4V implant surface. This technique is one of the several ways to develop functional graded coatings. Figure 6.5 shows the SEM micrograph of

≈ 12–15 μm thick electrophoretically deposited 1YSZ–2BG coating onto the YSZ-coated Ti–6Al–4V after heat treatment at 900 °C for 3 h. The nanoscale YSZ particles were chemisorbed onto the matrix of the BG microparticles. The dominant contribution of the YSZ nanoparticles surface charge is attributed to their higher surface to volume ratio in comparison to the BG microparticles. Interlocked agglomerated particles with appreciable internal porosity can be observed. Thus, titanium alloys can be functionalized with 1YSZ–2BG coatings are promising candidate materials for dental and orthopedic implant applications.

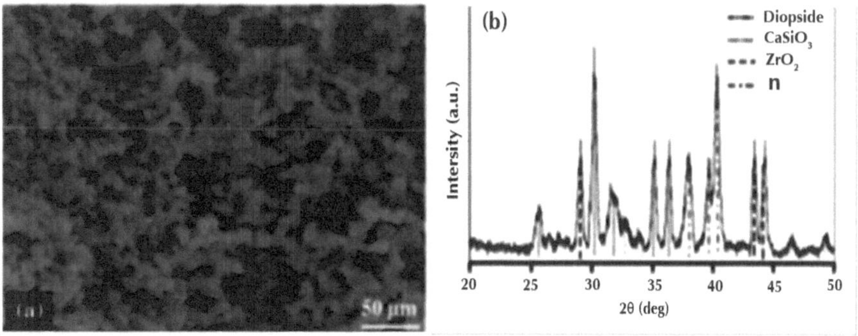

FIGURE 6.5 SEM micrograph (a) and XRD patterns (b) for the EPD 1YSZ–2BG coatings on Ti–6Al–4V after sintering at 900 °C for 3 h.[80]

Bilayer composite matrices are potential scaffolds. They can be prepared by combining two polymers along with a bioceramic polyvinyl alcohol (PVA)/HAP/Poly vinyl chloride (PCL) ceramics. Brittle scaffolds are formed when bead formation takes place during loading of HAP in PVA scaffolds accompanied by a reduction in crystallinity of PVA in the composite scaffold. Good porosity is a characteristic property of the bilayer composites and usually ranges from 60–70% in scaffold when compared to other control scaffolds. Suitable hydrophilicity of bilayer scaffold and optimum pore diameter above 500 nm aid the adherence of cells and allow better growth and proliferation.

On the other hand, the processing inconvenience due to the use of nanopowders has to be considered due to the higher agglomeration tendencies and oxidation of the reinforcing powders. Agglomeration hurdles can be overcome by preparing suspension of the powders, using a suitable dispersant and the solvent medium, thereby allowing for homogenous nanocomposites to be complete. Solvent selection is very critical as it should not cause the reinforcing powders to be oxidized and a dispersant selected should disperse each of the powders uniformly. Lubrizol 2155 id one of the effective dispersants with hexane as the solvent.[83] Composites materials made from micron sized powders, ZrO_2-50%TiC and ZrO_2-50%TiN, were prepared and sintered using Lubrizol 2155 and hexane to produce a homogenous distribution of the powders in a sintered sample. SEM analysis indicates that Lubrizol was a flourishing dispersant. The micron sized composites were used as a comparison for the nanocomposites.

6.8.3 SCAFOLDS AND ITS ROLE IN TISSUE ENGINEERING

Tissue engineering (TE) is an interdisciplinary fields that exploits a combination of living cells, engineering materials, and appropriate biochemical growth factors using various techniques to imitate the natural process and improves, replace, restores, maintains or enhances living tissues and the organs as a whole.[84–86] Tissue engineering endeavor is at resolving the difficulties faced by surgeons today, such as donor shortage and immune rejection faced by transplantation. Sergey has comprehensively reviewed the applications of calcium orthophosphates along with TE.[86] The first priority is to prepare the scaffolds by engineering a combination of polymers and bioceramics results in the formation of scaffolds. This along with TE using cell culture techniques would help us to develop a new generation of biomedical devices.

Tissue engineering explores the approach to develop scaffold made of nanobiomaterials surrounded by active regenerating cells.[87] This is an impertinent clinical challenge as far as bone is concerned, especially when the bones break into pieces and a load-bearing implant cannot be placed. One of the solutions is to use autologous bone grafting method to repair such bone defects and they are based on its osteogenic and osteoinductive potential.[87,88] The challenges in this regard include preparing the scaffold to match the exact size and shape of the bone and also obtaining the correct cell line for the patient with blood compatibility and immunological issues. The other simple method is to do bone grafting with allografts as this would lead to immunological problems and transmission of any unidentified disease, which the donor may be having.[89] To overcome these issues confronting us tissue engineering plays a predominant role were the scaffolds contain osteoconductive bioceramics materials along with growth factors that would take care of the osteogenic process along with the cells to complete the biomineralization process. The selection of the bioceramics would determine the bioactive, bioinert, and bioresorbable properties along with the surface roughness, pore volume, structure, and biodegradability. The resulting bone growth should be able to form a strong bond with the scaffold and should be well interconnected.

The number of bone graft procedures is increasing year upon year and in United States it is estimated that the number of bone graft procedures performed annually is more than 500,000 with more than half of which dedicated to spinal fusion cases.[90] The spinal system, rib cage, fingers in our hands and legs are a few places where we cannot replace them with metallic parts, especially when they are broken to parts. The current medical systems demands us to also address the problems that pertain to traditional bone graft systems such as implant failure due to lack of tissue regeneration around the implant surface, resulting in poor bone remodeling, and loosening of implants. In recent years, tissue engineering has revolutionized the direction of research for orthopedic applications because of the success of nanotechnological advancements in creating nouveau fabrication techniques for nanoscale materials, such as nanofibers and nanofibrous scaffolds. By mimicking the structural properties of natural tissues, the subtleties of extracellular matrix (ECM) proteins, particularly collagen fibrils in bone, enhanced absorption of biomolecules such as vitronectin is achievable on nanoscale materials due to their high surface area to volume ratio.[91] This would guide us to a more constructive atmosphere for cellular connections (cell-to-cell and cell material) to cement the biomineralization process. In order to optimize the functional performance of bone graft materials, one must envisage the chemical and structure properties of native bone. Bone in itself encompasses mainly nanohydroxyapatite (n-HA) and type I collagen. The collagenous nanofibrils are the main structural proteins, contributing up to 30% of the dry weight of bone and 90–95% of the organic, nonmineral components in bone other noncollagenous proteins for example, osteopontin, osteocalcin, and osteonectin constitute part of the composition of bone.[92]

Principally, synthetic polymers are utilized in attempt to mimic the main ECM protein in bone, type I collagen. Owing to the composition of bone, HAP is often incorporated in material substrates via various modes of processing methods. Ngiam et al. have successfully deposited bone-like nanoapatite on polymeric electrospun PLGA and PLGA/Col nanofibers using a biomimetic Ca–P treatment, without the need for pretreatment such as alkaline treatment in order to hydrolyze the surfaces for initial apatite nucleation.[93] The presence of collagen in the nanofibers expedited the deposition of n-HAP on the scaffolds. It was emblematic of the types of functional groups present in collagen, that is, scilicet carboxyl groups and carbonyl groups.[94] These functional groups served as nucleation sites for apatite formation and consequently, uniform distribution of n-HAP was apparent on the outer and inner surfaces of the PLGA/Col nanofibers.

Scaffolds (artificial extracellular matrices) have critical roles in tissue engineering. Nanofibrous poly(l-lactic acid) scaffolds under the hypothesis that synthetic nanofibrous scaffolding was prepared, mimicking the structure of natural collagen fibers that could create a more favorable microenvironment for cells.[95] The nanofibrous architecture built in three-dimensional scaffolds improved the features of protein adsorption, which mediates cell interactions with scaffolds. The nanofibrous architecture selectively enhances protein adsorption including fibronectin and vitronectin, even though both scaffolds were made from the same poly(l-lactic acid) material. These results demonstrate that the biomimetic nanofibrous architecture serves as superior scaffolding for tissue engineering. The apprehensions of pathogens/bacterial transmission, mechanical properties, methods of handling, and nature of degradation products formed in vivo. The adsorbed proteins mainly from serum proteins or those secreted by the cells medicate the polymer surfaces toward the growth and attachment of cells and its migration.

One of the principal methods in tissue engineering involves growing the relevant cell(s) *in-vitro* on required three-dimensional (3-D) tissue or organ and this is achieved by seeding the cells onto porous matrices, known as scaffolds, to which the cells attach and colonize.[96] The scaffold provides structure for cell attachment and subsequently proliferation. The successful cell seeding on scaffolds depend on the source and type of cells and on the extracellular components of the scaffold. The cells are grown on the scaffold that functions to mechanically support cells and regulate the functions of cells in a manner analogous to ECM of mammalian tissue. The fibroblast cells are common in the connective tissue that synthesize and continuously secretes precursors of the ECM in mammalian tissues. Fibroblasts are actively dividing cells and provide structural framework (stroma, cytoskeleton) for many tissues. Fibroblasts make collagens, glycosaminoglycans, glycoproteins, reticular, and elastic fibers found in the ECM and play an important role in the regeneration of new tissue. It has been reported that the presence of fibroblasts in a skin substitute stimulated epidermal differentiation and dermal regeneration. The fibroblasts accelerate growth of tissue cells by secreting several growth factors and ECM.[97,98] To prepare an ideal tissue engineered scaffold, it is important that seeded cells proliferate and migrate throughout the matrix and organize as a homogenously distributed ECM.

Bone marrow stromal cells are a potential source of osteoblasts and chondrocytes and can be used to regenerate damaged tissues using a tissue-engineering approach. These strategies require the use of an appropriate scaffold architecture that can support the formation de novo of either bone and cartilage tissue, or both, as in the case of osteochondral defects. The later has been attracting a great deal of attention since it is considered a difficult goal to achieve. A novel hydroxyapatite/chitosan (HA/CS)-bilayered scaffold has been developed by combining a sintering and a freeze-drying technique to show the potential of such type of scaffolds for being used in TE of osteochondral defects.[99] Therefore, on the basis of the osteochondral approaches, the development of bilayered osteochondral scaffolds combining both HA and chitosan (CS) layers thus seems a good approach but there is a need for more simple and reliable strategies to manufacture bilayered scaffolds for osteochondral applications.[100]

The pore size is critical to have vascular integration to take place. Hydroxyapatite provides a bioactive surface for osseointegration process to take place.[87] On the other hand, bioglasses have a bioactive surface with amorphous structure that enables the bone to grow as it slowly dissolves and redeposits over a period of time. The mesenchymal (MSC) stem cells are derived from bone marrow are commonly used as they are capable of auto renewal, form a strong bond, nonhematopoietic cells and they are able to differentiate into several phenotypes, including bone, adipocytes, and cartilage.

After their isolation and extension in tissue culture, osteoconductive scaffolds and osteoprogenitor cells are the two main factors that are involved in bone tissue regeneration.[101] The main challenge during the repair and reconstruction of bone defects is the search for biocompatible and functionally proven graft materials. The ability of a gelatin/nanobioglass scaffold to support the differentiation and viability of rMSCs and repair of critical cranial bone defects have been reported.[87] The bioglass nanocomposite scaffolds are biodegradable, osteoconductive, and biocompatible, and perform the role of a temporary matrix for cells.

To achieve improvement of implant osseointegration, the most promising approaches is to bestow a nano-HAP/collagen surface to mimic bone on bone-contacting implants.[102] Such a surface would probably make use of speciality from each one of nano-HAP and collagen in the interactions at the implant-bone interface, and thereby, probably produce a good bone tissue reaction and a bone-bonding interface. Nano-HAP sol and collagen gel were mixed and coated on different titanium surfaces by the deposition method. The cell responses to nano-HAP, nano-HAP/collagen, native, and anodized titanium surfaces were evaluated by cell culture studies. The in vitro studies showed that porous structures produced by anodic oxides on titanium served as positive anchorage sites for cell filopodia to connect, and nano-HAP decreased cell attachment of osteoblasts and induced well-developed long filopodia and broad lamellipodia, thereby enhancing cellular motility. Collagen involvement enhanced cell adhesion to nano-HA. Cell reactions to nano-HA, nano-HA/collagen, native, and porous titanium surfaces provide some guidance for an optimal osseointegration by their application in surface modifications for implants.

Santos et al. have evaluated the relative role of the calcium phosphate surface chemistry and surface topography on human osteoblast behavior.[103] Highly dense phosphate ceramics (single-phase HAP and β-tricalcium phosphates TCP) presenting two distinct nanoroughnesses were produced. The phosphate chemistry was responsible for changes in adhesion, proliferation, and cell differentiation.[104–106] On TCP, it was shown that the main influent parameter was surface chemistry, which negatively affected the initial cell adhesion but positively affected the subsequent stage of proliferation and differentiation. On HAP, the main influent parameter was surface topography, which increased cell differentiation but lowered proliferation.

The potential techniques for obtaining thinner and resorbable HAP coatings is by biomimetic and electrochemical deposition are two of the most promising new processes. Moreover, these processes can produce nano-hydroxyapatite (nano-HAP) on the metallic substrate surface.[107,108] Barrère et al. have reported that a biomimetic process can produce an apatite coating within 72 h.[128] The process has been applied to deposit the CaP coating on a roughened surface to develop a 25 μm coating containing octacalcium phosphate (OCP) and HAP in the coating.[109] OCP has been identified as one of the Ca–P that participates in the early stage of biomineralization of calcified tissues. The biodegradability and osteoconductivity of OCP have been demonstrated in vivo.[103]

An attractive strategy for fabricating mimics of different types of composite biomaterials is to electively grow apatite on polymers with control of structure, mechanical properties, and function.[110] Silk/apatite composites were prepared by growing apatite on functionalized nanodiameter silk fibroin fibers prepared by electrospinning. The functionalized fibers were spun from an aqueous solution of silk/polyethylene oxide (PEO) (78/22 wt/wt) containing poly(L-aspartate) (poly-Asp), which was introduced as an analogue of noncollageous proteins normally found in bone. Apatite mineral growth occurred preferentially along the longitudinal direction of the fibers, a feature that was not present in the absence of the combination of components at appropriate concentrations. The results suggest that this approach can be used to form structures with potential utility for bone-related biomaterials based on the ability to control the interface wherein nucleation and crystal growth occur on the silk fibroin. With this level of inorganic–organic control, coupled with the unique mechanical properties, slow rates of biodegradation, and polymorphic features of this type of proteins, new opportunities emerge for utility of biomaterials. Many organic polymers have been studied as substrates for hydroxyapatite nucleation and crystal growth including electrospun silk fibers are used as templates for growth of apatite as a route to generate new composite biomaterials with a broader range of properties. The electrospinning process offers an alternative approach to protein fiber formation that can generate nanometer diameter fibers, a useful feature in some biomaterial and tissue engineering applications due to the increased surface area for cell interactions and tissue ingrowth.[110–112]

Many papers have reported on the importance of silicon on the bone formation, the growth and development of bone, teeth, and some invertebrate skeletons.[113,114] Si element was introduced into hydroxyapatite ceramic by different techniques to further improve osseointegration. Increasing evidences have shown that the presence of Si contributes to the enhanced bioactivity of some bioactive ceramics in vitro. Silicon-substituted hydroxyapatite (Si-HAP) coatings with 0.14–1.14 wt.% Si on pure titanium were prepared by a biomimetic process.[113] The prepared Si-HAP coatings and HAP coating were only partially crystallized or in nanoscaled crystals. The introduction of Si element in HAP significantly reduced P and Ca content, but densified the coating. Both the HAP coating and the Si-HAP coatings demonstrated a significantly higher cell growth rate than the uncoated pure titanium in all incubation periods while the Si-HAP coating exhibited a significantly higher cell growth rate than the HAP coating. The synthesis mode of HAP and Si-HAP coatings in simulated body environments contribute significantly to good cell biocompatibility in the biomimetic process. Silicon-substituted hydroxyapatite ($Ca_{10}(PO4)_6\text{-}x(SiO_4)x(OH)_2\text{-}x$, Si-HAP) composite coatings on a bioactive titanium substrate were prepared by electrophoretic deposition technique with the addition of triethanolamine (TEA) to enhance the ionization degree of Si-HAP suspension.[115] The depositing thickness and the images of Si-HA coating can be changed with the variation of deposition time. The bioactive TiO_2 coating formed may improve the bond strength of the coatings. The interaction of Ti/Si-HAP coating with bovine serum albumin is much greater than that of Ti/HAP coating, suggesting that the incorporation of silicon in HAP is significant to improve the bioactive performance of HAP.

Modification of the chemistry and surface topography of nanophase ceramics was used to provide biomaterial formulations designed to direct the adhesion and proliferation of human mesenchymal stem cells (HMSCs).[116]

HMSC adhesion was dependent upon both the substrate chemistry and grain size, but not on surface roughness or crystal phase. The results demonstrated the potential of nanophase ceramic surfaces to modulate functions of HMSCs, which are pertinent to biomedical applications such as implant materials and devices.

Graphene as a biomaterial has augmented the biological research due to its amazing chemical and mechanical stability, thermal, and electrical conductivity.[117] Graphene was traditionally used in electronic devices as it possess good electrical conductivity and thermal properties. Now it is employed in biomedical devices for measuring the cell potential and as a substrate for conductive cell culture devices and biosensors. The flat monolayers of carbon atoms on the graphene are closely packed into a two-dimensional honeycomb crystal structure to act as a platform for the interaction of biomolecules through covalent, electrostatic, and hydrogen bonding.[118] The oxidized form of graphene is known as graphene oxide (GO) has rich oxygen functional groups (hydroxy, epoxy, ketone, carboxyl, and diols) on its basal plane. This assists in the functionalization on its exterior surface to attract biological molecules like nucleic acids, proteins, peptides, etc. Graphene and its oxides have inspired researches due to its ease of production, which can be synthesized in the lab itself along with the low cost. Elasticity, flexibility, and adaptability are a few special properties of graphene based nanomaterials that play a key role in cell proliferation, adhesion, and differentiation due to their mechanical strength with single atom thickness.[119–121] Sastry and his group are working on functionalized graphene oxide (FGO)/GO are used as a matrix to fabricate HAP particles on its surface and it is in vitro studies were carried out to find out its efficacy as an osteoinductive material. [117,122] They were able to demonstrate the formation of HAP on FGO/GO matrix and characterization studies have confirmed the same. These types of implants may be used in the defects of nonload bearing bones. FGHAP has further enhanced the formation of ALP and calcium levels, and MTT assay have shown the biocompatible nature of FGHAP compared to that of FGO and GOHAP. FGHAP possessed better osteoinductive potentials than FGO and it is a promising, composite for orthopedic applications.

Glass ionomer cements (GICs) are the most accepted and commercially available dental cements owing to their adhesive nature and fluoride releasing property.[123] The poor mechanical properties along wear resistance are a cause of concern when they have to be used for posterior restorations. Recently several studies on have focused on the biomedical applications of carbon nanotubes (CNTs) on stem cells, human gingival fibroblast, and osteosarcoma cell lines. [124–126] Glass ionomer cements reinforced with Multi

Walled Carbon Nano Tubes (MWCNTs) have been fabricated and their different properties have been studied. The interaction of MWCNT with the polymer matrix was characterized for the vibrational properties using infrared spectroscopy and the crystal patterns with X-ray diffraction (XRD), A shift in the peak corresponding to O—C=O vibrations and delicate changes in the diffraction patterns were observed, respectively. The surface morphology of the samples showed that aggregation of MWCNT resulted at higher concentrations and it influenced the properties of the set cement. The MWCNTs neither interfered nor influenced the cross-linking and neutralization of the polyacrylic acid. The hardness values of the cement were found to increase in hardness values with increase in concentration of MWCNT but at the same time its compressive strength was inferior when compared to that of conventional glass ionmer cements (GIC), which was attributed to the nonuniform distribution and aggregation of the nanotubes. The improved hardness values of the MWCNT reinforced cements indicate their potential applications as posterior restorative materials and for core build up.

6.8.4 POLYMER NANOCERAMICS-BASED COMPOSITES

Nanoceramics/nanocomposites are defined as novel bulk materials or coatings with microstructural architecture, characterized by at least one of the ceramic phases having a length scale between 1 and 100 nm. The major drive for wider interest in nanoceramics and its composites has been the fact that one can potentially achieve better and in a few cases unusual material properties by manipulating the length scale in the nano range. Therefore, better performance and newer applications of the materials have now been made possible.[127]

The preparation of suspensions is an essential feature for the successful production of nanoceramics. Agglomeration is a common problem encountered with these materials and this has to be avoided, especially with powders. Nanosized powder particles have a large surface area and are prone to agglomeration to minimize the total surface area or interfacial area of the system. The attractive van der Waals forces acting between the particles is the root cause that compel agglomeration to occur.[128] This tendency generates a major obstacle in the formation of a homogenous mixture of nanosized particles. The formation of a stable suspension of the high quality nanoceramic were all the particles repel each other would enable us to overcome by using dispersing agents and maintaining the pH.[83] The optimum pH or amount of dispersant to create a stable suspension can be experimentally found through rheological measurements. Rheology is the science of deformation and flow characteristics of matter. During rheological measurements the flow behavior is monitored in a response to an applied shear stress. A key rheological parameter is the viscosity of suspensions. Rheological measurements are necessary to characterize the properties of colloidal suspensions; specifically such measurements are used in order to determine the optimal amount of dispersant required to stabilize a suspension. This is done by measuring the viscosity of a suspension against varying dispersant concentrations.[129]

Polyethylene glycols (PEG) are polymers from oxyethylene polymerization, which are amphipathic and biocompatible polyethers widely used for biomedical research and applications.[130] A significant number of studies have reported that the presence of PEG can modify or control the surface of the nanometer crystal, moreover, can act as the dispersing agent of the nanometer crystal in the process of synthesis. According to this correlation, the spherical nano-hydroxyapatite, whose size is about 30–50 nm, can be synthesized in the presence of a certain concentration of the PEG by biomimetic method. The interaction between Ca^{2+} and PEG possibly is an important factor of the nucleation and growth of the spherical HAP crystal.

Polymer based synthetic bone graft materials have attracted the attention of many scientists due to the wide choice of materials available.[130] The success of polymer/ceramic composites in treating bone defects have been reported by several in vitro and in vivo studies.[131,132] These composite scaffolds present several advantages. Firstly, during degradation of the polymer by lowering of the pH is prevented by the calcium phosphate, which acts as a buffer and thus, dampening the acidic polymer degradation. Secondly, the incorporation of the ceramic filler particles is usually associated with a slight increase in stiffness and strength when compared to the polymer alone. Thirdly, the presence of calcium phosphates facilitates direct bone apposition at the scaffold-bone interface (osteoconduction), preventing the unfavorable fibrous encapsulation of the scaffold.[133]

Sangeetha and coworkers are working on sulphonated poly ether ether ketone (SPEEK), which is a rather new polymer that is being only recently considered for biomedical applications.[134] Poly ether ether ketone (PEEK) on the other hand has been demonstrated to perform exceptionally well as a load bearing prosthetic biomaterial. While PEEK is amenable only for injection moulding technique, sulphonated PEEK (SPEEK) can be prepared as membranes and nanofiber mats. The group had reported on the in vitro studies based on SPEEK beads and SPEEK/ poly (methyl methacrylate)/ SiO_2 composite membranes for orthopedic applications. The results of the *in vivo* studies suggested that only SPEEK/HA showed favorable host-implant response with minimal inflammatory response and good osseointegration, which was comparable with the commercial biphasic ceramic bone graft material. However, it was noted that the extent of healing was lesser when compared with the commercial graft material. Nonetheless, these studies throw up the possibility of incorporating proteins such as rhBMP withinSPEEK/HAP graft materials and further studying their bone healing response. The results of SPEEK/PMMA/SiO_2 composites were disappointing as they elicited an unfavorable host response leading ultimately to graft rejection.

6.8.5 NANOCOMPOSITE FIBERS

Nanofibers are one of the current applications of polymers as implant materials. It is composed of two terms where "nano" indicates the size of the polymeric molecules and "fiber" its shape and mechanical properties like strength and support. These fibers are naturally present in our human body as muscles, connective tissues, organs etc. In the recent past, "nano" is used for describing various physical quantities within the scale of billionth as nanometer and the fibers have diameter range in nanometer. Polymers are found to occur naturally as fibers in plants, animals and humans as silk, chitosan, hyaluronic acid, gelatin, and elastin to name a few. Electrospinning is one of the most common technique to prepare nanofibers using polymers like PCl, PLGA, etc. PLA (Poly Lactic Acid), PEVA (Polyethylene vinyl acetate), PCL (Poly-caprolactone), PLGA (Poly lactic-co-glyolic acid), PVA (Polyvinyl alcohol), PET (Poly ethylene terephthalate), and also by incorporating inorganic compounds like hydroxyapatite, titanium dioxide etc.,

Electrospinning is a simple and versatile method for generating nanofibers and ultrathin fibers, under the effect of an electric field, from a variety of materials including polymers. Seeram Ramakrishna and his group are the global pioneers in the field of electrospinning of nanofibers for more than a decade and have been working on all its aspects to commercializing the process.[110,130] Electrospun nanofibers have led to new approaches in the development of scaffolds for tissue engineering and delivery systems for bioactive molecules. Polymeric nanofibers have proved to be attractive materials for a wide range of applications because of their unique properties, e.g. very high surface area to volume ratio, the possibility of surface functionalization, superior mechanical properties, and the similarity in structural morphology of spun meshes to the fibrillar extracellular matrix.[112]

Incorporation of proteins in electrospun fibers has been studied and applied for delivery of various types of proteins, enzymes, growth factors, etc.

Subbu's group is working extensively on this topic and has reported that in most cases involving the electrospinning of protein-loaded fibers, the protein is loaded in the core only as an additive to another hydrophilic polymer, which makes up most of the core material.[111] The reason could be either the inability of the protein solution to be electrospun due its low viscosity, or the susceptibility of the molecule to the electrospinning process.[135] This leads to low overall loading amount of the protein in the fibers. Another limitation of electrospinning of protein-loaded monolithic fibers for example is the burst release owing to the segregation of the protein toward the fiber surface (due to noncompatibility with the fiber–polymer matrix), and loss of activity due to either the organic solvents used to dissolve the polymer in which the protein is dispersed and/or the high electric field applied during the electrospinning process. Jiang et al. have reported on modulation of protein release from biodegradable core–shell structured fibers prepared by coaxial electrospinning.[136] However, the limitation of loading capacity still remains due the prerequisite of using an additional polymer as additive to achieve the minimum viscosity of the core solution required for viscous drag by the shell solution being drawn by the electrostatic force.

Nanofiber is used for a wide range of applications from medical to consumer products and industrial to high-tech applications. Researchers worldwide are investigating the use of electrospun nonwoven mesh for a variety of applications, such as in environment technology, biotechnology, energy, aerospace, capacitors, transistors, drug delivery systems, battery separators, energy storage, fuel cells, defense and security, and in healthcare. Currently in the medical field, there are reported experiments on the use of the electrospun mesh as skin grafts, vascular grafts, cartilage grafts, nerve grafts, cardiac grafts, cornea grafts, bone grafts, musculoskeletal, ligament grafts, liver tissue, and kidney tissue.[137] They are also used as scaffolds in tissue engineering, drug delivery devices, wound dressing, enzyme mobilization, etc., It is also used in various other fields like filtration, barrier, wipes, personal care, composite, garments, insulation, and energy storage.

6.8.6 NANOCOMPOSITES

They are materials made and blended with nanoparticles and are ideal for implant application in the skeletal system. Generally, two critical factors involved in producing nanocomposites with bone-like properties are a good interfacial adhesion between organic polymers and inorganic HAP, and the uniform dispersion of HAP at nanolevel in the polymer matrix. The former contributes to the improved mechanical properties of nanocomposites and the latter is responsible for the enhanced protein adsorption and cell adhesion/proliferation due to the larger specific surface area. Lee et al. attempted to improve these properties by modifying the surface of HAP nanocrystals with a diverse class of coupling agents and polymers.[116] The effect of surface-modified hydroxyapatite (HAP) nanocrystals on the biocompatibility of a new-type nanocomposite consisting of poly(e-caprolactone) (PCL) and HAP was studied. PCL-grafted HAP in nanocomposites provided more favorable environments for protein adsorption, positive effects on adhesion and proliferation of fibroblasts compared with HAP ceramic.

Bilayered composite scaffolds can be prepared by combining two polymers along with a bioceramic. PVA/HAP/PCL ceramics are commonly prepared. Brittle scaffolds are formed when bead formation takes place during loading of HAP in PVA scaffolds accompanied by a reduction in crystallinity of PVA in the composite scaffold. Good porosity is a characteristic property of the bilayer composites and usually ranges from 60–70% in scaffold when compared to other control scaffolds. Suitable hydrophilicity of bilayer scaffold and optimum pore diameter above 500 nm aid the adherence of cells and allow better growth and proliferation.[138–140] Bilayer composite matrices are potential scaffold materials for tissue engineering applications.

Graphitic nanoreinforced cementitious (GNRC) composites show great assurance as an improved class of structural material with superior mechanical properties that include enhanced microcrack control capabilities resulting in better damage tolerance, fracture toughness and improved flexural strength, and stiffness.[165] The enhanced bone related cell attachment and arrangement of these novel composites largely depends on the nanomaterial dispersion properties and chemical bonding characteristics within the cementitious matrix.[141–142] The chemical functionality of the aqueous suspensions is modified by dispersing the highly aggregating carbonaceous nanofibers that lead to the nanoreinforcement of surfaces. The resulting functional groups act as chemical binding sites for the metal ions in the calcium oxide/silica/alumina matrix of cement, thus, providing key compatibility attributes to the composite system.[1] The interface between the nanoreinforcement and the surrounding cementitious matrix plays a crucial role in enhancing the mechanical properties. Innovative characterization tools would be needed to understand and probe the material compatibility between graphitic nanoreinforcement and cementitious matrices.

6.9 FUTURE PROSPECTS

The demand for load bearing orthopedic and dental implants is increasing day by day globally due to accidents from sports and elderly citizens. Metals and alloys would continue to play their dominance in the field of orthopedic implants for load bearing applications. Surface engineering, bioceramics, and functionally graded coatings are the promising techniques to battle corrosion of biomaterials. The immense growth in bioceramics would bring further research on nanobioceramics coupled with polymers and glass would result in development of composites with good mechanical properties matching bone structure. On the other hand, nanobioceramics and composite coatings on orthopedic implants and coupled with tissue engineering would lead to the development of new biomedical devices with the better understanding of the structure–property relationship. Modification of biomaterial surface properties through control of the characteristic length scale is one promising approach to modulate select cell functions. In the case of nanoscale ceramics, the length scale of the material approaches the length scale of the proteins, which mediate cell material interactions. This aspect affects the type, amount and conformation of adsorbed proteins and in turn modulates select cell functions. Biomimetic coatings and scaffolds would pave the way for advanced future research in this area of nanobiomaterials. On the other hand a more systematic and multinational research approach is needed to develop and standardize protocols for evaluation of tissue engineered products. The cost of these products should be made affordable to the larger population.

Cellular behavior largely depends on the physical and chemical characteristics of materials surface. The attachment of an implant to the surrounding tissue is affected by a variety of factors including design, stability, biocompatibility, surface chemistry, mechanical properties, tissue properties, etc. Open porosity in the implant would allow the cells responsible for bone remodeling to grow in the material, and for blood vessels to supply the cellular activity. To date, however, there is no synthetic material that exhibits mechanical and remodeling behavior similar to those of the bone. Functionally graded nanocoatings, three-dimensional printing of devices and scaffold with tissue engineering including gene therapy would lead to the design and development of a new generation of devices for the orthopedic appliances market.

ACKNOWLEDGMENT

The author thanks Dr. S. Sriman Naryanan, Professor and Head, Department of Analytical Chemistry, University of Madras for his constant support and encouragement and valuable suggestions in preparation of this chapter. Thanks are also due to the contributors, who kindly agreed to permit reprinting the figures and tables.

KEYWORDS

- **metallic implants**
- **nanobioceramics**
- **hydroxyapatite**
- **surface coatings**
- **scafolds**

BIBLIOGRAPHY

1. Williams, D.F. *Mater. Sci. Tech.* **1987**, *3*, 797.
2. Wrekstrom, J.K. *J. Mater*. **1996**, *1* (2), 366.
3. Mudali, U.K.; Sridhar, T.M.; Raj, B. Sādhanā, (Academy Proceedings in Engineering Sciences, Indian Academy of Sciences), **2003**, 28, Parts 3 and 4, 601–637.
4. Mudali, U.K.; Sridhar, T.M.; Eliaz, N.; Raj, B. *Corr. Rev*. **2003**, *21*, 231–267.
5. Patterson, S.P.; Daffner, R.H.; Gallo, R.A. Electrochemical Corrosion of Metal Implants. *Am. J. Roentgenol.* (AJR), **2005**, *184*, 1219–22.
6. Hansen, D.C. Metal Corrosion in the Human Body: The Ultimate Bio-Corrosion Scenario, the Electrochemical Society Interface. *Summer* **2008**, *17*, 31–34.
7. Gautam, S.; Chou, C.F.; Dinda, A.K.; Potdar, P.D.; Mishra, N.C.; *Mater. Sci. Eng*. C. **2014**, *34* 402–409.
8. Nithya, R.; Natarajan, T.S.; Rajiv, S. *Adv. Polym. Technol.*; **2013**, *32*, 21348.
9. Maheshwari, S.U.; Samuel, V.K.; Nagiah, V. Fabrication and Evaluation of (PVA/HAp/PCL) Bilayer Composites as Potential Scaffolds for Bone Tissue Regeneration Application. *Ceram. Int*. **2014**, *40*, 8469–8477.
10. Galperin, A.; Oldinsk, R.A.; Florczyk, S.J.; Bryers, J.D.; Zhang, M.; Ratner, B.D.; Integrated Bi-Layered Scaffold For Osteochondral Tissue Engineering, *Adv. Healthcare Mater*. **2013**, *2*, 872–883.
11. Wong, J.Y.; Bronzino, J.D. Biomaterials. 2007, Boca Raton: CRC Press. 1 v. (various pagings).

12. Dubok, V.A. Bioceramics—Yesterday, today, tomorrow. Powder Metallurgy and Metal Ceramics, **2000**, *39*(7–8): p. 381–394.
13. Rusin, R.P.; Fischman, G.S.; S. American Ceramic, Bioceramics. *Mater. Appl.* 1995: ACS.
14. Bioceramics: Engineering in medicine. Biomedical Materials Symposium. 1972, New York: Interscience Publishers. xi, 484 p.
15. Park, J. Bioceramics [electronic resource]: Properties, Characterizations, and Applications. 2009, New York, NY: Springer-Verlag New York. Digital.
16. Ducheyne, P.; Lemons, *J. Ann. N. Y. Acad. Sci.* **1988**, *523,* 64.
17. Sridhar, T.M., Rajeswari, S. Biomaterials Corrosion, *Corros. Rev*. Spl Issue, **2009**, 27, 287–332.
18. Raemdonck, W.V.; Ducheyne, P.; Meester, P.D. "Calcium Phosphate Ceramics", In Metal and Ceramic Biomaterials: Strength and Surface, P. Ducheyne and G.W. Hastings (Eds.), Vol II, CRC Press, Boca Raton, FL, 1984, 144.
19. Van Haaren, E.H.; Smit, T.H.; Phipps, K.; Wuisman, P.I.; Blunn, G.; Heyligers, I C. Tricalcium-Phosphate and Hydroxyapatite Bone-Graft Extender for Use in Impaction Grafting Revision Surgery. An In Vitro Study on Human Femora. *J. Bone Joint Surg.* (Br) **2005**, *87* (2), 267–71.
20. Walschot, L.H.; Schreurs, B.W.; Verdonschot, N.; 1,2, Buma, P. The Effect of Impaction and a Bioceramic Coating on Bone in Growth in Porous Titanium Particles A Bone Chamber Study in Goats. *Acta Orthop*. **2011**, *82* (3), 372–377.
21. Black, J.; Hastings, G.W.; Knovel. Handbook of Biomaterial Properties [electronic resource]. 1st ed. 1998, London; New York: Chapman and Hall. xxvi, 590 p.
22. American Ceramic, S., Progress in Bioceramics. Progress in ceramic technology series book. 2004, Westerville, Ohio: The American Ceramic Society. ix, 342 p.
23. Sridhar, T.M.; Praveen, R.; Shanmugaraj, S.; Srinivas, S.K.; Das, D. Crystallographic Changes of Bioceramic Composition of Cadever Bone on Heat Treatments. Key Engineering Materials. "Bioceramics 20", **2008**, Vol. 361–363 p 219–222. ISBN-13: 978-0-87849-457-6.
24. Holand, W., Biocompatible and Bioactive Glass-Ceramics—State of the Art and New Directions. *J. Non-Cryst. Solids*. **1997**. *219*: p. 192–197.
25. Suominen, E.; Turun, Y. Bioactive Ceramics in Reconstruction of Bone Defects: Studies with Bioactive Glasses, Glass-Ceramics, Hydroxyapatite and their Composites. Turun yliopiston julkaisuja = Annales Universitatis Turkuensis. Sarja—Ser. D, Medica—Odontologica, 2310355–9483. 1996, Turku: Turun Yliopisto. 1 v. (various pagings).
26. Ylanen, H.O., Bioactive Glasses [electronic resource]: Materials, Properties and Applications. 2011, Cambridge: Woodhead Publishing Ltd. 288 p.
27. Hashmi, M.U.; Shah, S.A. Bioactive Glass Ceramics for Orthopedic Applications: Significance of CaO/MgO Ratio for Mechanical, Structural and Biological Properties. 2012, Saarbrücken: Lambert Academic Publishing. 72 p.
28. Kokubo, T. et al. Mechanical-Properties of a New Type of Apatite-Containing Glass Ceramic for Prosthetic Application. *J. Mater. Sci.* **1985**, *20*(6), 2001–2004.
29. Bembey, A.K. Micro-Mechanical Properties and Composite Behaviour of Bone. 2008. 215 leaves.
30. International Symposium on Ceramics in, M., G. Daculsi, and P. Layrolle, Bioceramics : Volume *20* : proceedings of the 20th International Symposium on Ceramics in Medicine, the Annual Meeting of the International Society for Ceramics in Medicine (ISCM),

Nantes, France, 24–26 October 2007. Key engineering materials, v. 361-3631013-9826. 2008, Stafa-Zuerich; UK: Trans Tech Publications.
31. Hench, L.L. *J. Am. Ceram. Soc.* **1991**, *74* 1487.
32. Yazdimamaghani, M.; Vashaee, D.; Assefa, S.; Walker, K.J.; Madihally, S.V.; Köhler, G.A.; Tayebi, L. Hybrid Macroporous Gelatin/Bioactive-Glass/Nanosilver Scaffolds with Controlled Degradation Behavior and Antimicrobial Activity for Bone Tissue Engineering. *J. Biomed Nanotechnol.* **2014**, *10*(6), 911–31.
33. Martın, A.; Encinas-Romero; Salvador Aguayo-Salinas; Santos, J.; Castillo, F. F.; Castillo′n-Barraza; Victor, M.; Castano. Synthesis and Characterization of Hydroxyapatite–Wollastonite Composite Powders by Sol–Gel Processing. *Int. J. Appl. Ceram. Technol.* **2008**, *5* (4) 401–411.
34. Sridhar, T.M. Nano Bioceramic Coatings for Biomedical Applications. *Mater. Technol.* **2010**, *25*, 184–195, (DOI 10.1179/175355510X12723642365449).
35. Ducheyne, P.; Lemons, J. *Ann. N. Y. Acad. Sci.* **1988**, *523*, 64.
36. Hench, L.L., ed. An Introduction to Bioceramics. Second edition/editor, Larry L. Hench, University of Florida, USA. ed. 2013, World Scientific Imperial College Press: Hackensack, New Jersey London. xix, 600 pages.
37. Kokubo, T.M. Institute of Materials, and Mining, Bioceramics and their Clinical Applications. Woodhead Publishing in Materials. 2008, Cambridge: Woodhead Publishing Ltd. xxiv, 760–784 p.
38. Kossler, W.; Fuchs, J.; Ebrary, I. Bioceramics : Properties, Preparation and Applications. 2009, New York: Nova Biomedical Books. xi, 299 p.
39. Šesták, J., J.i.J. Mareš, and P. Hubík, Glassy, amorphous and Nano-Crystalline materials: Thermal Physics, Analysis, Structure and Properties. Hot topics in thermal analysis and calorimetry. 2010, Dordrecht; London: Springer. xvii, 380 p.
40. Japan Medical Materials. Bioceramics and their Clinical Applications. Cambridge: Woodhead Publishing Limited, 2008. pp. 28–30, 243–263.
41. Narasimha, R.; Sastry, T.P. Biointegration to Bone—Current Concepts and Perspectives. *Trends Biomater. Artif. Organs.* **2014**, *28*(3), 113–120.
42. Noiri, A.; Hoshi, F.; Murakami, H.; Sato, K.; Kawai, S.; Kawai, K.; Ganki, **2002**, *53*(6), 476–480.
43. Thamaraiselvi, T.V.; Rajeswari, S. "Biological Evaluation of Bioceramic Materials—A Review". *Trends Biomater. Artif. Organs.* **2004**, *18*, 9–17.
44. Piconia, G.; Maccaurob. Biomaterials 20 (1999) 1–25, Zirconia as a ceramic biomaterial.
45. Ramesh, T.R.; Gangaiah, M.; Harish, P.V.; Krishnakumar, U.; Nandakishore, B. Zirconia Ceramics as a Dental Biomaterial—An Overview. *Trends Biomater. Artif. Organs.* **2012**, *26*(3), 154–160.
46. Balamurugan, A.; Balossier, G.; Kannan, S.; Michel, J.; Faure, J.; Rajeswari, S. Electrochemical and Structural Characterisation of Zirconia Reinforced Hydroxyapatite Bioceramic Sol–Gel Coatings on Surgical Grade 316L SS for Biomedical Applications. *Ceram. Int.* **2007**, *33*, 605–614.
47. Chevalie, J. What Future for Zirconia as a Biomaterial? *Biomaterials.* **2006**, *27*(4), 535–543.
48. Köbel, S. Mechanische Eigenschaften von TZP BIO HIP, Technical Report, Metoxit AG, 19.10.2006.
49. Manivasagam, G.; Dhinasekaran, D.; Rajamanickam, A.; Biomedical Implants: Corrosion and its Prevention—A Review. Recent Patents on Corrosion Science, **2010,** 2, 40–54 1877–6108.

50. Breme, J.; Biehl, V.; Hoffmann, A. *Adv. Eng. Mater.* **2000**, *2*(5), 270.
51. Lakes, R. "Composite Biomaterials." The Biomedical Engineering Handbook: 2nd ed. Joseph D. Bronzino Boca Raton: CRC Press LLC, 2000.
52. Yang, Y.; Kim K.H.; Ong, J.L. *Biomater* **2005**, *26*, 327–37.
53. de Senaa, L.A.; Rochab, N.C.C.; Andradec M.C.; Soaresa, G.A. *Surf. Coat. Tech.* **2003**, *166*, 254–258.
54. Kim, H.M. *J. Curr. Opin. Solid State Mater. Sci.* **2003**, *7*, 289–299.
55. Sridhar, T.M.; Kishen, A.; Shanmugaraj, S.; Praveen, R.; Srinivas, S.K.; Das, D.; Subbiya, A. Biomineralization Studies in Age Induced Human Teeth Key Enginering Materials, "Bioceramics 20", Vol. 361–363 (2008) 893–896. ISBN-13: 978-0-87849-457-6.
56. Wang, H.; Eliaz, N.; Xiang, Z.; Hsu, H.-P.; Spector, M.; Hobbs, L.W. *Biomaterials* **2006**, *27*, 4192–4203.
57. Balamurugan, A.; Balossier, G.; Kannan, S.; Rajeswari, S. *Mater. Lett.* **2006**, *60*, 2288–2293.
58. Sridhar, T.M.; Eliaz, N.; Kamachi Mudali, U.; Raj, B. Electrophoretic Deposition of Hydroxyapatite Coatings and Corrosion Aspects of Metallic Implants. *Corros. Rev.* **2002**, *20*, 255–293.
59. Hijon, M.V. Cabanas, I. Izquierdo-Barba, and M. Vallet-Regi: *Key Eng.Mater.* **2004**, *363*, 254–256.
60. Abdeltawab, A.A.; Shoeib, M.A.; Mohamed, S.G. Electrophoretic Deposition of Hydroxyapatite Coatings on Titanium from Dimethylformamide Suspensions. *Surf. Coat. Tech.* **2011**, *206*, 43–50.
61. Mahmoodi S.; Sorkhi, L.; Farrokhi-Rad, M.; Shahrabi, T. Electrophoretic Deposition of Hydroxyapatite–Chitosan Nanocomposite Coatings in Different Alcohol. *Surf. Coat. Tech.* **2013**, *216*, 106–114.
62. Boccaccini AR1, Keim S, Ma R, Li Y, Zhitomirsky I. Electrophoretic Deposition of Biomaterials. *J. R. Soc. Interface.* **2010**, *7*, S581–613.
63. Sridhar, T.M.; Mudali, U.K.; Subbaiyan, M. Preparation and Characterisation of Electrophoretically Deposited Hydroxyapatite Coatings on Type 316L Stainless Steel. *Corros. Sci.* **2003**, *45*, 237–252.
64. Sridhar, T.M.; Mudali, U.K. Development of Bioactive Hydroxyapatite Coatings on Type 316L Stainless Steel by Electrophoretic Deposition for Orthopaedic Applications. *T. Indian. I. Metals.* 2003, *56*, 221–230.
65. Sridhar, T.M.; Mudali, U.K.; Subbaiyan, M. Sintering Atmosphere and Temperature Effects on Hydroxyapatite Coated Type 316L Stainless Steels. *Corros. Sci.* **2003**, *45*, 2337–2359.
66. Sridhar, T.M.; Mudali, U.K., Depth Profile and Interface Characteristics of Hydroxyapatite Coatings on Type 316L SS, In Surface Modification Technologies XXIII, Edited by Sudarshan, T. S.; Mudali, U.K.; Raj, B. **2010**, 507–513, ISBN: 978-81-910570-0-2.
67. Cordero-Arias L.; Cabanas-Polo, S.; Gilabert, J.; Goudouri, O.M.; Sanchez, E.; Virtanen, S.; Boccaccini, A.R. Electrophoretic Deposition of Nanostructured TiO2/Alginate and TiO2-Bioactive Glass/Alginate Composite Coatings on Stainless Steel. *Adv. Appl. Ceram.* **2014,** *113* NO 1, 42–49.
68. Mohan, L.; Durgalakshmi, D.; Geetha, M.; Sankara, T.; Narayanan; Asokamani, R. 'Electrophoretic Deposition of Nanocomposite (HAp z TiO2) on Titanium Alloy for Biomedical Applications'. *Ceram. Int.* **2012**, *38*, 3435–3443.

69. Kwok, C.T.; Wong, P.K.; Cheng, F.T.; Man, H.C. *Appl. Surf. Sci.* **2009**, *255*, 6736–6744.
70. Yasuda, K.; Schmuki, P. *Electrochim. Acta.* **2007**, *52*, 4053–4061.
71. Eliaz, N.; Sridhar, T.M.; Mudali, U.K.; Raj, B. Electrochemical and Electrophoretic Deposition of Hydroxyapatite for Orthopaedic Applications. *Surf. Eng.* **2005**, *21*, N0. 3 238–242.
72. Eliaz, N.; Sridhar, T.M. Electrocrystallization of Hydroxyapatite and its Dependence on Solution Conditions. *Cryst. Growth. Des.* **2008**, *8*, N0. 11 3965–3977 (DOI 10.1021/cg800016h).
73. Tsuchiya, H.; Macak, J M.; Muller, L.; Kunze, J.; Muller, F.; Greil, P.; Virtanen, S.; Schmuki, P. *J. Biomed. Mater. Res.* **2006**, *77A*, 534–541.
74. Song, W.H.; Jun, Y. K.; Han, Y.; Hong, S. H. *Biomater.* **2004**, *25*, 3341–3349.
75. Kunze, J.; Muller, L.; Macaka, J. M.; Greil, P.; Schmuk, P.; Muller, F. A. *Electrochim. Acta.* **2008**, *53*, 6995–7003.
76. Narayanan, R.; Seshadri, S.; Kwon, T.; Kim, K. *Scr. Mater.* **2007**, *56*, 229–232.
77. Narayanan, R.; Kwon, T-Y.; Kim, K-H. *Mater. Sci. Eng.: C.* **2008**, *28*, 1265–1270.
78. Lee, E.J.; Lee, S.H.; Kim, H.W.; Kong, Y.M.; Kim, H.E. *Biomater* **2005**, *26*, 3843–51.
79. Cheng, K.; Zhang, S.; Weng, W.; Zhang, X. *Surf. Coat. Tech.* **2005**, *198*, 242–246.
80. Ananth, K.P.; Suganya, S.; Mangalaraj, D.; Ferreira, J.M.F.; Balamurugan, A. Electrophoretic Bilayer Deposition of Zirconia and Reinforced Bioglass System on Ti6Al4V for Implant Applications: An In Vitro Investigation. *Mater. Sci. Eng.: C.* **2013**, *33*, 4160–4166.
81. Okazaki, Y.; Gotoh, E. Comparison of Metal Release from Various Metallic Biomaterials *In Vitro. Biomaterials.* **2005**, *26,* 11–21.
82. Luiz de Assis, S.; Wolynec, S.; Costa, I. Corrosion Characterization of Titanium Alloys by Electrochemical Deposition. *Electrochim. Acta.* **2006**, *51*, 1815–1819.
83. Vythilingum, V. Zirconia Nanocomposites for Biomedical Applications, University of the Witwatersrand, Faculty of Engineering and the Built Environment, School of Chemical and Metallurgical Engineering, 2012.
84. Nanotechnology and Tissue Engineering—the Scafold, C.T. Laurencin and L.S. Nair ed., 2008, 163, CRC Press, Boca Raton.
85. Griffith, L.G.; Naughton, G. *Science* **2002**, *295*, 1009–14.
86. Sergey V.; Dorozhkin: *Biomater.* 2009, 2010 Mar;31(7):1465–85.
87. Aghozbeni E.A.H; Fooladi A.A.I; Koudehi M.F; Amiri A; Nourani, M.R. Use of Nano-Bioglass Scaffold Enhanced with Mesenchymal Stem Cells for Rat Calvarial Bone Tissue Regeneration. *Trends Biomater. Artif. Organs.* **2014**, *28*(1), 8–18.
88. Azami, M.; Rabiee, M.; Moztarzadeh, F. Glutaraldehyde crosslinked Gelatin/Hydroxyapatite Nanocomposite Scaffold, Engineered via Compound Techniques. *J. Polym. Compos.* **2010**, *31*, 2112–20.
89. D Ben-David, T A. Kizhner, T Kohler, R Muller, Livne E, Srouji S. Cell-scaffold Transplant of Hydrogel Seeded with Rat Bone Marrow Progenitors for Bone Regeneration. *J. Cranio. Maxillofac. Surg.* **2011**, 364–371.
90. Greenwald, A.S.; Boden, S.D.; Goldberg, V.M.; Khan, Y.; Laurencin, C.T.; Rosier, R.N. *J. Bone Joint Surg. Am.* **2001**, *83*, S98–S103.
91. Woo, K.M.; Chen, V. J.; Ma, P/ X. *J. Biomed Mater. Res.* **2003**, *67A*, 531–7.
92. 121. M. Vallet-Regi: *J. Chem. Soc. Dalton Trans.* **2001**, *100*, 97–108.
93. Ngiam, M.; Liao, S.; Patil, A. J.; Cheng, Z.; Chan, C. K.; Ramakrishna, S. *Bone.* **2009**, *45*, 4–16.

94. Zhang, W.; Huang, K.L.; Liao, S.S.; Cui, F.Z. *J. Am. Ceram. Soc.* **2003**, *86*, 1052–4.
95. Woo, K.M.; Chen, V.J.; Ma, P.X. *J. Biomed Mater. Res.* **2003**, *67A*, 531–537.
96. Naveen Kumar, Anil K. Gangwar, Dayamon D. Mathew, Sameer Shrivastava, Mamta Negi, Himani Singh, Ashok K. Sharma, Swapan K. Maiti, Remya Vellachi, Sonal and Manoj P. Singh, Development of Collagen Based Decellularized Biomaterials as 3-D Scaffold for Tissue Engineering of Skin. *Trends. Biomater. Artif. Organs.* **2014**, *28*(3), 83–91.
97. Smith, M.; McFetridge, P.; Bodamyali, T.; Chaudhuri, J.B.; Howell, J.A. Porcine-Derived Collagen as a Scaffold for Tissue Engineering. *Trans. Inst. Chem. Eng.* **2000**, *78*, 19–24.
98. Xian, W.; Schwertfeger, K.L.; Vargo-Gogola, T.; Rosen, J.M. Pleiotropic Effects of FGFR1 on Cell Proliferation, Survival, and Migration in a 3D Mammary Epithelial Cell Model. *J Cell Biol.* **2005**, *171* (4), 663–66.
99. Al-Munajjed, A. A.; Plunkett, N. A.; Gleeson, J.P.; Weber, T.; Jungreuthmayer, C.; Levingstone, T.; Hammer, J.; O'Brien, F. J. Development of a Biomimetic Collagen-Hydroxyapatite Scaffold for Bone Tissue Engineering using a SBF Immersion Technique. *J. Biomed. Mater. Res.* **2009**, *90B*: 584–591.
100. Uematsu, K.; Hattori, K.; Ishimoto, Y.; Yamauchi, J.; Habata, T.; Takakura, Y.; Ohgushi, H.; Fukuchi T.; Sato, M. *Biomater* **2005**, *26*, 4273–9.
101. Sila-Asna, M.; Bunyaratvej, A.; Maeda, S.; Kitaguchi, H.; Bunyaratavej, N. Osteoblast Differentiation and bone formation gene expression in Strontium Inducing Bone Marrow Mesenchymal Stem Cell. *Kobe J. Med. Sci.* **2007**, 25–35.
102. Songa, W.H.; Juna, Y.K.; Hana, Y.; Honga, S.H. *Biomater.*, **2004**, *25*, 3341–3349.
103. dos Santos, E.A.; Farina, M.; Soares, G. A; Anselme, K. *J Biomed Mater Res.* **2009**, *89A*, 510–520.
104. Rokusek, D.; Davitt, C.; Bandyopadhyay, A.; Bose, S.; Hosick, H.L. *J Biomed Mater Res A.*, **2005**, *75*, 588–594.
105. Dorozhkin, S.V.; Epple, M. *Angew. Chem. Int. Ed. Engl.* **2002**, *41*, 3130–3146.
106. Ehara, A.; Ogata, K.; Imazato, S.; Ebisu, S.; Nakano, T.; Umakoshi, Y. *Biomater* **2003**, *24*, 831–836.
107. Smith, I.O.; McCabe, L.R.; Baumann, M.J. *Int. J. Nanomedicine.* **2006**, *1*, 189–94.
108. Chen, F.; Lam, W.M.; Lin, C.J.; Qiu, G.X.; Wu, Z.H.; Luk, K.D.; Lu, W.W. *J. Biomed Mater. Res. B Appl. Biomater.* **2007**, *82*, 183–91.
109. Barrère, F.; Layrolle, P.; Van Blitterswijk, C.A.; Groot, K. De. *J. Mater. Sci. Mater. Med.* **2001**, *12*, 529–34.
110. Seeram Ramakrishna, Fujihara, K; Teo, W-E; Lim, T-C; Ma, Z. "An Introduction to electrospinning and Nanofibers". World Scientific, Singapore (2005)
111. Tiwari, S. K.; Venkatraman, S. Polymer International, **2012**, *61* (10), 1549–1555.
112. Pillai C.K.S and Chandra P. Sharma, *Trends. Biomater. Artif. Organs.* **2009**, *22* 179–201.
113. Zhang, E.; Zou, C.; Yu, G. *Mater. Sci. Eng.* **2009**, C *29*, 298–305.
114. Tian, T.; Jiang, D.; Zhang, J.; Lin, Q. *Mater. Sci. Eng.* **2008,** C *28*, 57–63.
115. Juan, F.; Ying, Z.; Jiang, Y. *Trans. Non Ferrous Met. Soc.* China, **2009**, *19*, 125–130.
116. Lee, H.J.; Kim, S.E.; Choi, H.W.; Kim, C.W.; Kim, K.J.; Lee, S.C.; *Eur. Poly. J.* **2007**, *43*, 1602–1608.
117. Deepachitra, R.; Chamundeeswari, M.; Santhosh kumar, B.; Krithiga, G.; Prabu, P.; Devi, M.P.; Sastry, T.P. Osteo mineralization of fibrin-decorated graphene oxide, C A R B ON 5 6 (2013) 6 4–7 6.

118. Zhao, X.; Zhang, Q.; Hao, Y.; Li, Y.; Fang, Y.; Chen, D. Alternate Multilayer Films of Poly(Vinyl Alcohol) and Exfoliated Graphene Oxide Fabricated via a Facial Layer-By-Layer Assembly. *Macromolecules*. **2010**, *43*(22), 9411–9416.
119. Lee, W.C.; Lim, C.H.; Shi, H.; Tang, L.A.; Wang, Y; Lim, C.T., et al. Origin of Enhanced Stem Cell Growth and Differentiation on Graphene and Graphene Oxide. *ACS Nano*. **2011**, *5*(9), 7334–7341.
120. Zhang, Y.; Nayak, T.R.; Hong, H.; Cai, W. Graphene: a Versatile Nanoplatform for Biomedical Applications. *Nanoscale*. **2012**, *4*(13), 3833–3842.
121. Nayak, T.R.; Andersen, H.; Makam, V.S.; Khaw, C.; Bae, S.; Xu, X.; et al. Graphene for Controlled and Accelerated Osteogenic Differentiation of Human Mesenchymal Stem Cells. ACS Nano **2011**, *5*(6), 4670–4678.
122. Muthukuamr, T.; Senthil, R.; Sastry, T.P. Synthesis and Characterization of Biosheet Impregnated with Macrotylomauniflorum Extract for Burn/Wound Dressings. Colloids Surf. B Biointerfaces. **2013**, (102), 694–699.
123. K.A. Bhat, R. N Raghavan, D. Sangeetha, S. Ramesh, Multi-walled Carbon Nanotube Reinforced Glass Ionomer Cements for Dental Restorations, *Trends. Biomater. Artif. Organs*. **2013**, *27*(4), 168–176.
124. Stout, D.A.; Webster, T.J. "Carbon nanotubes for stem cell control", *Mater. Today*. **2012**, *15*, 7–8, 312–319.
125. Cicchetti, R.; Divizia, M.; Valentini, F.; Argentin, G. "Effects of Single-Wall Carbon Nanotubes in Human Cells of the Oral Cavity: Geno-Cytotoxic Risk", Toxicol. In Vitro. **2011**, *25*, 1811–1819.
126. E. Hirata, T. Akasaka, M. Uo, H. Takita, F. Watari, A. Yokoyama, "Carbon nanotube-coating accelerated cell adhesion and proliferation on poly (L-lactide)". *Appl. Surf. Sci.* **2012**, *262*, 24–27.
127. Basu, B. Nanoceramics and Nanocomposites. *Curr. Sci.* **2008**, *95*, 570–571.
128. Edelstein, A.S Cammarata, R.C. Nanomaterials: Synthesis, Properties, and Applications. Abingdon : Taylor and Francis Group, **1996**. pp. 57–59.
129. Rahaman, M.N. Ceramic Processing and Sintering 2nd ed. 2nd New York: Marcel Dekker, 2003. pp. 230-233, 828, 778–779.
130. Ramakrishna, S.; Mayer, J.; Wintermantel, E.; Leong, K.W. Biomedical Applications of Polymer-Composite Materials: A Review. *Compos. Sci. Technol.* **2001**, 61, 1189–1224.
131. Ai, J.; Rezaei-Tavirani, M.; Biazar, E.; Heidari, K.S.; Jahandideh, R. Mechanical Properties of Chitosan-Starch Composite Filled Hydroxyapatite Micro-and Nanopowders. *J. Nanomater*. **2011**, Article ID 391596.
132. Lei, L.; Li, L.; Zhang, L.; Chen, L,D.; Tian, W. Structure and performance of nano-hydroxyapatite filled biodegradable poly ((1, 2-propanediolsebacate)- citrate) elastomers, *Polym. Degrad. Stab.* **2009**, *94*, 1494–1502.
133. Kroeze, R.J.; Helder, M.N.; Govaert, L.E.; Smit, T.H. Biodegradable Polymers in Bone Tissue Engineering. *Materials* **2009**, *2*, 833–856.
134. Aravind, K.; Shanmuga, S.S.; Sangeetha, D. In Vivo Studies of Sulphonated Polyether Ether Ketone Based Composite Bone Graft Materials, *Trends. Biomater. Artif. Organs*. **2014**, *28*(2), 52–57.
135. Dror, Y.; Kuhn, J.; Avrahami, R.; Zussman, E. *Macromolecules*. **2008**, *41*, 4187–4192.
136. Jiang H, Hu Y, Zhao P, Li Y, Zhu K. *J Biomed. Mater. Res. Part B: App. Biomater*. **2006**, *79B*, 50–57.
137. Maleki, M.; Latifi, M.; Tehran, A.M.; Mathur, S. *Poly. Eng. Sci.* **2013**, *53*, 1770–79.

138. Anuradha, S.; Maheswari, U.K.; Swaminathan, S.; *BioMed Res. Int.* (2013), http://dx.doi.org/10.1155/2013/390518.
139. Timothy M. O'Shea and Xigeng Miao. Bilayered Scaffolds for Osteochondral Tissue Engineering; Tissue Engineering Part B: Reviews. December **2008**, 14(4): 447–464.
140. Maheshwari, S.U.; Samuel, V.K.; Nagiah, N. Fabrication and Evaluation of (PVA/HAp/PCL) Bilayer Composites as Potential Scaffolds for Bone Tissue Regeneration Application. *Ceram. Int. 40*, Issue 6, July **2014**, Pages 8469–8477.
141. Aich, N.; Zohhadi, N.; Khan, A.I.; Matta, F.; Ziehl, P.; Navid, B. Saleh, Applied TEM Approach for Micro/Nanostructural Characterization of Carbon Nanotube Reinforced Cementitious Composites, *J. Res. Updates Polym. Sci.* **2012**, *1*, 14–23.

PART IV

Nano-Hybrid Materials for Advanced Engineering Applications

CHAPTER 7

GREEN NANOHYBRID SILVER-EPOXY ECO-FRIENDLY COATINGS

R. MANJUMEENA[1,*], D. DURAIBABU[2], and S. ANANDA KUMAR[2]

[1]*Centre for Advanced Studies in Botany, University of Madras, Chennai 600025, India. *E-mail: manjumeena1989@gmail.com*

[2]*Department of Chemistry, Anna University, Chennai 600025, India.*

CONTENTS

ABSTRACT

TGBAPB epoxy resin was reinforced with 1, 3, and 5 wt% of surface functionalized AgNPs (F-AgNPs), which were synthesized using *Rosa indica* wichuriana hybrid leaves extract with a view of augmenting the corrosion control property of the epoxy resin and also imparting antimicrobial activity to epoxy coatings on mild steel. Corrosion resistance of these coatings was evaluated by EIS, potentiodynamic polarization studies, and cross scratch tests. Results showed that the corrosion resistance increased at 1 wt% of F-AgNPs. The F-AgNPs/TGBAPB coatings also offered manifold antimicrobial protection to the steel surfaces by inhibiting the growth of bacteria like *Pseudomonas aeruginosa*, *Bacillus subtilis*, and *Escherichia coli.*

7.1 INTRODUCTION

Nanotechnology represents an emerging technology that has the potential to have an impact on an incredibly wide number of industries, such as the medical, the environmental, and the pharmaceutical industries. Nanoparticles (NPs) are clusters of atoms in the size range of 1–100 nm. The use of nanoparticles is gaining impetus in the present century as these particles possess defined chemical, optical, and mechanical properties.[1-3] There is a growing need to develop an environmentally friendly process for corrosion control that does not employ any toxic chemicals. Generally, nanoparticles are prepared by a variety of chemical and physical methods,[4–7] which are not environmentally friendly. Nowadays, green chemistry procedures using plant extract[8,9] for the synthesis of nanoparticles are commonly employed. In the present study, a new approach was made to synthesize silver nanoparticles (AgNPs) from *Rosa indica* wichuriana hybrid leaf extract.

Owing to their high surface-to-volume ratio, surface energy, spatial confinement, and reduced imperfections, metal nanoparticles have characteristic physical, chemical, electronic, electrical, mechanical, magnetic, thermal, dielectric, optical, and biological properties as opposed to bulk materials.[10,11] Silver is a metal known for its broad spectrum of antimicrobial activity against Gram-positive and Gram-negative bacteria, fungi, protozoa, and certain viruses. In recent years, a rapid increase in microbes that are resistant to conventional antibiotics has been observed.[12] Especially, the frequency of infections provoked by opportunistic fungal strains has increased drastically. In such scenarios, nanoscale materials have emerged as novel antimicrobial agents.[13,14]

Due to the excellent mechanical properties and low cost, mild steel is extensively used as a constructional material in many industries. However, when exposed to the corrosive industrial environment, it is easily corroded. Paint coating is regarded as one of the most economical and widely used methods of protecting metal and steel surfaces. The coating layer acts as a barrier that isolates the metal and steel surfaces from the corrosive environment that causes deterioration of material. Thus, the coating layer enhances the lifetime of the metal and steel surfaces.[15] Epoxy is one of the most well-known binders that are used for a wide variety of protective coatings because of its excellent adhesion, good mechanical properties, and tremendous corrosion resistance performance.[16] Corrosion and wear resistant coatings can be used in a variety of industries, such as in automobile, power generation, utility, aerospace, defense, optical equipment, magnetic storage devices and bearings, engine parts, and seals, etc.[17,18] The epoxy coatings are the common victim of surface abrasion and poor resistance to the initiation and propagation of cracks.[19] Such processes propagate defects in the coating and impair their appearance and mechanical strength. The defects can act as pathways accelerating the diffusion of water, oxygen, and corrosive species onto the metallic substrate, resulting in its localized corrosion. Due to limited hydrophobicity, epoxy coatings experience large volume shrinkage upon curing and can absorb water from surroundings.[20] The presence of pores in the cured epoxy coating can assist in the migration of absorbed water and other species to the epoxy–metal interface leading to the initiation of corrosion of the metallic substrate and to the delamination of the coating.[21] It has been reported[22] that higher cross-linking density increases the anticorrosion behavior by decreasing the free volume and segmental mobility in the coating. It is difficult for the aggressive molecules to penetrate through the coating. The barrier performance of epoxy coatings can be enhanced by the incorporation of a second phase that is miscible with the epoxy polymer, by decreasing the porosity and restraining the diffusion path for deleterious species. For instance, inorganic filler particles at nanometer scale can be dispersed within the epoxy resin matrix to form an epoxy nanocomposite coating. The incorporation of nanoparticles into epoxy resins offers environmentally benign solutions to enhance the integrity and durability of coatings, since the fine particles dispersed in coatings can fill cavities[23–25] and cause crack bridging, crack deflection, and crack bowing.[26] Apparently, the demand for coatings with superior technical characteristics have induced the use of nanoparticles, such as TiO_2, Fe_2O_3, ZnO, SiO_2, and $CaCO_3$ either as fillers or as accessory ingredients in coatings.[27–29]

The main problem to the wide use of nanoparticles in coatings is the stable dispersion of nanoparticles because of their high surface activities. This could be overcome by functionalization of nanoparticles. Functionalization of nanoparticles can also prevent epoxy disaggregation during curing, resulting in a more homogenous coating. Nanoparticles tend to occupy small hole defects formed from local shrinkage during curing of the epoxy resin and act as a bridge interconnecting more molecules. This results in a reduced total free volume as well as an increase in the cross-linking density.[30,31] In addition, epoxy coatings containing nanoparticles offer significant barrier properties for corrosion protection[32,33] and reduce the trend for the coating to blister or delaminate. In recent years, the use of antimicrobial agents in surface coatings has been increasing to control the growth of bacteria and fungus in biomedical coatings and adhesives. Recently, considerable attention has been paid to the utilization of epoxy resin as antimicrobial coatings[34–37] for steel substrate. Epoxy antimicrobial coatings are of great interest for protection of surfaces, since survival of microorganisms on surface environment can be detrimental to the materials. Therefore, the development of epoxy nanostructured coatings with antimicrobial properties is essential.

Owing to the advantages of green synthesis of nanoparticles and the properties exhibited by nanosilver, we chose green synthesized nanosilver as reinforcing additive for our present study. This work is an attempt to formulate green nanoepoxy coatings by reinforcing a very low content of surface functionalized green synthesized AgNPs (F-AgNPs) in TGBAPB epoxy resin and study the anticorrosion and antimicrobial behavior of such eco-friendly coatings. This work may provide proper guidance for the design of high functionality TGBAPB epoxy resin green nanostructured coatings that are seldom explored for corrosion and microbial protection of steel substrates. This epoxy-green nanohybrid coating formulation also offers manifold antimicrobial protection to the steel surfaces by inhibiting the growth of bacteria like *Pseudomonas aeruginosa*, *Bacillus subtilis*, the most common human pathogen *Escherichia coli*. This dual behavior exhibited by surface functionalized green nanosilver is corroborated by experimental evidences. This approach interfaces nanotechnology with corrosion control science, which would eventually increase the longevity of steel and metal surfaces.

7.2 EXPERIMENTAL

7.2.1 BIOSYNTHESIS OF SILVER NANOPARTICLES

Aqueous extract of *Rosa indica* wichuriana hybrid was prepared using freshly collected leaves (10 g). The leave's surface was cleaned using running tap water, followed by distilled water, and boiled with 100 mL of distilled water at 60°C for 10 min. This extract was filtered through nylon mesh, followed by Millipore filter (0.45 μm). Filtered leaf extract was stored at −4°C for further use, being usable for 1 week.[38]

7.2.2 SYNTHESIS OF AGNPS FROM LEAVES EXTRACT

Total, 60 mL of aqueous solution of silver nitrate was reduced using 2.5 mL of leaves extract at room temperature for 10 min. This setup was incubated in dark (to minimize the photoactivation of silver nitrate) under static condition resulting in a brown-yellow liquid indicating the formation of silver (AgNPs).

7.2.3 SYNTHESIS OF TGBAPB VIA BAPB

The TGBAPB resin was synthesized using EPC and BAPB with 40% NaOH solution. A pale brown colored TGBAPB resin obtained (yield 80%) was purified and preserved for further use. The reaction sequence is illustrated in Scheme 7.1.[39]

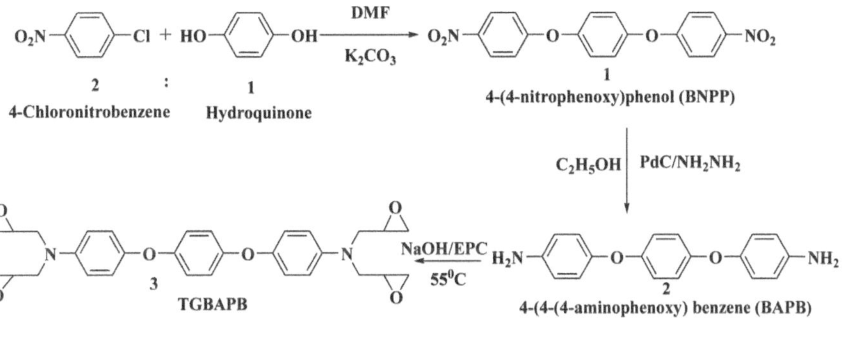

SCHEME 7.1 Synthesis of TGBAPB epoxy resin.

7.2.4 SURFACE FUNCTIONALIZATION OF AGNPS

The introduction of reactive NH_2 group onto the surface of AgNPs was achieved through the reaction between 3-aminopropyltriethoxysilane and the hydroxyl groups on the surface of AgNPs. Typically, 2.0 g AgNPs and 2 mL 3-aminopropyltriethoxysilane in 40 mL o-xylene were kept at 150°C for 3 h under stirring and argon protection. After that, the amine functionalized AgNPs were collected by filtration and rinsed thrice with acetone. Finally, the amine functionalized AgNPs were dried under vacuum for 12 h.[39] The reaction sequence is given in Scheme 7.2.

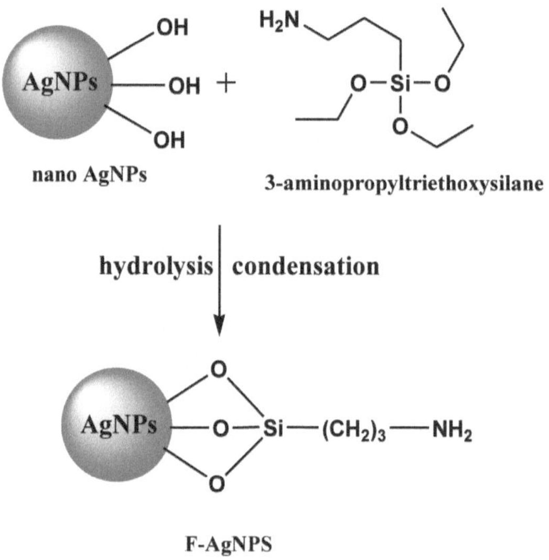

SCHEME 7.2 Surface functionalization of F-AgNPs.

7.2.5 SURFACE PREPARATION OF THE MILD STEEL SPECIMENS

The mild steel specimens of size 30 × 15 × 3 mm and composition Cr 17.20%, Ni-12.60%, Mo 2.40%, Mn 1.95%, N 0.02%, C 0.03%, Fe balance were polished by silicon carbide (SiC) papers up to 1200 grit. After polishing, the sample surface was rinsed with distilled water followed by acetone and finally dried at 40 °C before coating.

7.2.6 FORMULATION OF TGBAPB/F-AGNPS EPOXY COATINGS

TGBAPB epoxy was mixed separately with varying wt% of F-AgNPs (1, 3, and 5 wt%) at 60 °C for 15 min, with constant stirring so as to disperse the F-AgNPs completely within the epoxy resin and stoichiometric amount of curing agent (Aradur 140) was added to fabricate TGBAPB/F-AgNPs coating. This formulation was applied on mild steel specimen using bar coater. The samples were kept for 7 days at room temperature to allow complete curing. After curing, coating thickness was confirmed by Mini test 600FN, EXACTO-FN type. The thickness of the coating was found to be approximately ± 100 μm. Finally, the coated specimens were tested for their anticorrosion and antimicrobial properties. The functionalized AgNPs (F-AgNPs) form a covalent bond with TGBAPB epoxy resin as shown in Scheme 7.3. The method of application is shown in Figure 7.1. The nomenclature of varying wt% of F-AgNPs reinforced TGBAPB coating systems is given in Table 7.1.

TABLE 7.1 Composition of TGBABP/F-AgNPs Epoxy Nanocomposites.

Systems	TGBAPB/%F-AgNPs	Curing agent
a	100/0	Aradur 140
b	100/1	Aradur 140
c	100/3	Aradur 140
d	100/5	Aradur 140

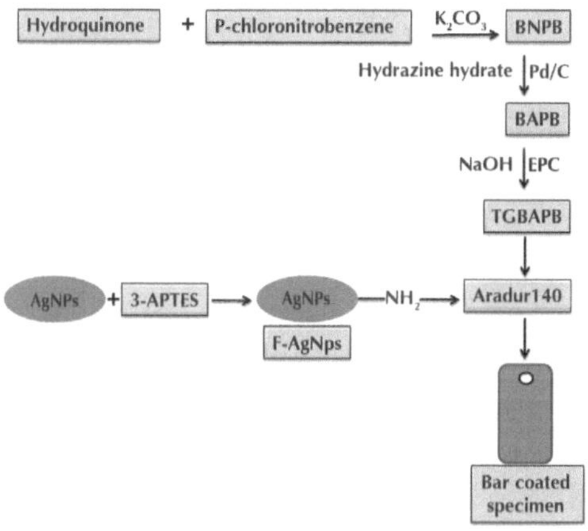

FIGURE 7.1 Flowchart involving the preparation of F-AgNPs-TGBAPB epoxy coating.

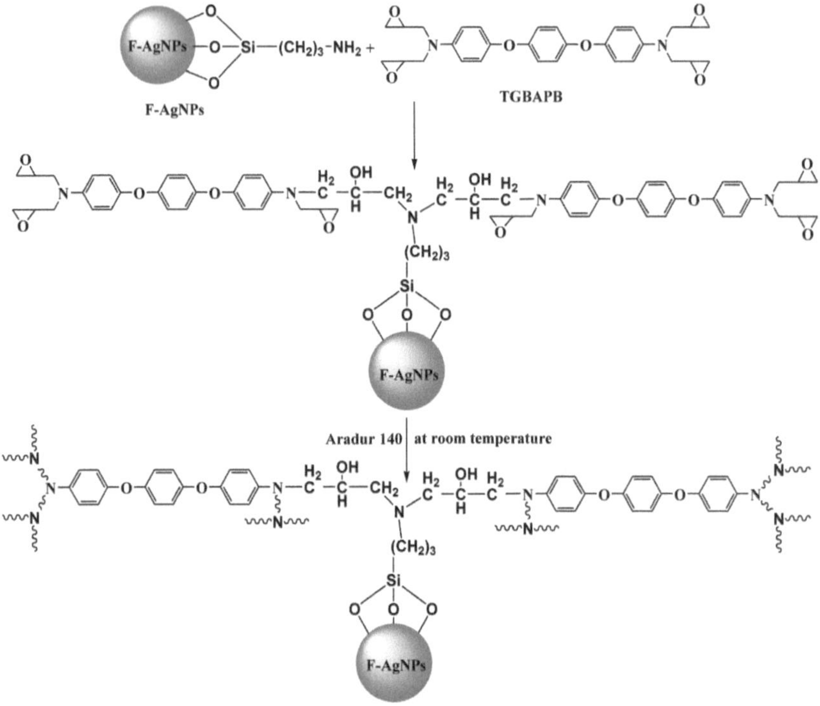

SCHEME 7.3 The reaction mechanism of TGBAPB with F-AgNPs nanoparticle.

7.2.7 EVALUATION OF ANTIBACTERIAL AND ANTIFUNGAL ACTIVITIES OF THE TGBAPB/F-AGNPS COATING

As 1 wt% F-AgNPs reinforced TGBAPB coating was found to be the best formulation with better anticorrosion properties, the same sample was tested for antibacterial and antifungal activities by an inhibition zone method. *Escherichia coli* (ATCC 8739), *Bacillus subtilis* (ATCC 6633), and *Pseudomonas aeruginosa* (ATCC 15442) were taken as the test bacteria. 100 mL Muller Hinton broth, 200 mL Muller Hinton agar, Petri dish, and the samples were autoclaved at 121°C, 15 psi for 15 min. A loop of the bacterial culture was inoculated from fresh colonies on agar plates into 100 mL Muller Hinton culture medium.

The culture was allowed to grow until the optical density reached 0.2 at 600 nm (OD of 0.2 corresponding to a concentration of 10^8 CFU mL^{-1} of medium). This indicates that the bacterial culture is in exponential or log phase of growth, which is ideal for the experiment. Then it was swabbed uniformly onto individual Mueller Hinton agar plates using sterile cotton swabs. The coated mild steel samples and control (only epoxy without any nanoparticle) were placed in the centre of the culture swabbed petriplate in such a manner that the coating is in contact with the culture. The plates were examined for possible clear zone formation after overnight incubation at 37 °C. The presence of clear zone around the steel samples on the plates was recorded as an inhibition against the test bacterial species.[39]

7.3 RESULTS AND DISCUSSION

7.3.1 POWDER X-RAY DIFFRACTION OF AGNPS

The crystalline nature of AgNPs was confirmed from X-ray diffraction (XRD) analysis. Figure 7.2 shows the XRD pattern of the synthesized AgNPs. The five diffraction peaks were observed at 38.2°, 44.4°, 64.6°, 77.6°, and 81.7° in the 2 range can be indexed to the (111) (200) (220) (331), and (222) reflection planes of face centered cubic structure of metallic silver nanopowders are shown in Figure 7.2. The mean size of the biosynthesized AgNPs was determined to be in the range of 18–20 nm. The calculated lattice parameter was found to be 4.08 Å.

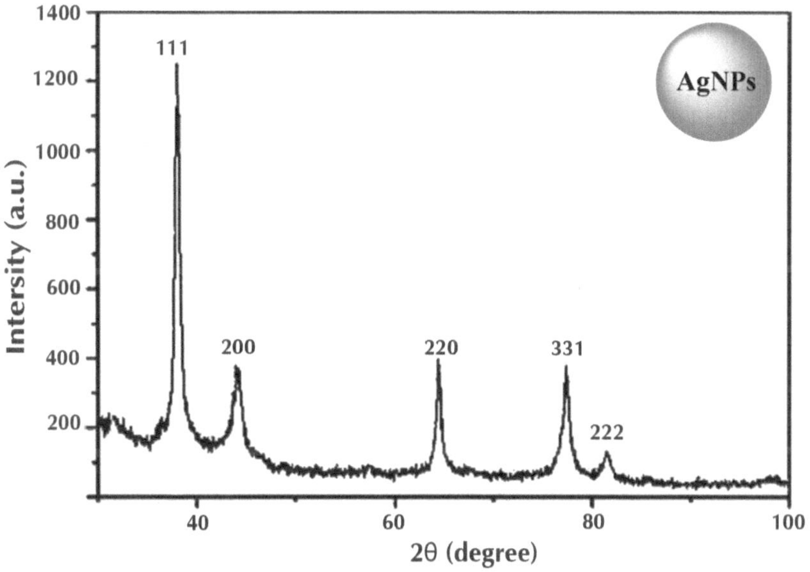

FIGURE 7.2 X-ray diffraction of AgNPs.

7.3.2 TEM ANALYSIS OF AGNPS

The TEM technique was used to visualize the size and shape of the synthesized AgNPs. Figure 7.3 shows the typical bright field TEM micrograph of the synthesized AgNPs. The observed morphology of nanoparticles in the micrograph is spherical. The TEM micrograph suggests that the size of the particles is around 15–25 nm.

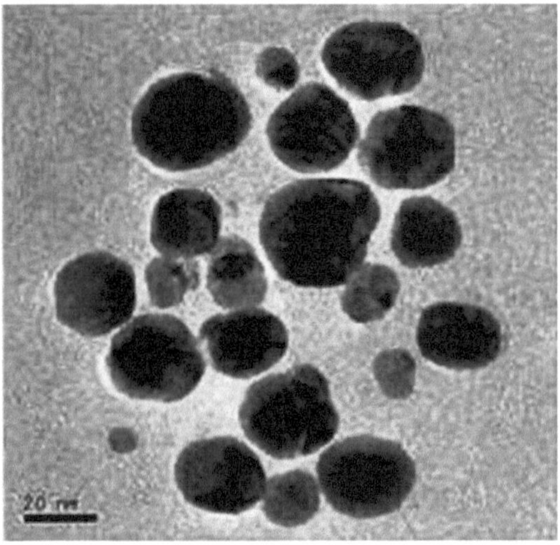

FIGURE 7.3 TEM of AgNPs.

7.3.3 CHARACTERIZATION OF AMINE FUNCTIONALIZED AGNPS

Figure 7.4(a) and (b) shows the FTIR spectrum of AgNPS and F-AgNPs (amine functionalized AgNPs). Broad absorption peaks at 2958 and 1236 cm^{-1} were observed, which are assigned to the symmetric methylene stretch (CH_2) and Si—C stretching respectively are presented in Figure 7.4(a). The presence of NH_2 could be confirmed by the appearance of a peak at 3234 cm^{-1}. Furthermore, peaks at 1392 and 1082 cm^{-1} correspond to methylene scissoring vibration and primary amine CN stretching, respectively. Thus, the spectrum confirmed that coupling agent has been successfully grafted onto the surface of AgNPs thereby surface functionalizing it is shown in Figure 7.4(b).

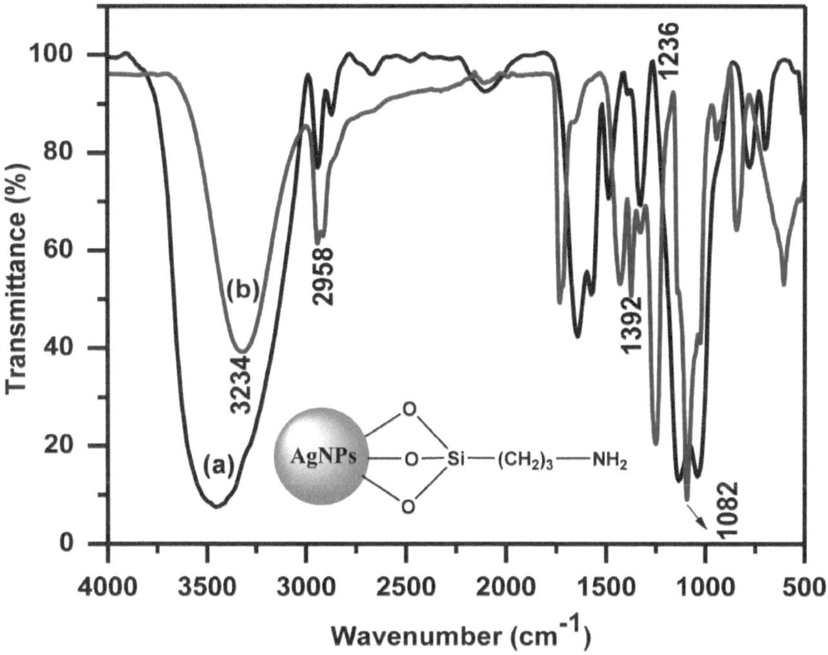

FIGURE 7.4 FTIR spectrum of AgNPs.

Figure 7.5(a) and (b) shows the diffused reflectance spectra of AgNPs and F-AgNPs. In the case of nanoscale metal particles, the absorption wavelength associated with s-p transitions depends on size and shape of the particles, which is due to the fact that the s-p electrons are largely free to move about throughout the nanoparticle and consequently their energies are therefore, sensitive to the size and shape of the particles. Furthermore, nonfunctionalized AgNPs show absorbance at longer wavelengths, while F-AgNPs absorb lower wavelengths. The absorption at shorter wavelength observed for F-AgNPs is due to the effective transfer of electron from the exited state APTES to the conduction band of AgNPs, whereas electron transfer is not possible in nonfunctionalized AgNPs. Hence, the absorption max remains unaltered. It was interesting to note that the UV absorption of virgin APTES (Fig. 7.5a) and AgNPs were found to be at 213 and 415 nm, respectively. However, the F-AgNPs showed an absorption max at wavelength of 314 nm (Fig. 7.5b). And this value falls in between the absorption max values of the virgin APTES and AgNPs. Therefore, this observation is further indicative of the coupling reaction that occurred between virgin APTES and AgNPs.

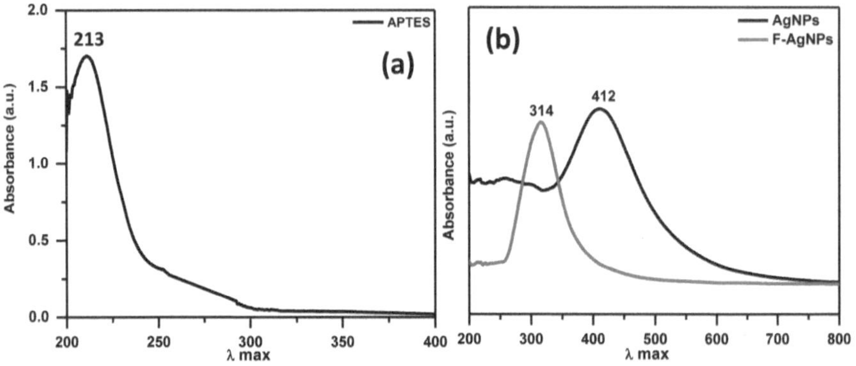

FIGURE 7.5 UV–vis spectrum of (a) APTES and (b) AgNPs and F-AgNPs.

7.3.4 FUNCTIONAL GROUPS ANALYSIS

The FTIR spectrum of TGBAPB epoxy and cured with DDM is shown in Figure 7.6(a) and (b). The absorption band at 913 cm^{-1} was due to the presence of oxirane ring of TGBAPB. The peaks at 2945, 1596, and 1500 cm^{-1} correspond to the aromatic C—H, C=C and C—O—C stretching of TGBAPB, respectively are depicted in Figure 7.6(a) Figure 7.6(b) shows the FTIR

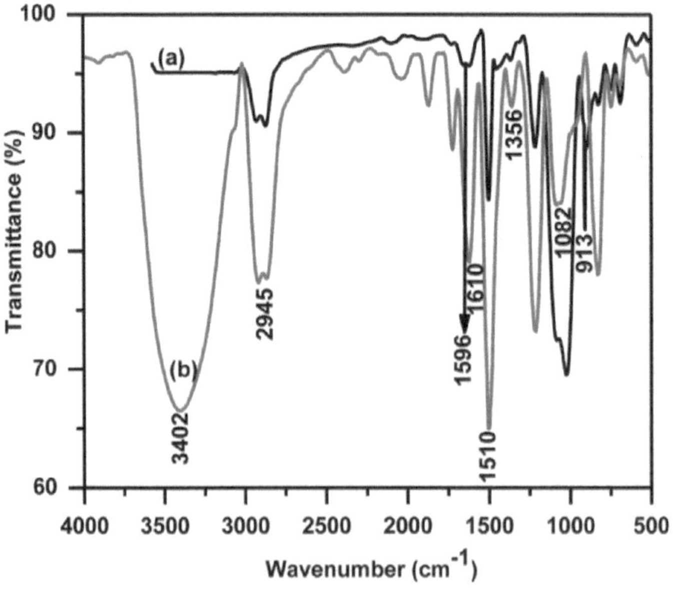

FIGURE 7.6 FTIR spectrum of TGBAPB epoxy and cured with DDM.

spectrum of Aradur 140 cured TGBAPB epoxy resin with F-AgNPs reinforcement. The disappearance of oxirane ring at 913 cm^{-1} and appearance of secondary OH at 3402 cm^{-1} and also confirmed the curing reaction of epoxy by Aradur 140. The remaining peaks at 1610, 1510, 1356, and 1082 cm^{-1} indicate that the F-AgNps have been successfully reinforced into the TGBAPB epoxy matrix, thus forming an inter-cross-linked network structure along with the curing agent.

7.3.5 EVALUATION OF CORROSION RESISTANCE BY POTENTIODYNAMIC POLARIZATION STUDIES

Figures 7.7 and 7.8 show typical potentiodynamic polarization curves for system "a," "b," "c," and "d" coated mild steel specimens in 3.5% NaCl. It was apparent that the polarization curves for mild steel coated with systems "b", "c", and "d" showed remarkable potential shifts to noble values compared to the mild steel coated with systems "a." The various polarization parameters, such as corrosion potential (E_{corr}) and corrosion current density (I_{corr}) obtained from cathodic and anodic curves by extrapolation of Tafel lines are given in Table 7.2. The polarization response of coating systems "b" and "c" after 1000 h of immersion in NaCl was toward anodic side indicating their superior corrosion resistance, whereas the responses of coating systems "a" and "d" were toward cathodic side. It is worth observing that the decrease in corrosion resistance of the system "d" is considerably lower than the neat system "a" coated sample after 1000 h immersion. This indicates that the corrosive electrolyte cannot easily permeate into the coating/metal interface when the coating includes F-AgNPs. For instance, the corrosion rate for mild steel coated with system "a" after 1000 h of immersion in 3.5% NaCl was found to be 1.941×10^{-8} and it was minimized by coating the metal with system "b" to a lower value of 3.8737×10^{-10} than the other coatings namely coating systems "c" and "d" whose values are, respectively 2.44×10^{-9}, and 4.88×10^{-9} (Table 7.2). It should be mentioned that the E_{corr} values decreased significantly for mild steel coated with the system "b". This observation clearly showed that mild steel coated with system "b" control both cathodic and anodic reactions.[50] Moreover, the corrosion rate (CR) of system "b" is found to be very minimum corrosion rate. It could be attributed to the considerable restriction offered by system "b" between the metal and the electrolyte when compared to other systems. The extent of restriction to the entry of corrosive species was more in the case of system "b" and least in the case of system "a".

The reasons for this behavior exhibited by system "b" may be due to the even distribution of F-AgNPs, which increases the coating resistance against corrosive species, thus, offering a barrier effect to the underlying mild steel substrate by interconnecting more molecules leading to the formation of an impeccable coating. Lower corrosion resistance of system "d" can be attributed to the decrease in cross-linking density between the F-AgNPs epoxy matrix and the tendency of the nanoparticles to form aggregates at high wt% loadings.[44]

TABLE 7.2 Data Resulted from Tafel Study After 0 Days and 50 Days of Immersion in 3.5% NaCl.

Immersion days	Coating systems	E_{corr} Mv	I_{corr} mA/cm^2	CR (mpy)
0 days	a	−200	$10^{-5.4}$	1.5421×10^{-10}
	b	−163	10^{-8}	3.8737×10^{-13}
	c	−269	$10^{-6.6}$	9.7305×10^{-12}
	d	−214	$10^{-5.9}$	4.8768×10^{-11}
50 days	a	−246	$10^{-3.3}$	1.941×10^{-8}
	b	−181	10^{-5}	3.8737×10^{-10}
	c	−283	$10^{-4.2}$	2.44×10^{-9}
	d	−275	$10^{-3.9}$	4.88×10^{-9}

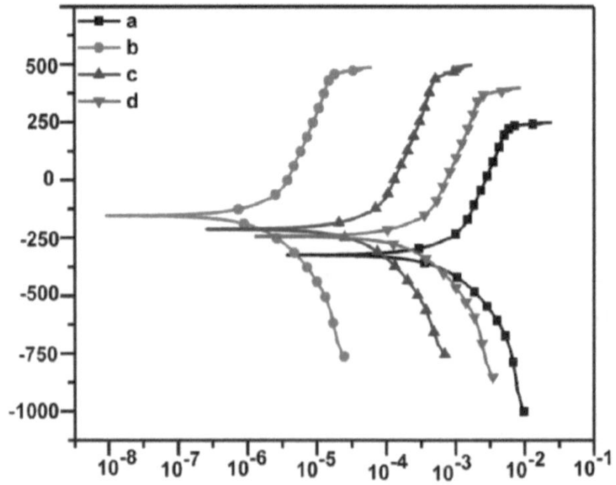

FIGURE 7.7 Polarization response of MS specimens coated with coating systems: (a) TGBAPB neat, (b) 1 wt% F-AgNPs/ TGBAPB, (c) 3 wt% F-AgNPs/ TGBAPB, and (d) 5 wt% F-AgNPs/ TGBAPB for 0 days of immersion in 3.5% NaCl.

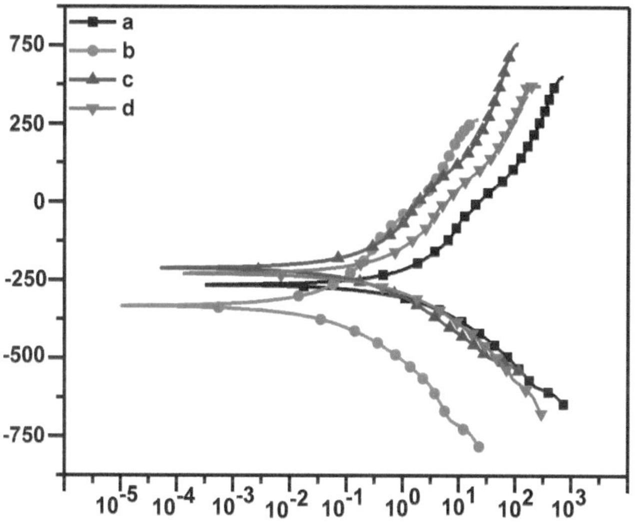

FIGURE 7.8 Polarization response of MS specimens coated with coating systems: (a) TGBAPB neat, (b) 1 wt% F-AgNPs/ TGBAPB, (c) 3 wt% F-AgNPs/ TGBAPB, and (d) 5 wt% F-AgNPs/TGBAPB for 50 days of immersion in 3.5% NaCl.

7.3.6 ELECTROCHEMICAL IMPEDANCE SPECTROSCOPIC STUDIES (EIS)

The impedance graphs obtained for MS specimen coated with systems "a," "b," "c," and "d" for 0 days and 50 days of immersion in 3.5 % NaCl solution are displayed in Nyquist (Figs. 7.9 and 7.10) and Bode plots (Figs. 7.11 and 7.12).

The corresponding impedance values are given in Table 7.3. On this basis, the greater semicircle in the Nyquist plot for coating system "b" indicates better corrosion resistance, which may be explained by the presence of F-AgNps at a low loading of 1 wt%. The higher value of impedance shows an augmentation in the barrier properties of neat system "a," which is due to the presence of F-AgNPs forming a layered framework by uniform and stagger distribution in the coating matrix, which significantly restrained the electrolyte penetration into the epoxy coating. It can be observed that the Nyquist plots of systems "b" and "c" derived from the EIS measurements show two prominent capacitive loops, in the high-frequency region, which were attributed to the high resistance and low capacitance offered by the systems "b" and "c" coating on the steel–electrolyte interface. It can be seen that the systems "a" and "d" coating has exhibited the least resistance to corrosion compared to their counterparts. It can also be noticed that systems "a" and "d" coating exhibit an incomplete semicircle (Figs. 7.9 and 7.10) in the high-frequency region, followed by a low frequency Warburg diffusion tail. The formation of an incomplete semicircle suggests that the sodium chloride solution has just started permeating through the coating system. The formation of a low-frequency diffusion tail in coating systems "a" and "d" confirmed that the corrosion process was a diffusion-controlled reaction. Both the systems "b" and "c" exhibited a capacitive behavior nevertheless system "b" exhibited the optimum capacitive behavior indicating its excellent corrosion resistance. The order of corrosion resistance could be put in the following order:

Systems "b" > "c" > "d" > "a."

The inferior corrosion resistance offered by system "d" coating might be due to the improper dispersion of F-AgNPs at a high loading of 5 wt% in the epoxy resin, which would have led to an agglomeration, leading to the entry of corrosive species through the metal coating interface. However, at a low loading of F-AgNPs covalent bond is formed with TGBAPB epoxy, which eventually leads to a well-dispersed TGBAPB/F-AgNPs epoxy coating with superior corrosion than the other systems of the present study. As observed in the case of potentiodynamic polarization study, the impedance value of coating systems "b" was better than that of coating systems "a" "c," and "d". This observation further supports the results of polarization study and accounts for the low loading of F-AgNPs in offering optimum corrosion resistance to mild steel surface.[39,49]

TABLE 7.3 Data Resulted from Impedance Studies After 0 Days and 50 Days of Immersion in 3.5% NaCl.

Days	Systems	Impedance: Ω. cm²	Frequency
0 day	a	10^6	10^3
	b	10^9	10^3
	c	10^8	10^3
	d	10^7	10^3
50 days	a	10^4	10^3
	b	10^8	10^3
	c	10^6	10^3
	d	10^5	10^3

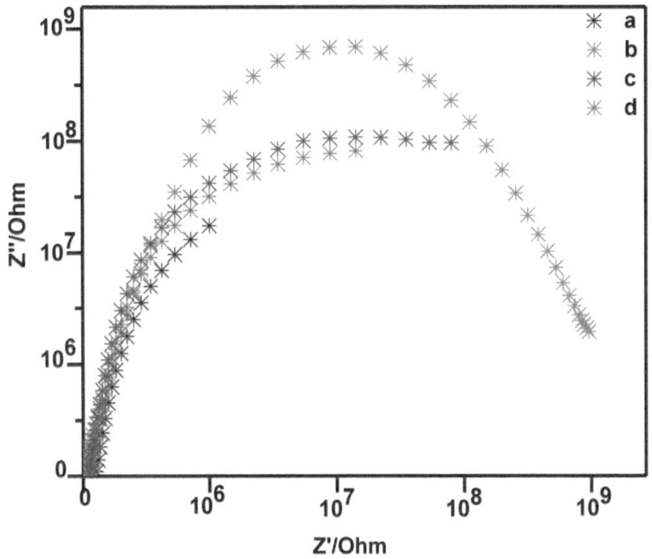

FIGURE 7.9 EIS Nyquist plot of MS specimens coated with coating systems (a) TGBAPB neat, (b) 1 wt% F-AgNPs/TGBAPB, (c) 3 wt% F-AgNPs/TGBAPB, and (d) 5 wt% F-AgNPs/TGBAPB for 0 days of immersion in 3.5% NaCl.

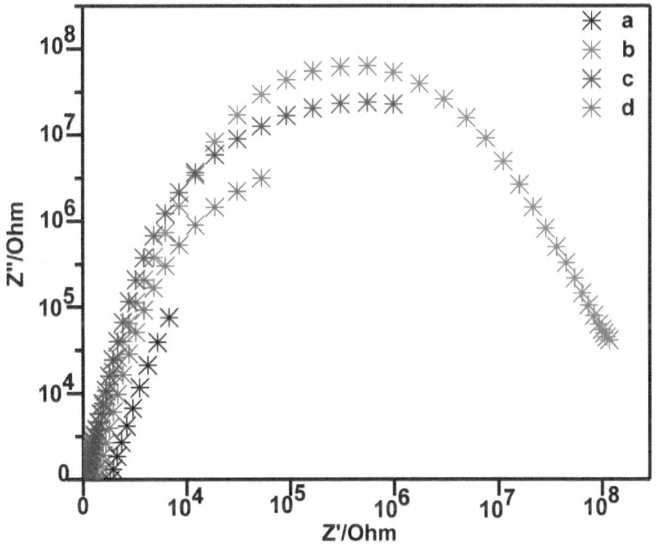

FIGURE 7.10 EIS Nyquist plot of MS specimens coated with coating systems (a) TGBAPB neat, (b) 1 wt% F-AgNPs/ TGBAPB, (c) 3 wt% F-AgNPs/ TGBAPB, and (d) 5 wt% F-AgNPs/TGBAPB for 50 days of immersion in 3.5% NaCl.

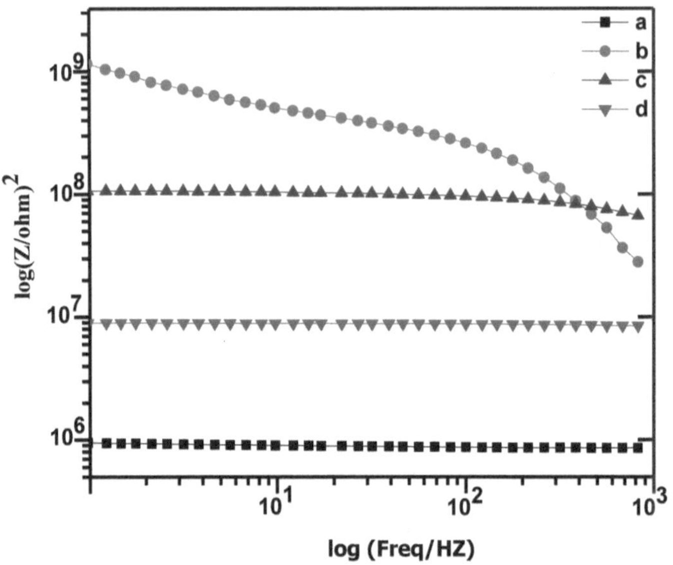

FIGURE 7.11 Bode plot of MS specimens coated with coating systems: (a) TGBAPB, neat, (b) 1 wt% F-AgNPs/TGBAPB, (c) 3 wt% F-AgNPs/TGBAPB, and (d) 5 wt% F-AgNPs/TGBAPB for 0 days of immersion in 3.5% NaCl.

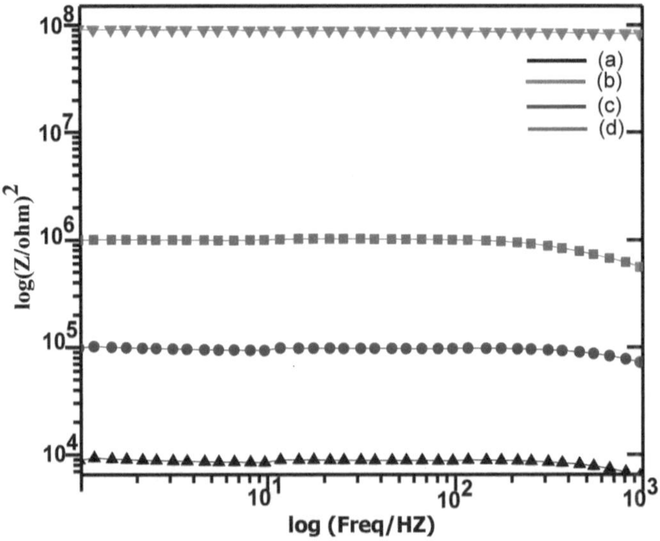

FIGURE 7.12 Bode plot of MS specimens coated with coating systems: (a) TGBAPB neat, (b) 1 wt% F-AgNPs/TGBAPB, (c) 3 wt% F-AgNPs/TGBAPB, and (d) 5 wt% F-AgNPs/TGBAPB for 50 days of immersion in 3.5% NaCl.

7.4 SURFACE MORPHOLOGY

7.4.1 TEM ANALYSIS

Figure 7.13 depicts the HRTEM micrograph of 1 and 5 wt% F-AgNPs reinforced TGBAPB resin, respectively. The micrograph of 1 wt% F-AgNPs reinforced TGBAPB resin manifests a homogeneous dispersion of AgNPs into skeletally modified epoxy matrix, which attributes an excellent compatibility between skeletally modified epoxy matrix and amine functionalized AgNPs. This may be due to the efficient adhesion and the influence of intermolecular interactions between skeletally modified epoxy matrix and amine functionalized AgNPs. The size of the F-AgNPs is consistent and there is no evidence of phase inversion (the particulate morphology is retained). Furthermore, there is no change in the morphology of the F-AgNPs at 1 wt% loading [Figure 7.13 (a)]. Contrary to this observation, at 5 wt% loading of F-AgNPs, the skeletally modified epoxy matrix is saturated with F-AgNPs, which apparently led to increased agglomeration as shown in Figure 7.13(b). F-AgNPs are not distorted but agglomerated due to the weak interfacial bonding between the reinforcement and the TGBAPB epoxy matrix. Apparently, the agglomerates reduce the reinforcing effects of F-AgNPs because they are acting as flaws in the epoxy matrix. The poorly dispersed F-AgNPs at 5 wt% loading reflects negatively on the corrosion behavior of the epoxy nanocomposite than the well dispersed ones at 1 wt% loading.

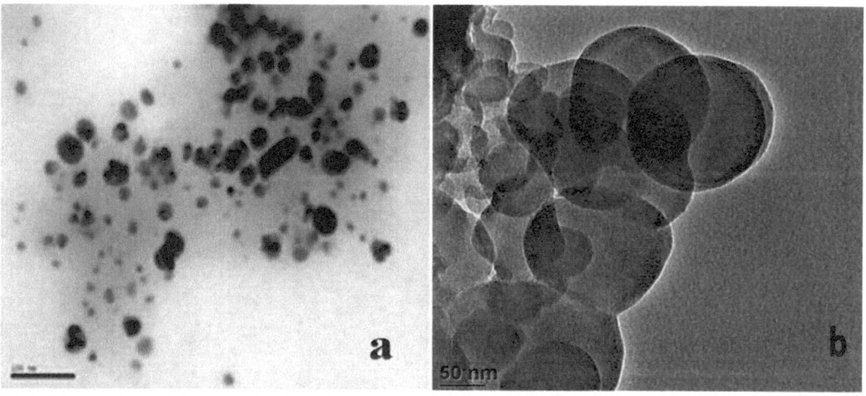

FIGURE 7.13 TEM analysis of (a) 1 wt% of F-AgNPs/TGBAPB and (b) 5 wt% of F-AgNPs/TGBAPB.

7.5 ANTIMICROBIAL STUDIES

7.5.1 EVALUATION OF ANTIBACTERIAL ACTIVITY OF THE TGBAPB/F-AGNPS COATING

Metallic silver and silver-based compounds have been investigated by several researchers as an antibacterial and antifungal agent over several decades. Silver is a metal known for its broad spectrum antimicrobial activity against Gram-positive and Gram-negative bacteria, fungi, protozoa, and certain viruses.[52,53] Exploiting these unique antimicrobial properties of AgNPs, an attempt was made to fabricate F-AgNPs reinforced TGBAPB epoxy coatings with good corrosion control property and also a significant antibacterial and antifungal activity.

The mild steel sample coated with system "b" showed significant antibacterial activity against all tested microorganisms, while the sample with only TGBAPB epoxy (system "a") coating did not show any antibacterial activity (Fig. 7.14). When the steel sample with system "b" coating was brought in contact with the test organism cultures in the petri plate, silver ions are released, which resulted in the formation of zone of inhibition. AgNPs penetrate

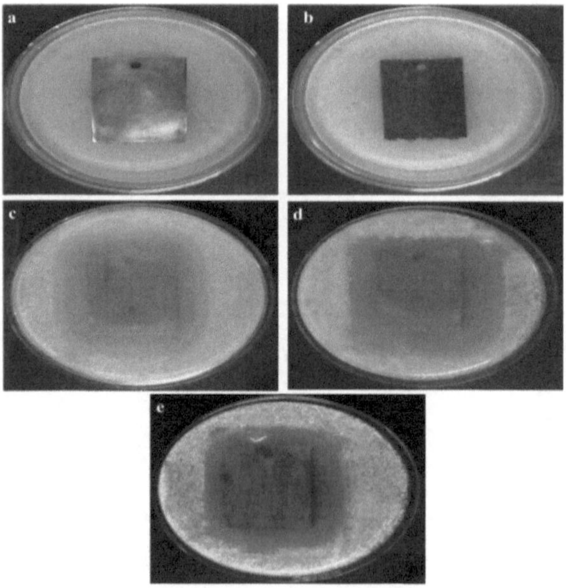

FIGURE 7.14 Antimicrobial activity of (a) control (MS without TGBAPB coating), (b) MS with only TGBAPB coating, (c) 1 wt% F-AgNPs/TGBAPB coating against *Pseudomonas aeruginosa*, (d) 1 wt% F-AgNPs/TGBAPB coating against *Bacillus subtilis*, (e) 1 wt% F-AgNPs/TGBAPB coating against *E. coli*.

and disrupt the membranes of microorganisms, which result in a massive loss of intracellular potassium and decreased ATP levels. The possible molecular targets for the AgNPs could be protein thiol groups in present in enzymes, such as NADH dehydrogenases and disrupt the respiratory chain, facilitating the release of active oxygen species, leading to oxidative stress, and resulting in significant damage to the cell structures and ultimate cell death.[54,55]

7.6 CONCLUSION

Green nano silver-epoxy coating for "corrosive and microbial safe" mild steel surfaces was formulated by reinforcing AgNPs, which were green synthesized using *Rosa indica* wichuriana hybrid leaves extract. The green-synthesized AgNps were surface functionalized using 3-aminopropyltriethoxysilane to achieve better adhesion and dispersion of the AgNps in the TGBAPB epoxy matrix. Different wt% of F-AgNPs (1, 3, and 5 wt %) were reinforced into the epoxy matrix to ascertain the corrosion resistance and antimicrobial behavior of the coatings. Of all the systems studied, system "b" (1 wt% F-AgNPs-reinforced TGBAPB coating) exhibited the maximum corrosion and microbial resistance. The antimicrobial behavior of this epoxy-green nanohybrid coating formulation is noteworthy as it offers manifold antimicrobial protection to the steel surfaces by inhibiting the growth of bacteria like *P. aeruginosa*, *B. subtilis*, the most common human pathogen *E. coli*. This dual behavior of corrosion and microbial resistance exhibited by surface functionalized green nanosilver TGBAPB coatings offer environmentally benign solutions by enhancing the corrosion control property of conventional epoxy coatings that are currently in use and imparting excellent antimicrobial activity leading to better performance and longevity of steel surfaces.

KEYWORDS

- **TGBAPB epoxy coating**
- **nanoparticle**
- **silver nitrate**
- **surface morphology**
- **microbial resistance**

BIBLIOGRAPHY

1. Debaditya, B.; Rajinder, G. Nanotechnology and Potential of Microorganisms. *Crit. Rev. Biotechnol.* **2005**, *25*, 199–204.
2. Lee, B.I; Qi, L.; Copel, T. Nanoparticles for Materials Design: Present and Future. *J. Ceram. Process. Res.* **2005**, *6*, 31–40.
3. Schultz, S.; Smith, D.R.; Mock, J.J.; Schultz, D.A. Single-Target Molecule Detection with Nonbleaching Multicolor Optical Immunolabels. *PNAS.* **2000**, *97*, 996–1001.
4. Yu, D.G. Formation of Colloidal Silver Nanoparticles Stabilized by Na^{+}–Poly(γ-Glutamic Acid)—Silver Nitrate Complex via Chemical Reduction Process. *Colloids. Surf. B.* **2007**, *59*, 171–178.
5. Mallick, K.; Witcombb, M.J.; Scurrella, M.S. Self-Assembly of Silver Nanoparticles in a Polymer Solvent: Formation of a Nanochain through Nanoscale Soldering. *Mater. Chem. Phys.* **2005**, *90*, 221–224.
6. Liu, Y.C.; Lin, L.H. New Pathway for the Synthesis of Ultrafine Silver Nanoparticles From Bulk Silver Substrates in Aqueous Solutions by Sonoelectrochemical Methods. *Electrochem. Commun.* **2004**, *6*, 1163–1168.
7. Smetuna, A.B.; Klabunde, K.J.; Sorensea, C.M. Synthesis of Spherical Silver Nanoparticles by Digestive Ripening Stabilization with Various Agents, and their 3-D and 2-D Super Lattice Formation. *J. Colloid. Interface. Sci.* **2005**, *284*, 521–526.
8. Badri Narayanan, K.; Sakthivel, N. Coriander Leaf Mediated Biosynthesis of Gold Nanoparticles. *Mater. Lett.* **2008**, *62*, 4588–4590.
9. Prathap Chandran, S.; Chaudhary, M.; Pasricha, R.; Ahmad, A.; Sastry, M. Synthesis of Gold Nanotriangles and Silver Nanoparticles using Aloe Vera Plant Extract. *Biotechnol. Prog.* **2006**, *2*, 577–583.
10. Schmid, G. Large Clusters and Colloids. Metals in the Embryonic State. *Chem. Rev.* **1992**, *92*, 1709–1727.
11. Daniel, M.C.; Astruc, D. Gold Nanoparticles: Assembly, Supramolecular Chemistry, Quantum-Size-Related Properties, and Applications Toward Biology, Catalysis, and Nanotechnology. *Chem. Rev.* **2004**, *104*, 293–346.
12. Goffeau, A. Drug Resistance: the fight Against Fungi. *Nature.* **2008**, *452*, 541–542.
13. Morones, J.R.; Elechiguerra, J.L.; Camacho, A.; Ramirez, J.T.; Yacaman, M.J. The Bactericidal Effect of Silver Nanoparticles *Nanotechnology.* **2005**, *16*, 2346–2353.
14. Rai, M.; Yadav, A.; Gade, A. Silver Nanoparticles as a new Generation of Antimicrobials. *Biotechnol. Adv.* **2009**, *27*, 76–83.
15. Mathivanan, L.; Radhakrishna, S. Protection of Steel Structures in Industries with Epoxy-Silicone Based Coatings. *Anti-Corros. Methods Mater.* **1998**, *45*, 301–305.
16. Armelin, E.; Pla, R.; Liesa, F.; Ramis, X.; Iribarren, J.I.; Aleman, C. Corrosion Protection with Polyaniline and Polypyrrole as Anticorrosive Additives for Epoxy Paints. *Corrosion. Sci.* **2008**, *50*, 721–728.
17. Tjong, S.C.; Haydn, C. Nanocrystalline Materials and Coatings. *Mat. Sci. Eng. R* **2004**, *45*, 1–88.
18. Veprek, S.; Argon, A.S.; Mechanical Properties of Superhard Nanocomposites *Surf. Coat. Technol.* **2001**, 146, 175–182.
19. Wetzel, B.; Haupert, F.; Qiu, Z.M. Epoxy Nanocomposites with High Mechanical and Tribological Performance. *Compos. Sci. Technol.* **2003**, *63*, 2055–2067.

20. Perreux, D.; Suri. C. A Study of the Coupling between the Phenomena of Water Absorption and Damage in Glass/Epoxy Composite Pipes. *Compos. Sci. Technol.* **1997**, *57*, 1403–1413.
21. Shi, X.; Nguyen, T.A.; Suo, Z.; Liu, Y.; Avci, R. Effect of Nanoparticles on the Anticorrosion and Mechanical Properties of Epoxy. Coating *Surf. Coat. Technol.* **2009**, *204*, 237–245.
22. Ayman, M.; Atta, N.O.; Shaker, N.E.; Maysour. Influence of the Molecular Structure on the Chemical Resistivity and Thermal Stability of Cured Schiff Base Epoxy Resins. *Prog. Org. Coat.* **2006**, *56*, 100–110.
23. Lam, K.; Lau, K.T. Localized elastic modulus distribution of nanoclay/epoxy composites by using nanoindentation *Compos. Struct.* **2006**, *75*, 553–558.
24. Shi, G.; Zhang, M.Q.; Rong, M.Z.; Wetzel, B.; Friedrich, K. Friction and wear of Low Nanometer Si_3N_4 Filled Epoxy Composites. *Wear.* **2003**, *254*, 784–796.
25. Hartwig, A.; Sebald, M.; Putz, D.; Aberle, L. Preparation, Characterisation and Properties of Nanocomposites Based on Epoxy Resins—An Overview. *Macromol. Symp.* **2005**, *221*, 127–136.
26. Dietsche, F.; Thomann, Y.; Thomann, R.; Mulhaupt, R. Translucent Acrylic Nanocomposites Containing Anisotropic Laminated Nanoparticles Derived from Intercalated Layered Silicates. *J. Appl. Polym. Sci.* **2000**, *75*, 396–405.
27. Zhou, S.X.; Wu, L.M. Preparation Technology and Product Development of Nanocomposite Coatings. *Mater. Rev.* **2002**, *16*, 41–43.
28. Stamataskis, P.; Palmer, B.R. Optimum Particle Size of Titanium Dioxide and Zinc Oxide for Attenuation of Ultraviolet Radiation. *J. Coat. Technol. Res.* **1990**, *62*, 95–102.
29. Hu, Z.S.; Dong, J.X.; Chen, G.X. Preparation and tribological properties of Nanoparticle Lanthanum Borate. *Wear.* **2000**, *243*, 43–47.
30. Huong, N. Improvement of bearing strength of laminated composites by nano clay and Z-pin reinforcement, Ph.D. Thesis, University of New South Wales, Australia, **2006**.
31. Becker, O.; Varley, R.; Simon, G. Morphology, Thermal Relaxations and Mechanical Properties of Layered Silicate Nanocomposites Based upon High-Functionality Epoxy Resins. *Polymer.* **2002**, *43*, 4365–4373.
32. Yang, L.H.; Liu, F.C.; Han, E.H. Effects of P/B on the properties of anticorrosive coatings with different particle size. *Prog. Org. Coat.* **2005**, *53*, 91–98.
33. Lamaka, S.V.; Zheludkevich, M.L.; Yasakau, K.A.; Serra, R.; Poznyak, S.K.; Ferreira M.G.S. Nanoporous Titania Interlayer as Reservoir of Corrosion Inhibitors for Coatings with Self-Healing Ability. *Prog. Org. Coat.* **2007**, *58*, 127–135.
34. Atta, A.M.; Shaker, N.O.; Abdou, M.I. Abdelfatah, M. Synthesis and Characterization of High Thermally Stable Poly(Schiff) Epoxy Coatings. *Prog. Org. Coat.* **2006**, *56*, 91–99.
35. Kaan Emregul, C.; Duzgun, E.; Atakol, O. The Application of some Polydentate Schiff Base Compounds Containing Aminic Nitrogens as Corrosion Inhibitors for Mild Steel in Acidic Media. *Corros. Sci.* **2006**, *48*, 3243–3260.
36. Monticelli, C.; Brunoro, G.; Frignani, A.; Marchi, A. Inhibitive Action of Some Schiff Bases and Amines Towards the Corrosion of Copper in an Aqueous Alcoholic Medium. *Surf. Coat. Technol.* **1986**, *27*, 175–186.
37. Patel, S.; Navin, P.; Patel, J.S.; Harshad, A.J. Study on Novel Epoxy Based Poly (Schiff Reagents). *Polym. Mater.* **2000**, *46*, 499–509.
38. Manjumeena, R.; Girilal, M.; Peter, M.; Sudha, J.; Kalaichelvan, P.T. Potential Antifungal and Antibacterial Activities of Green Synthesized Silver Nanoparticles using

Rosa indica-wichuriana Hybrid Leaf Extract Against Selected Pathogenic Strains. *Int. J. Curr.Sci.* **2013**, *8*, 1–8.

39. Duraibabu, D.; Ganeshbabu, T.; Manjumeena, R.; Ananda kumar, S.; Priya, D.; Unique Coating Formulation for Corrosion and Microbial Prevention of Mild Steel. *Prog. Org. Coat.* **2014**, *77*, 657–664.
40. Madhankumar, A.; Rajendran, N.; Nishimura, T. Influence of Si Nanoparticles on the Electrochemical Behavior of Organic Coatings on Carbon Steel in Chloride Environment *J. Coat. Technol. Res.* **2012**, *9*, 609–620.
41. Sinebryukhov, S.L.; Gnedenkov, A.S.; Mashtalyar, D.V.; Gnedenkov, S.V. PEO-Coating/Substrate Interface Investigation by Localised Electrochemical Impedance Spectroscopy. *Surf. Coat. Technol.* **2010**, *205*, 1697–1701.
42. Abraham, R.; Thomas, S.P.; Kuryan, S.; Isac, J.; Varughese, K.T.; Thomas, S. Mechanical Properties of Ceramic-Polymer Nanocomposites. *Express. Polym. Lett.* **2009**, 3, 177–189.
43. Behzadnasab, M.; Mirabedini, S.M.; Kabiri, K.; Jamali, S. Corrosion Performance of Epoxy Coatings Containing Silane Treated ZrO_2 Nanoparticles on Mild Steel in 3.5% NaCl Solution. *Corros. Sci.* **2011**, *53*, 89–98.
44. Ramezanzadeh, B.; Attar, M.M.; Farzam, M. A Study on the Anticorrosion Performance of the Epoxy–Polyamide Nanocomposites Containing ZnO Nanoparticles. *Prog. Org. Coat.* **2011**, *72*, 410–422.
45. Shi, X.; Nguyen, T.A.; Suo, Z.; Liu, Y.; Avci, R. Effect of nanoparticles on the anti-corrosion and mechanical properties of epoxy coating *Surf. Coat. Technol.* **2009**, *204*, 237–245.
46. Kanimozhi, K.; Devaraju, S.; Vengatesan, M.R.; Selvaraj, V.; Alagar, M. Studies on Synthesis and Characterization of Surface-Modified Mullitefibre-Reinforced Epoxy Nanocomposites. *High. Perform. Polym.* **2013**, *25*, 658–667.
47. Zainuddin, S.; Hosur, M.V.; Zhou, Y.; Alfred Narteh, T.; Ashok, K.; Jeelani, S. Experimental and Numerical Investigations on Flexural and Thermal Properties of Nanoclay–Epoxy Nanocomposites. *Mat. Sci. Eng. A.* **2010**, *527*, 7920–7926.
48. Tianxi, L.; Wuiwui, C.T.; Yuejin, T.; Chaobin, H.; Sok, S.G.; Tai, S.C. Morphology and Fracture Behavior of Intercalated Epoxy/Clay Nanocomposites. *J. Appl. Polym. Sci.* **2004**, *94*, 1236–1244.
49. Ananda Kumar, S.; Balakrishnan, T.; Alagar, M.; Denchev, Z. Development and Characterization of Silicone/Phosphorus Modified Epoxy Materials and their Application as Anticorrosion and Antifouling Coatings. *Prog. Org. Coat.* **2006**, *55*, 207–217.
50. Arthananareeswari, M.; Sankara Narayanan, T.S.N.; Kamaraj, P.; Tamilselvi, M. Polarization and Impedance Studies on Zinc Phosphate Coating Developed Using Galvanic Coupling. *J. Coat. Technol. Res.* **2012**, *9*, 39–46.
51. Kavitha, C.; Priya Dasan, K. Nanosilver/Hyperbranched Polyester (HBPE): Synthesis, Characterization, and Antibacterial Activity. *J. Coat. Technol. Res.* **2013**, *10*, 669–678.
52. Sondi, I.; Salopek-Sondi, B. Silver Nanoparticles as Antimicrobial Agent: A Case Study on *E. coli* as a Model for Gram-Negative Bacteria. *J. Colloid Interface Sci.* **2004**, *275*, 177–182.
53. Li, Y.; Leung, P.; Yao, L.; Song, Q.; Newton, E. Antimicrobial Effect of Surgical Masks Coated with Nanoparticles. *J. Hosp. Infect.* **2006**, *62*, 58–63.

54. Yamanaka, M.; Hara, K.; Kudo, J. Bactericidal Actions of a Silver Ion Solution on *Escherichia coli*, Studied by Energy-Filtering Transmission Electron Microscopy and Proteomic Analysis. *Appl. Environ. Microbiol.* **2005**, *71*, 7589–7593.
55. Kim, K.; Woo, S.S.; Bo, K.S.; Seok-Ki, M.; Jong-Soo, C.; Jong, G.K.; Dong, G.L. Antifungal Activity and Mode of Action of Silver Nanoparticles on *Candida albicans. Biometals.* **2009**, *22*, 235–242.

CHAPTER 8

BINDING INTERACTION AND MORPHOLOGY STUDIES OF 3-AMINOPROPYLTRIETHOXYSILANE TREATED NANOPARTICLES

P. SARAVANAN[1], K. JAYAMOORTHY[1], and S. ANANDA KUMAR[2*]

[1]*Department of Chemistry, St. Joseph's College of Engineering, Chennai 600119, India*

[2]*Department of Chemistry, Anna University, Chennai 600025, India.*
**E-mail: sri_anand_72@yahoo.com*

CONTENTS

ABSTRACT

Binding interaction studies of 3-aminopropyltriethoxysilane (3-APTES) with zinc oxide (ZnO), titania (TiO_2), zirconia (ZrO_2), alumina (Al_2O_3), silica (SiO_2), and iron oxide (Fe_2O_3) nanocrystals have been carried out by UV–visible and fluorescence spectral studies. 3-APTES acts as a host for ZnO and TiO_2 nanocrystals that enhance its fluorescence due to the electron transfer, whereas ZrO_2, Al_2O_3, SiO_2, and Fe_2O_3 do not affect the fluorescence intensity of 3-APTES. 3-APTES gets adsorbed on the surface of all nanoparticles and the surface treatment of nanoparticles by 3-APTES, its mechanism is discussed in detail. Scanning electron microscopy (SEM) and Fourier transform-infrared spectra (FT-IR) spectral studies are carried out to support the biding interaction between 3-APTES with nanoparticles.

8.1 INTRODUCTION

Surface properties govern the behavior and overall performance of materials. Surface modifications thereby provide unique possibilities to control the subsequent surface interaction and response, which are required for a particular application. By tailoring the material surface, a wide portfolio of additional functionalities can be enabled to overcome material deficiencies while maintaining its bulk material properties. As a consequence, the functionalization of surfaces has become pivotal for academic research as well as industrial product development. Nowadays, innumerable materials and commercial products with specific surface functionalities are readily available and their usage in technical field is predominant. For all these reasons, surface functionalization is considered to be of paramount importance for the development and design of new materials, as well as for prospective engineered systems. In this chapter, strategies for surface functionalization of inorganic material with a special focus on ZnO, Al_2O_3, ZrO_2, TiO_2, SiO_2, and Fe_2O_3 are discussed. These materials are widely used in biomedical and bioengineering applications as biomedical implants and devices, lab-on-chips, chromatographic supports, bioreactors, and clean water management systems, and are thus not only of great academic but also of steadily increasing human and commercial interest. The development and application of novel materials for tissue engineering, diagnostic tools, and sensoring, for instance, are under intensive investigation. Over the years, many surface functionalization strategies have been developed and optimized. They can be divided into three main categories: physical, chemical, and biological.

The current trend consists of the application of multiple surface modification approaches and/or several surface modifiers on the same surface.

Nanocrystalline materials may be considered as the challenge of this age. Intensive investigations have been stimulated for several applications for these new classes of materials. Metal oxide nanoparticles can be composed of variety of materials, including titanium, zinc, cerium, aluminum, and iron oxides.[1] Nanoparticles possess different chemical properties when compared to bulk types of similar chemical composition.[2] Furthermore, the size of such particles is one of the major causes responsible for the changes in their fundamental physical and chemical properties yielding completely new and different physiochemical properties. Zinc oxides of particle size in nanometer range have been paid more attention for their unique properties. They are widely used for solar energy conversion, nonlinear optics, catalysis, pigments, gas sensors, cosmetics, etc.[3–9] ZnO being a wide band-gap semiconductor has been widely studied in varistors, transparent conductors, transparent UV protection films, chemical sensors, and so on.[10,11] The most commonly available form was crystalline in nature, namely corundum or α-aluminum oxide, which is good electrical insulator.[12] The hardness of the material makes it suitable for using as an abrasive. Aluminum oxide offers resistance to metallic aluminum, when exposed to weathering. Different methods like mechanical and chemical methods have been developed to solve the agglomeration of nanoparticles.

Herein we focus the surface treatment and binding interaction of nanoparticles by 3-APTES. The binding of 3-APTES on the surface of ZnO, that is, functionalized ZnO (f-ZnO) and nonfunctionalized ZnO (n-ZnO), is discussed elaborately. Absorption and emission spectra have been carried out to explain interaction of 3-APTES with ZnO, TiO_2, ZrO_2, Al_2O_3, SiO_2, and Fe_2O_3 nanocrystals and it gives surprising results.

8.2 MATERIALS AND METHODS

8.2.1 SURFACE TREATMENT OF VARIOUS NANOPARTICLES WITH 3-APTES

The surface treatment with 3-APTES as a coupling agent is the most popular chemical method to modify the surface structure of various nanoparticles and it is widely used nowadays. The 3-APTES contains functional groups that could react with inorganic fillers. The 3-APTES normally contains two types of $–NH_2$ and $–CH_2$ groups.

Silanol molecule chains are able to form oligomer structures and form hydrogen bonds on the surface of inorganic nano-size fillers. Moreover, additional condensation reactions will also occur between 3-APTES, silanol groups, and the surface hydroxyls of inorganic fillers. Further condensation and dehydration reactions can also be obtained by drying the nanofillers after surface treatment with silane. The inorganic nanosize fillers with organic functional groups, that attach to their surface by strong chemical bounds, can be finally obtained, as shown in Scheme 8.1.

8.2.2 PREPARATION OF NANOPARTICLES SOLUTION

Various concentrations of different nanoparticles were synthesized by dispersing the nanoparticles with ethanol and trace amount of ethylene glycol to increase the dispersion stability. This solution was sonicated before the measurement of UV–visible and fluorescence spectral studies. After the sonication it has been noted that the nanoparticles had not settled even for two days.

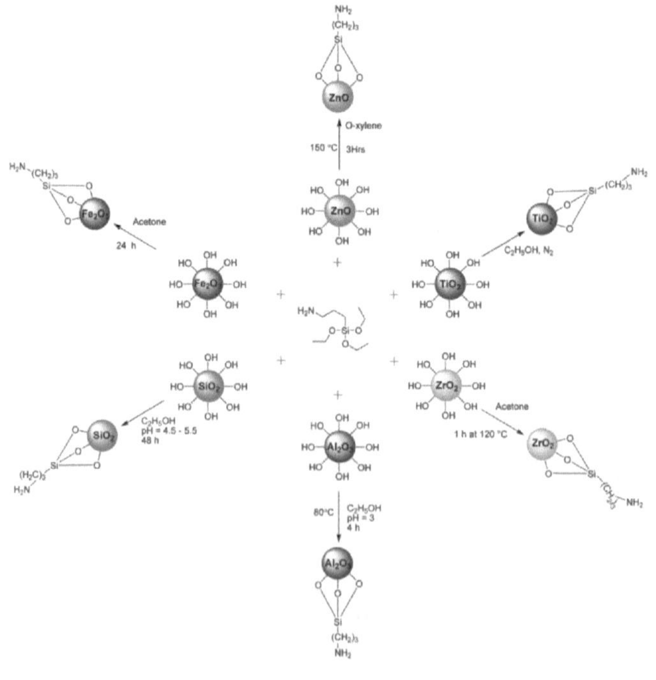

SCHEME 8.1 Surface treatment of different nanoparticles by 3-APTES.

8.2.3 SURFACE TREATMENT OF NANO ZnO PARTICLES

The introduction of reactive NH_2 group onto the surface of ZnO nanoparticles was achieved through the reaction between 3-APTES and the hydroxyl groups on the ZnO nanoparticle surface. Typically, 2.0 g ZnO nanoparticles and 2 mL 3-APTES in 40 mL O-xylene were kept at 150°C for 3 h under ultrasonic bath stirring and argon protection. The reaction mixture was refluxed for 24 h. Rotary evaporator was used to remove the solvent from the 3-APTES-modified ZnO. After that, the ZnO nanoparticles were collected by filtration and rinsed three times with acetone. Afterwards, the 3-APTES-functionalized ZnO nanoparticles were dried under vacuum for 12 h.

8.2.4 SURFACE TREATMENT OF NANO TiO_2 PARTICLES

A total of 0.5 g of TiO_2 nanoparticles was dispersed in 50 mL deionized water by ultrasonication for 10 min. Then, the silane coupling agent 3-APTES (5 g) were added in the dispersion. The mixture was kept refluxing at different reaction conditions. After that, dispersed particles were separated from solvent by centrifuge (10 min at 10,000 rpm) followed by washing with ethanol and water alternatively for at least 2 cycles to remove excessive silanes. To redisperse the centrifuged particles in fresh solvent, they were put in ultrasonic bath for more than 10 min to make sure that a visually well-dispersed suspension appeared before centrifugation. Once the process was finished, the modified particles were dried in an oven at 100°C for 24 h and cooled in a vacuum chamber for 1 h at room temperature.

8.2.5 SURFACE TREATMENT OF NANO-ZrO_2 PARTICLES

In the first stage, 10 g of ZrO_2 nanoparticles were kept in a vacuum chamber for 1 h at 120°C and then dispersed in 30 mL acetone via stirring at 300 rpm for 1 h at ambient temperature and sonicated for 20 min under power of 70 W, 20 kHz frequency, and 0.7 s pulse on and 0.3 s pulse off. In the second step, 50 wt.% (5 g) 3-APTES was gradually added to the dispersion and stirred for further 24 h in ambient temperature. Finally, it was centrifuged (6000 rpm) and the residue was washed with acetone. The washing procedure was repeated for three times and the remaining precipitate was dried in a vacuum oven at 50°C for 48 h.

8.2.6 SURFACE TREATMENT OF NANO AL_2O_3 PARTICLES

A three-necked flask was charged with Al_2O_3 (5 g) and C_2H_5OH (120 mL). The solution was stirred for 30 min under ultrasonic resonance. KH550 (0.15 g) and deionized water (4 g) were placed in a beaker and the pH of the mixture was adjusted to 3 through adding glacial acetic acid. Then, the obtained solution was put into the three-necked flask containing nano Al_2O_3, stirred continuously for 4 h under heating in a water bath (80°C). The obtained nano-powders were washed and filtrated repeatedly, dried for 12 h at 80°C in a vacuum oven. Finally, the nano alumina (Al_2O_3), which was modified by KH550 coupling agent, was prepared and grinded for further use.

8.2.7 SURFACE TREATMENT OF NANO SiO_2 PARTICLES

To improve the nanoparticle dispersion, 3-APTES was used as a coupling agent. The surface treatment of SiO_2 nanoparticles with silane coupling agent has been carried out. The detail process is discussed below: Nano SiO_2 particles (5g) are dispersed into 100 mL ethanol solution by using ultrasonic mixing method for 1 h. A mixture solution that contains 95% ethanol and 5% of water is adjusted to a pH value between 4.5 and 5.5 by using acetic acid. A total of 50 mL of silane solution is added into the mixture solution by magnetic stirring for 15 min. The mixture solution is then added into the ethanol solution that contains nano SiO_2 particles to form silanol group. The mixture is then mixed for another 48 h to wet the particles with silanol group. The mixture is then separated by using a centrifuge. The particles are washed in hexane solution and then separated by using a centrifuge.

8.2.8 SURFACE TREATMENT OF γ-Fe_2O_3 NANOPARTICLES

Surface modification of γ-Fe_2O_3 was performed as follows. Two grams of γ-Fe_2O_3 nanoparticles was kept in a vacuum chamber at 110°C for 75 min and then dispersed in acetone by stirring for 1 h at ambient temperature and finally was sonicated for 20 min. Then, 1 g 3-APTES (50 wt.%) was gradually added to the dispersed solution and stirred for further 24 h at ambient temperature. Finally, it was centrifuged and the residue is washed with acetone. The washing procedure was repeated for three times and the remaining precipitate was dried in a vacuum oven at 50°C for 72 h.

8.3 RESULTS AND DISCUSSION

8.3.1 CHARACTERIZATION OF NANO ZnO

The X-ray diffraction (XRD) spectrum of ZnO nanoparticles is depicted in Figure 8.1. A series of characteristic peaks 2.8140 (100), 2.6026 (002), 2.4751 (101), 1.9106 (102), 1.6245 (110), 1.4767 (103) 1.4081 (200) 1.3781 (112), and 1.3580 (201) observed are in accordance with the zincite phase of ZnO (International Center for Diffraction Data, JCPDS 65-3411). No peaks of impurity are observed, suggesting that the high-purity nano ZnO was obtained. In addition, the peak is widened, implying that the particle size is so small with an average crystallite size D 26 nm calculated using the Debye–Scherrer formula.

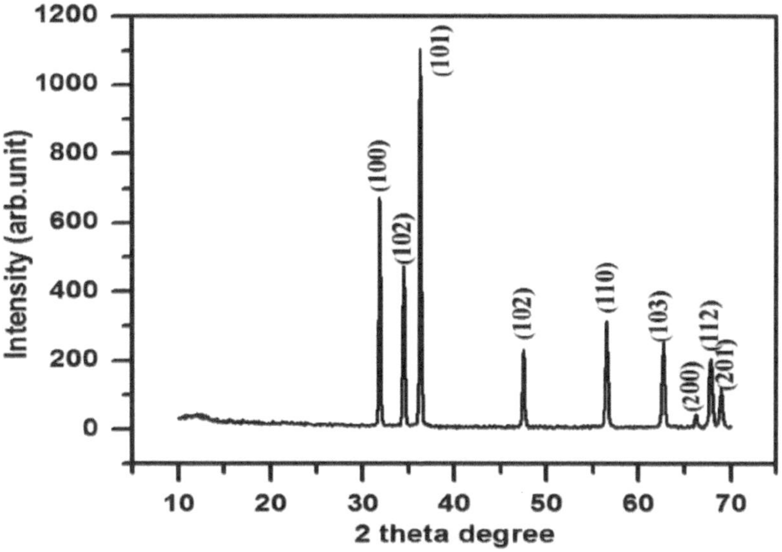

FIGURE 8.1 The XRD spectrum of ZnO nanoparticles.

Figure 8.2(a) and (b) depicts the SEM image of the n-ZnO and f-ZnO nanoparticles. It can be clearly seen from the SEM image that most of the f-ZnO particles exhibit spherical morphology and monodispersity than n-ZnO[13] with an average diameter of about 20 nm. This result is in good accordance with the value calculated from the XRD method. The SEM results show no large agglomerations in f-ZnO and more agglomerations in n-ZnO samples.

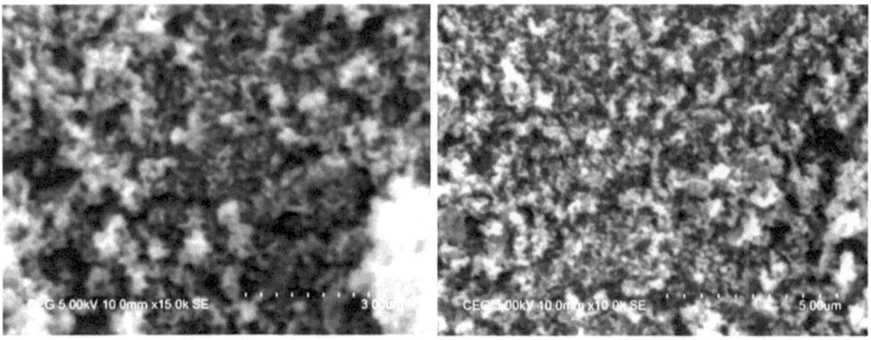

FIGURE 8.2 (a) The SEM image of n-ZnO. (b) The SEM image f-ZnO nanoparticles.

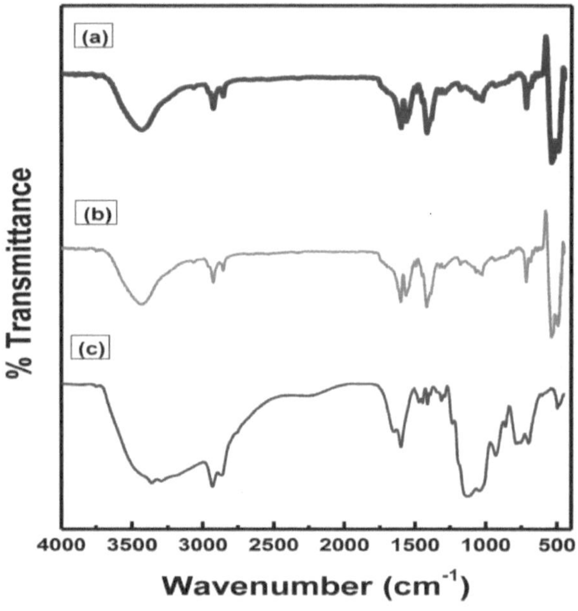

FIGURE 8.3 FT-IR spectrum of (a) n ZnO, (b) f ZnO, and (c) 3-APTES.

The FT-IR spectra of n-ZnO, f-ZnO, and 3-APTES is given in Figure 8.3. FT-IR spectrum of f-ZnO with 3-APTES gave a broad absorption band located at 3419 cm^{-1}, which was attributed to -OH and $-NH_2$ groups. The peaks at 2928 and 1016 cm^{-1} were assigned to the symmetric methylene stretch ($-CH_2$) and the Si–O stretch, respectively. FT-IR illustrated in Figure 8.3 confirmed that coupling agent has been successfully grafted onto the surface of nano ZnO.

8.3.2 ABSORPTION STUDIES

The absorption spectra of ZnO, TiO_2, ZrO_2, Al_2O_3, SiO_2, and Fe_2O_3 nanoparticles modified with and without 3-APTES are displayed in Figure 8.4. Nanoparticles enhance the absorbance of 3-APTES with slight shifting of its absorption maximum due to the effective transfer of electron from the excited state of 3-APTES to the conduction band of semiconductor nanoparticles. Alumina is an insulator nanoparticle, so electron transfer is not possible in alumina, and hence absorption intensity does not alter. This indicates that the nanoparticles slightly modify the excitation process of 3-APTES. The enhanced absorption and slight modification of excitation process is mainly due to the adsorption of 3-APTES on nanoparticles.

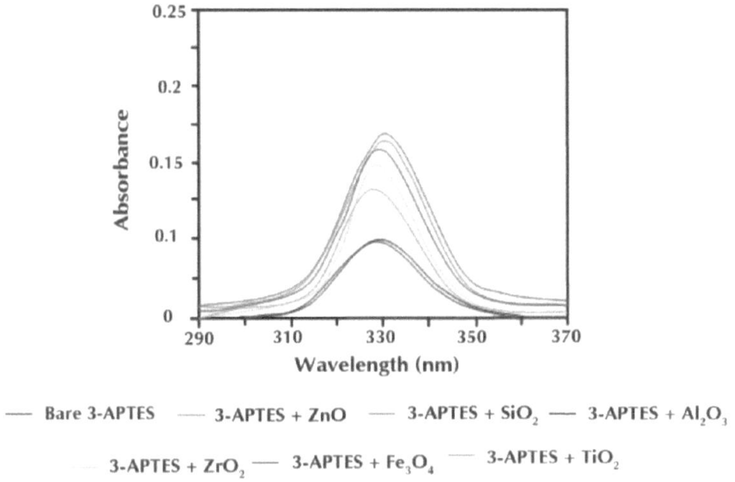

FIGURE 8.4 Absorbance spectra of 3-APTES in the presence and absence of various nanoparticles.

8.3.3 FLUORESCENCE STUDIES

The fluorescence spectra of TiO_2, ZnO, ZrO_2, Al_2O_3, SiO_2, and Fe_2O_3 nanoparticles modified with and without 3-APTES are displayed in Figure 8.5. The TiO_2 and ZnO nanoparticles enhance the emission of 3-APTES with shifting of its emission maximum. Fluorescence enhancing arises due to formation of 3-APTES nanoparticles complex. ZrO_2 Fe_2O_3, and SiO_2 quench the fluorescence of 3-APTES remarkably without shifting its emission maximum. Insulator alumina does not modify the fluorescence intensity.

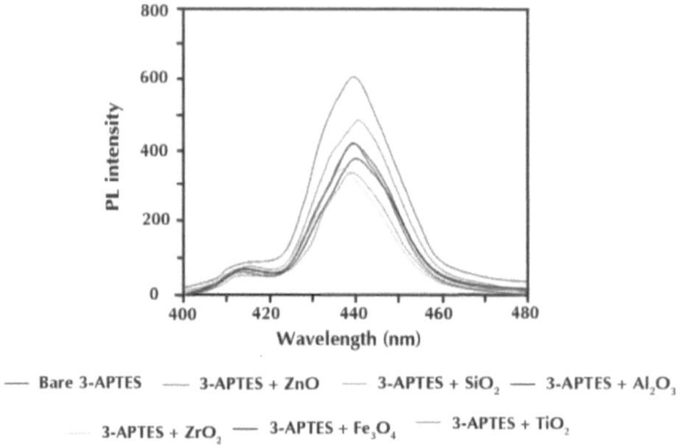

FIGURE 8.5 Fluorescence spectra of 3-APTES in the presence and absence of various nanoparticles.

8.3.4 DIFFERENT CONCENTRATION OF NANOPARTICLES WITH 3-APTES

The effect of increasing concentration of ZnO and TiO_2 nanoparticles at various concentrations on the fluorescence spectrum of 3-APTES is displayed in Figure 8.6(a) and (b). Addition and increasing the concentration of nanoparticles to the solution of 3-APTES resulted in the enhancing of fluorescence emission of ZnO and TiO_2.

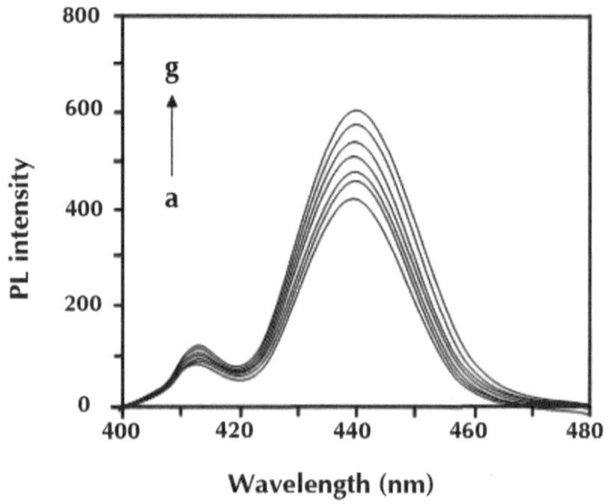

FIGURE 8.6(a) Fluorescence spectra of 3-APTES in the presence and absence of various concentrations of ZnO. ((a) Bare 3-APTES, (b)–(g) 3-APTES and ZnO 1×10^{-5} M to 4.5×10^{-5} M, respectively).

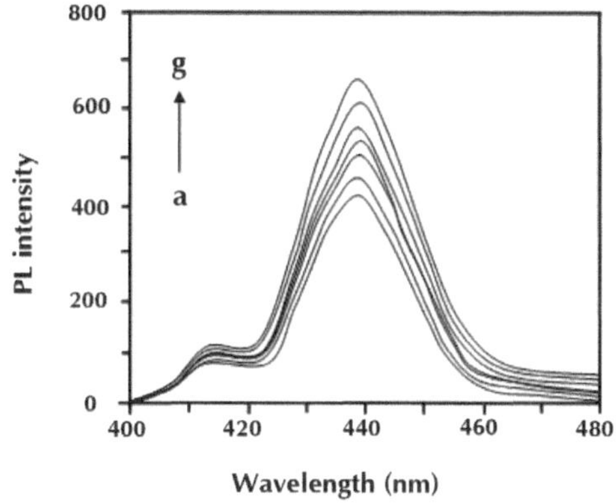

FIGURE 8.6(b) Fluorescence spectra of 3-APTES in the presence and absence of various concentration of TiO_2 ((a) Bare 3-APTES, (b)–(g) 3-APTES and TiO_2 1×10^{-5} M to 4.5×10^{-5} M, respectively).

The apparent association constants (K_{app}) have been obtained (ZnO = -2.209×10^{-7}, $TiO_2 = -1.859 \times 10^{-7}$) from the fluorescence sensing data using the following equation:

$$1/(F_0 - F) = 1/(F_0 - F) + 1/\, K_{app}\, (F_0 - F)\ [\text{nanoparticles}] \qquad (1)$$

where F_0 is the initial fluorescence intensity of the 3-APTES and F is the fluorescence intensity of the 3-APTES adsorbed on nanoparticles. The fluorescence quenching behavior is usually described by Stern–Volmer relation. Linear S–V plot for steady-state fluorescence quenching and enhancing of 3-APTES by nanoparticles is displayed in Figure 8.7(a) and (b).

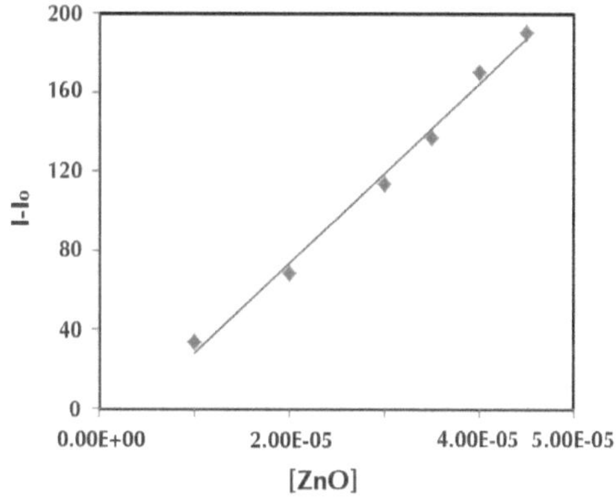

FIGURE 8.7(a) Linear S–V plot for steady-state fluorescence quenching and enhancing of 3-APTES by ZnO nanoparticles

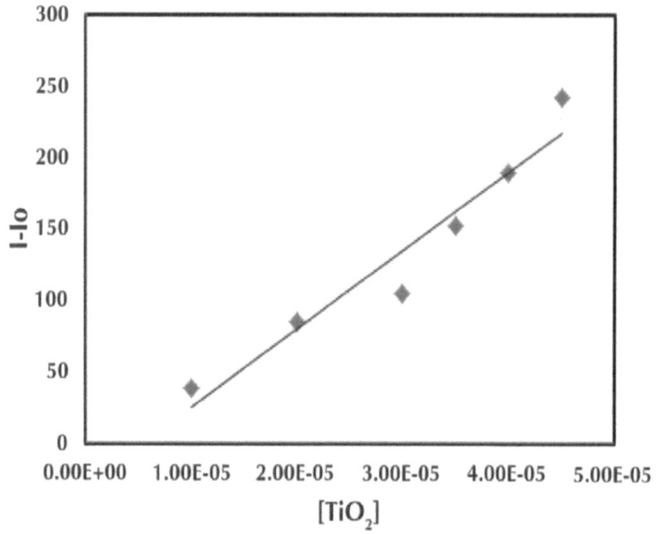

FIGURE 8.7(b) Linear S–V plot for steady-state fluorescence quenching and enhancing of 3-APTES by TiO_2 nanoparticles

8.3.5 ENERGETICS

The fluorescence spectra of 3-APTES in presence of TiO_2 nanoparticles dispersed at different concentrations are shown in Figure 8.6(b). TiO_2 nanoparticles enhance the emission of 3-APTES with slight shifting of its emission maximum at 440 nm. This indicates that the TiO_2 nanocrystals slightly modify the excitation process of 3-APTES. The enhanced emission is due to adsorption of 3-APTES on semiconductor surface. This is because of emission due to effective transfer of electron from the excited state of 3-APTES to the conduction band of TiO_2 nanoparticles.

The electron in the highest occupied molecular orbital (HOMO) of the excited molecule is of higher energy than the conduction band of TiO_2. This should lead to transfer of electron from lowest unoccupied molecular orbital (LUMO) of the excited molecule to the conduction band of rutile TiO_2, thereby quenching the fluorescence of 3-APTES. However, contrary to the expectations, enhancement of fluorescence is observed in presence of

TiO_2 nanocrystals. This may be because of the lowering of the HOMO and LUMO energy levels of 3-APTES due to adsorption on rutile TiO_2 nanoparticles. Rutile TiO_2 is of primitive tetragonal pattern. It consists of deformed TiO_2 octahedra connected differently by corners and edges. In rutile, two octahedral edges are shared to form linear chains along the 0 0 1 direction and the TiO_2 chains are linked to each other through corner shared bondings. The cross section of the void space in rutile is ~ 4.6 Å, which is the unit cell length of rutile phase. This void space permits seating of 3-APTES comfortably. Ducking or the perfect seating of 3-APTES in the void space of rutile allows binding of each 3-APTES to TiO_2 molecules. This explains the different mode of adsorption of 3-APTES on the TiO_2 and ZnO nanocrystals.

On irradiation at 330 nm both 3-APTES and TiO_2 are excited. Duel emission is expected due to LUMO → HOMO and CB (conduction band) → VB (valence band) electron transfer. Also possible is electron jump from the excited 3-APTES to the nanocrystal; the electron in the LUMO of the excited 3-APTES is of higher energy compared to that in the CB of TiO_2 nanocrystals. The polar TiO_2 surface enhances the delocalization of the π electrons and lowers the HOMO and LUMO energy levels due to adsorption. The emission intensity of 3-APTES bound to TiO_2 is far larger than that of the isolated molecule; the excited 3-APTES emits fluorescence at 440 nm. With 3-APTES adsorbed on TiO_2 nanomaterial the semiconductor is also excited on irradiation. The recombination of the electron in the conduction band with the hole in the valence band results in emission at 380 nm. In addition, emission from the LUMO of 3-APTES to the conduction band of rutile TiO_2 is possible by 3-APTES adsorbed on rutile TiO_2. This emission could be at about 440 nm if the HOMO and conduction band are not widely separated. Due to the additional path opened for emission the emission intensity has been increased.

On the basis of the relative positions of 3-APTES and ZnO energy levels, the interfacial electron injection would be thermodynamically allowed from the excited singlet of 3-APTES to the conduction band of ZnO. HOMO-LUMO orbital pictures of 3-APTES are shown in Figure 8.8. HOMO is located on the amino group, whereas LUMO is located on the entire molecule. On excitation at 330 nm both 3-APTES and nano-semiconductor are excited. Dual emission is expected due to LUMO → HOMO and CB → VB electron transfer. Also possible is electron jump from the excited 3-APTES to the nanocrystals; the electron in the LUMO of the excited 3-APTES is of higher energy compared to that in the CB of ZnO nanocrystals. This should lead to quenching of fluorescence in 3-APTES. However, contrary to the expectations, enhancement of fluorescence is observed in presence of

ZnO nanocrystals. This may be because of the lowering of the HOMO and LUMO energy levels of 3-APTES due to adsorption on ZnO nanocrystals. The polar ZnO surface enhances the delocalization of the π electrons and lowers the HOMO and LUMO energy levels due to adsorption.[14] The excited state energy of 3-APTES is larger than the conductance band energy levels of nano-semiconductors.[15] This makes possible the energy transfer from the excited state of 3-APTES to the nanoparticles.

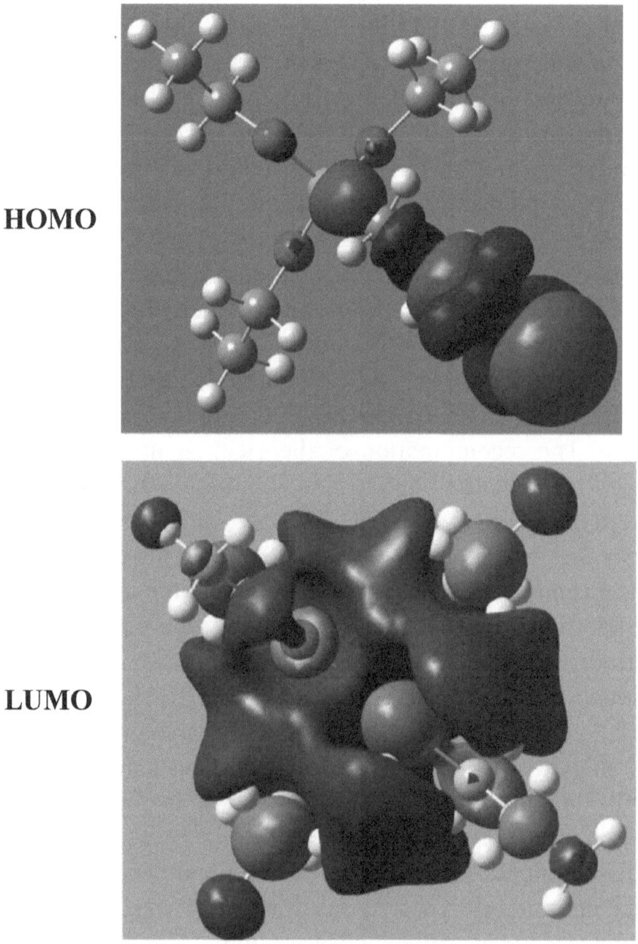

FIGURE 8.8 HOMO-LUMO orbital picture view of 3-APTES.

According to Forster's nonradiative energy transfer theory,[16] the energy transfer efficiency is related not only to the distance between the acceptor and donor (r_0), but also to the critical energy transfer distance (R_0). That is,

$$E = R^6_0/(R^6_0 + r^6_0) \tag{2}$$

where, R_0 is the critical distance when the transfer efficiency is 50%.

$$R^6_0 = 8.8 \times 10^{-25} K^2 N^{-4} \varphi J \tag{3}$$

where, K^2 is the spatial orientation factor of the dipole, N is the refractive index of the medium, φ is the fluorescence quantum yield of the donor and J is the overlap integral of the fluorescence emission spectrum of the donor and the absorption spectrum of the acceptor. The value of J can be calculated (J = 380 nm) by using eq 4,

$$J = \int F(\lambda)\, \varepsilon(\lambda)\, \lambda^4 d\lambda / F(\lambda) d\lambda .... \tag{4}$$

where $F(\lambda)$ is the fluorescence intensity of the donor, $\varepsilon(\lambda)$ is molar absorptivity of the acceptor. The parameter J can be evaluated by integrating the spectral parameters in eq 4. Under these experimental conditions, the value of R_0 calculated from eq 3 is found to be about 1.70 nm in all cases; the values of K^2 (2/3) and N (1.3467) used are from the literature[17] and the φ value (0.15) is from the present study. Obviously, the calculated value of R_0 is in the range of maximal critical distance.[18] This is in accordance with the conditions of Forster's nonradiative energy transfer theory,[18] indicating the static quenching interaction between nanoparticles and 3-APTES. The value of r_0 obtained using eq 3. This suggests that nonradiative energy transfer occurs between the nanoparticles and 3-APTES with high probability.[19] The fact that the value of r_0 is larger than that of R_0 in the present study also reveals the operation of static type of quenching mechanism.[20-26]

8.4 CONCLUSION

Binding interaction between 3-APTES with ZnO, TiO_2, ZrO_2, Al_2O_3, SiO_2, and Fe_2O_3 were discussed. Electron transfer from excited state 3-APTES to semiconductor ZnO and TiO_2 nanocrystal is studied. Fluorescence intensity of 3-APTES does not alter by insulator Al_2O_3 nanocrystals. Fluorescent enhancement has been explained based on photoinduced electron transfer mechanism. Whereas ZrO_2, Fe_2O_3, and SiO_2 nanoparticles quiches the fluoresce intensity of 3-APTES. The silane group is adsorbed on the surface of semiconductor nanoparticles. The inorganic nanosize fillers with organic functional groups that attach to their surface can be finally obtained. The interaction increases the surface tension of inorganic nanoparticles results f-ZnO.

ACKNOWLEDGMENTS

Instrumentation facility provided under FIST-DST and DRS-UGC to Department of Chemistry, Anna University, Chennai are gratefully acknowledged. Two of the authors Mr. P. Saravanan and Dr. K. Jayamoorthy thank the management of St. Joseph's College of Engineering, Chennai for the infrastructure and moral support. The authors sincerely thank Ms. S. Aparana (PG student), Department of Chemistry, Anna University, Chennai-25, for her timely help in proof correction.

KEYWORDS

- **surface treatment**
- **binding interaction**
- **nanoparticle**
- **fluorescence**
- **nanocrystals**

BIBLIOGRAPHY

1. Johnston, H.J.; Hutchison, G.R.; Christensen, F.M.; Peters, S.; Ankin, S.; Stone, V. *Part. Fibre Toxicol.* **2009**, *17*, 6–33.
2. Warheit, D.B. *Toxicol. Sci.* **2008**, *101*, 183–185.
3. Sakohapa, S.; Tickazen, L.D.; Anderson, M.A. *J. Phys.Chem.* **1992**, *96*, 11086–11091.
4. Harada, K.; Asakura, K.; Ueki, Y.; Toshina, N. *J. Phys.Chem.* **1992**, *96*, 9730.
5. Lee, J.; Hwang, J.H.; Mashek, T.T.; Mason, T.O.; Miller, A.E.; Siegel, R.W. *J. Mater Res.* **1995**, *10*, 2295.
6. Hara, K.; Horiguchi, T.; Kinoshita, T.; Sayama, K.; Sugihara, H.; Arakawa, H. *Sol. Energy Mater. Sol. Cells.* **2000**, *64*, 115.

7. Ko, S.C.; Kim, Y.C.; Lee, S.S.; Choi, S.H.; Kim, S.R. *Sens. Actuators, A, Phys.* **2003**, *103*, 130.
8. Sberveglieri, G.; Groppelli, S.; Nelli, P.; Tintinelli, A.; Giunta, G. *Sens. Actuators B.* **1995**, *25*, 588.
9. Shishiyance, S.T.; Shishiyance, T.S.; Lupan, O.I. *Sens. Actuators B.* **2005**, *107*, 379.
10. Cao, H.; Jy Xu.; Zhang, D.Z. *Phys. Rev. Lett.* **2000**, *84*, 5584.
11. Bagnall, D.M.; Chen, Y.F.; Shen, M.Y.; Zhu, Z.; Goto, T.; Yao, T.J. *Cryst. Growth* **1998,** *184/185,* 605.
12. Sadiq, I.M.; Chowdhury, B.; Chandrasekaran, N.; Mukherjee, A. *Nanomedicine* **2009**, *5*, 282.
13. Hong, R.Y.; Li, J.H.; Chen, L.L.; Liu, D.Q.; Li, H.Z.; Zheng, Y.; Ding, J. *Pow. Techn.* **2009**, *189*, 432.
14. Mallakpour, S.; Bull, M.M. *Mater. Sci.* **2012**, *35*, 333.
15. Yongjun He. *Pow. Techn.* **2004**, *147*, 59.
16. Cheng, H.M.; Hsich, W.F. *Nanotech.* 2010, 21, 485202.
17. Lin, B.; Fu, Z.; Ji, Y. *Appl. Phy. Lett.* 1991, *79*, 943.
18. Maity, P.; Kasisomayajula, S.; Parameswaran, V.; Basu, N. *IEEE Trans.* **2008**, *15*, 63.
19. Gurumurthy, C.; Hui, C. *J. Adhes.* **2006**, *82*, 239.
20. Shin, E.J.; Kim, D.; *J. Photochem. Photobiol. A: Chem.* **2002**, *152*, 25–31.
21. Murov, S.L.; Carmichael, I.; Hug, G.L. *Handbook of Photochemistry*, 2nd ed., M.Dekker, Inc. New York, 1993; pp 269–273.
22. Lin, B.; Fu, Z. Ji, Y. *Appl. Phy. Lett.* **1991**, *79*, 943.
23. Cyril, L.; Earl, J.K.; W.M. Sperry, *Biochemists Handbook*, E & F.N. Spon, London, 1961.
24. Chen, G.Z.; Huang, X.Z.; Xu, J.G.; Wang, Z.B.; Zhang, Z.Z. *Method of Fluorescent Analysis*, 2nd ed., Science Press, Beijing, 1990; p. 123, 126
25. He, W.Y.; Li, Y.; Xue, C.X.; Hu, Z.D.; Chen, X.G.; Sheng, F.L. *Bioorg. Med. Chem.* **2005**, *13*, 1837.
26. Hu, Y.J.; Liu, Y.; Zhang, L.X. *J. Mol. Struct.* **2005**, *750*, 174.

CHAPTER 9

SURFACE MORPHOLOGICAL, THERMAL, AND DIELECTRIC BEHAVIOR OF TOUGHENED TETRAGLYCIDYL EPOXY-POSS NANOCOMPOSITES

D. DURAIBABU[1], S. ANANDA KUMAR[1,*], and R. MANJUMEENA[2]

[1]*Department of Chemistry, Anna University, Chennai 600025, India.*
**E-mail: sri_anand_72@yahoo.com*

[2]*Centre for Advanced Studies in Botany, University of Madras, Chennai 600025, India*

CONTENTS

ABSTRACT

The objective of the present work is to synthesize N,N'-Tetraglycidylbis amine-phenoxy benzene (TGBAPB) epoxy resin using 1,4-bis (4-amine-phenoxy) benzene (BAPB) and epichlorohydrin. The molecular weight was determined by gel permeation chromatography (GPC) and equivalent weight by means of epoxy equivalent weight (EEW) titration. The amino-functionalized polyhedral oligomeric silsesquioxane (POSS) was synthesized via octa(nitrophenyl)silsesquioxane (ONPS) method and its molecular structure has been confirmed by Fourier Transform-Infra Red spectra and nuclear magnetic resonance spectra (NMR). The TGBAPB epoxy resin was further reinforced using POSS with varying weight percentages (1–5 wt.%) and cured with diaminodiphenylmethane (DDM). Thermomechanical behavior of TGBAPB epoxy matrix and nanocomposites were examined by dynamic mechanical analysis (DMA), thermo gravimetric analysis (TGA), and differential scanning calorimetry. The surface morphology of the epoxy nanocomposites was investigated by X-ray diffraction (XRD), transmission electron microscopy (TEM), scanning electron microscopy (SEM), and atomic force microscope (AFM) studies.

9.1 INTRODUCTION

Epoxy resins are widely used in many industrial applications in the form of adhesives, sealants, and matrices for advanced composite, owing to their high tensile strength and modulus, good chemical, corrosion resistance, and excellent dimensional stability, etc.[1–3] However, the major drawback of diglycidyl epoxy resins is that in the cured state they are brittle in nature and exhibited inferior weather resistance. This inherent brittleness causes the poor damage tolerance to impact of the composites made from epoxy resin and poor peeling and shear strength of epoxy-based matrices.[4] To overcome the above said problems, the modifiers with flexible segments, namely, carboxyl-terminated butadiene nitrile rubber, hydroxyl-terminated butadiene nitrile rubber, amine-terminated butadiene nitrile rubber, polyurethanes, and silicones have been used to improve their properties suitable for high performance applications.[5,6] Furthermore, a new type of multifunctional epoxy resins has been developed by chemical and physical modifications.

The multifunctional epoxy resins are widely used because of their higher cross-link density and excellent adhesion, which can be exploited for high performance applications.[7,8] Although modification of multifunctional epoxy resin improved their impact characteristics, it led to a decrease in mechanical properties. The recent development of several families of functional hybrid materials based on POSS affords a tremendous potential for the modification of organic–inorganic hybrid polymers.[9,10] However, polymer-layered silicate nanocomposites that are reinforced with well-defined, nanosized inorganic clusters have been reported with improved stiffness, strength, fracture toughness, fire resistance, barrier properties, and dimensional stability.[11–13] The nanoscaled distribution of reinforcing agents can optimize the interactions between different molecular components and can afford materials with improved properties.[14,15] Generally, incorporation of inorganic or organometallic segments (POSS cages) to polymers is performed to afford the improved properties through the formation of covalent bonds between the POSS cages and the organic polymer matrices.[16,17] The objective of the present work is to develop new ether-linked skeletal-modified tetraglycidyl epoxy resin TGBAPB with toughness and stiffness. To achieve both toughness and stiffness to a significant extent, TGBAPB may be reinforced with POSS of varying percentages (1, 3, and 5 wt.%). The curing and thermal behavior, thermomechanical and mechanical properties, dielectric properties, and water absorption behavior of resulting TGBAPB epoxy-based nanocomposites have been studied and compared with those of neat TGBAPB epoxy matrix.

9.2 EXPERIMENTAL

9.2.1 SYNTHESIS OF TGBAPB EPOXY RESIN

The TGBAPB resin was synthesized using electrochemical pseudo supercapacitors and BAPB with 40% NaOH solution. A pale brown-colored TGBAPB resin obtained (yield 80%) was purified and preserved for further use. The reaction sequence is illustrated in Scheme 9.1. Fourier transform-infrared (FT-IR) spectra, NMR spectrum, and electron spray ionization-mass spectroscopy [M+] the molecular mass of TGBAPB epoxy groups has already been confirmed and reported elsewhere.[15] TGBAPB epoxy resin is synthesized in the laboratory as per the procedure reported in the literature.[18]

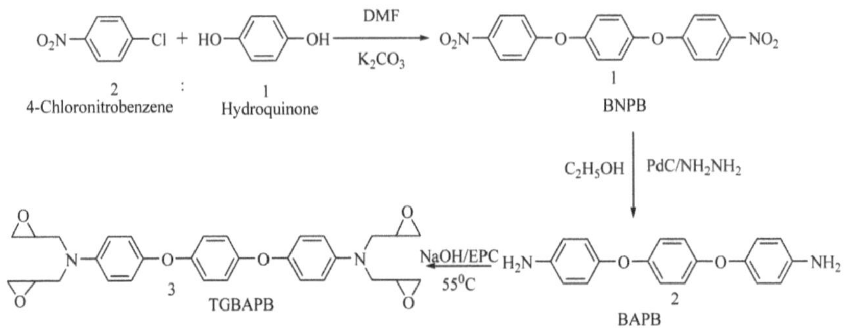

SCHEME 9.1 Schematic representation of synthesis of TGBAPB.

9.2.2 SYNTHESIS OF OCTAPHENYLSILSESQUIOXANE (OPS)

The OPS was synthesized in a manner reported by Huang et al.[19] 12.4 mL (10.9 g, 0.05 mol) of phenyl trichlorosilane was dissolved in 75 mL of benzene and 150 mL of water, the mixture was stirred for 5 h at 25°C. The acid layer formed was removed and the benzene layer was washed with water until pH reached neutral. It was then dried with anhydrous magnesium sulfate 1.2 mL (3 mmol.) subsequently 40% benzyl trimethyl ammonium hydroxide/methanol solution was added to this organic layer and the whole mixture was refluxed for 4 h, and then allowed to stand for 4 days. The mixture was refluxed again for 24 h, cooled and then filtered, which resulted in a white microcrystalline powder (Scheme 9.2). The product thus obtained was extracted using dry benzene in a Soxhlet extractor to remove the soluble resin. It was further dried overnight in vacuum at 70°C.[20] The yield was 6.5 g (yield 92.2%).

9.2.3 SYNTHESIS OF OCTA (NITROPHENYL) SILSESQUIOXANE (ONPS)

The ONPS was synthesized following the procedure reported by Laine's et al.[21] A total of 90 mL of fuming nitric acid was added in small portions to 10 g (9.7 mmol) of OPS and placed under stirring in an ice bath (0°C) for half an hour. It was then allowed to stir at ambient temperature for 20 h (Scheme 9.2). After filtration through glass wool, the solution was then poured over 100 g of ice to precipitate the product. It was then filtered to obtain a light yellow color precipitate, which was washed with water until the pH reached

6 and was finally washed with ethanol. The residual solvent was removed by drying the powder under vacuum at room temperature. The yield was 11.5 g (yield 94.8%) of ONPS.

9.2.4 SYNTHESIS OF OCTA (AMINO PHENYL) SILSESQUIOXANE (OAPS)

A total of 10.0 g (7.16 mmol) of ONPS and 5 wt.% Pd/C (1.22 g) (0.574 mmol.) was taken in a 500 mL round-bottom flask equipped with a condenser under nitrogen, distilled tetrahydrofuran (THF) (200 ml) was poured into it and stirred at room temperature. Through an addition funnel, 16 mL of formic acid (80%) was added in drops. Then the mixture was heated to 60°C (Scheme 9.2). Bilayer formation was observed after an hour. It contained a top layer and a bottom black suspension. After cooling the mixture to room temperature 30 mL of THF was added to the flask to dissolve the remaining slurry. The suspension was filtered again. The mixture was then subjected to ethyl acetate and was water washed consecutively. A total of 15 g of anhydrous magnesium sulfate was added to the organic layer to dry it. The product was precipitated by adding 2 liters of hexane. It was filtered and dissolved in 30:50 THF/ethyl acetate and reprecipitated using hexane. The powder was dried under vacuum to yield 6 g of OAPS (yield 70.5%).

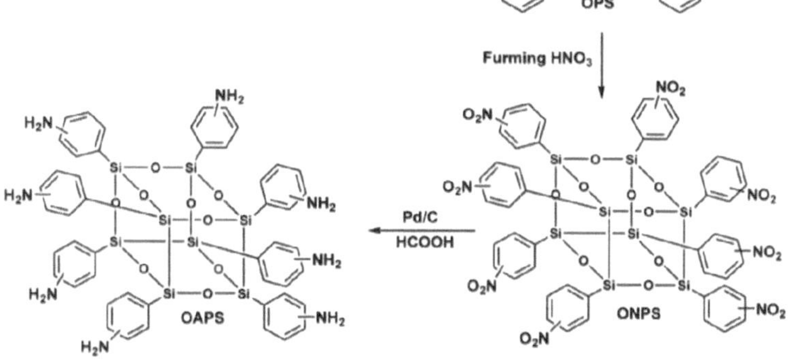

SCHEME 9.2 Synthesis of OAPS.

9.2.5 PREPARATION OF POSS-REINFORCED EPOXY NANOCOMPOSITES

In order to prepare the composites of epoxy resins with OAPS, the latter was first dissolved in the smallest possible amount of THF and the solution was mixed with the desired amount of TGBAPB epoxy resin at 25°C. The mixture obtained was heated to 60°C with continuous stirring to evaporate the majority of solvent and then degassed under vacuum at 60°C to remove the residual solvent. Subsequently, the stoichiometric amount of curing agent (DDM) was added (Table 9.1) into the molten epoxy at 120°C and stirred continuously until a homogeneous, transparent solution was formed. After degassing, the mixture was poured into a Teflon coated iron mold. It was thermally cured: first at 140°C (4 h) before the temperature was raised slowly to 200°C and then maintained at this temperature (3 h) for the curing reaction to reach completion which is shown in Scheme 9.3.

TABLE 9.1 Nomenclature of TGBAPB Epoxy Nanocomposites.

System	TGBAPB/% POSS	Curative
a	100/0	DDM
b	100/1	DDM
c	100/3	DDM
d	100/5	DDM

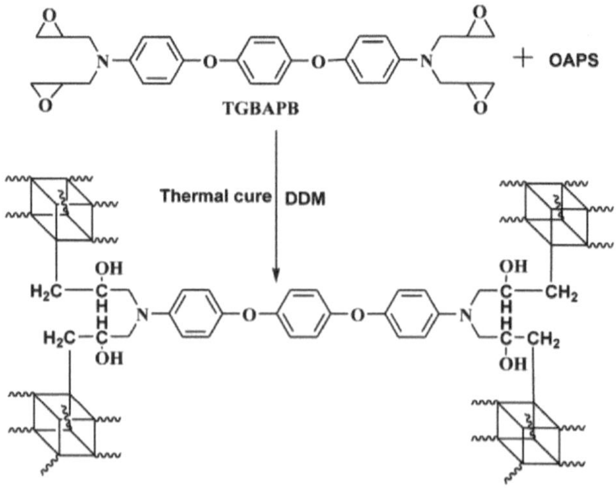

SCHEME 9.3 Schematic representation of POSS-reinforced epoxy nanocomposites.

9.3 RESULTS AND DISCUSSION

9.3.1 FT-IR SPECTROSCOPY

The peaks appearing at 1597 and 1115 cm^{-1} correspond to —Si—O—Si— of OPS are shown in Figure 9.1(a) of FT-IR spectra. FT-IR spectra of ONPS depicted in Figure 9.1(b) and OAPS in Figure 9.1(c), indicated a complete conversion of nitro groups to amino groups. In the FTIR spectrum of OAPS, the peak at 1090 cm^{-1}, indicates the presence of (Si—O—Si) and peaks at 1350 cm^{-1}, 1527 cm^{-1} indicate the presence of (NO_2). The disappearance of peak at 1527 cm^{-1} and the formation of peak at 3360 cm^{-1} (N-H) indicated the formation of amino-functionalized POSS. ^{1}HNMR spectra of OAPS, the appearance of chemical shifts at around δ = 7.6–6.5 ppm (b, 32H) and δ = 5.0–3.9 ppm (b, 16H) confirmed the presence of aromatic protons and are presented in Figure 9.2. In Figure 9.3(a), the band at 912 cm^{-1} appears due to the presence of oxirane of the tetraglycidyl epoxy ring. The bands appearing at 2975 cm^{-1} and 1650 cm^{-1} correspond to the aromatic rings of TGBAPB and 821 cm^{-1} for the C—N stretching of TGBAPB. Figure 9.4 shows the GPC graph of TGBAPB epoxy with a number average molecular weight (Mn) 600, weight average molecular weight (Mw) 886, and poly dispersity index (Mw/Mn) 1.47, respectively.

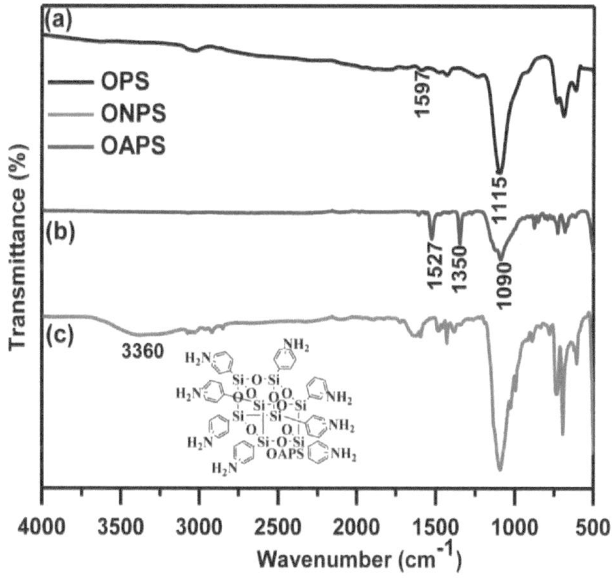

FIGURE 9.1 FT-IR spectra of OPS, ONPS, and OAPS.

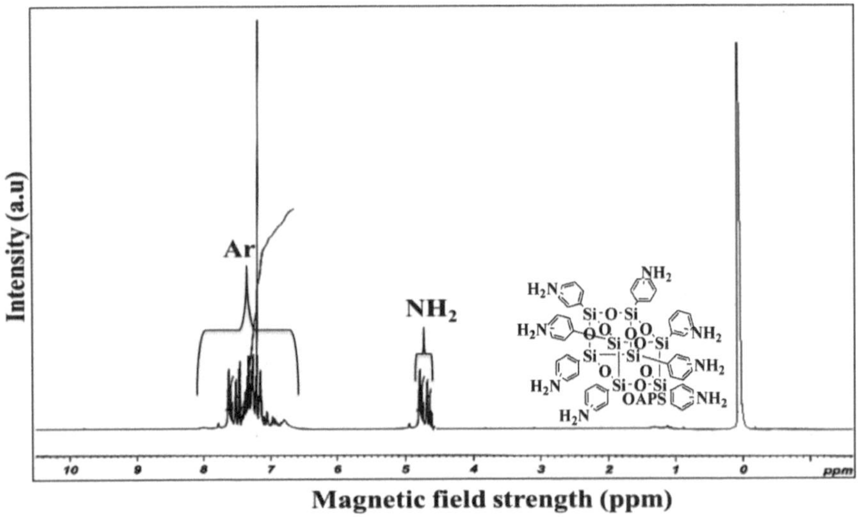

FIGURE 9.2 ^{1}HNMR spectra of OAP.

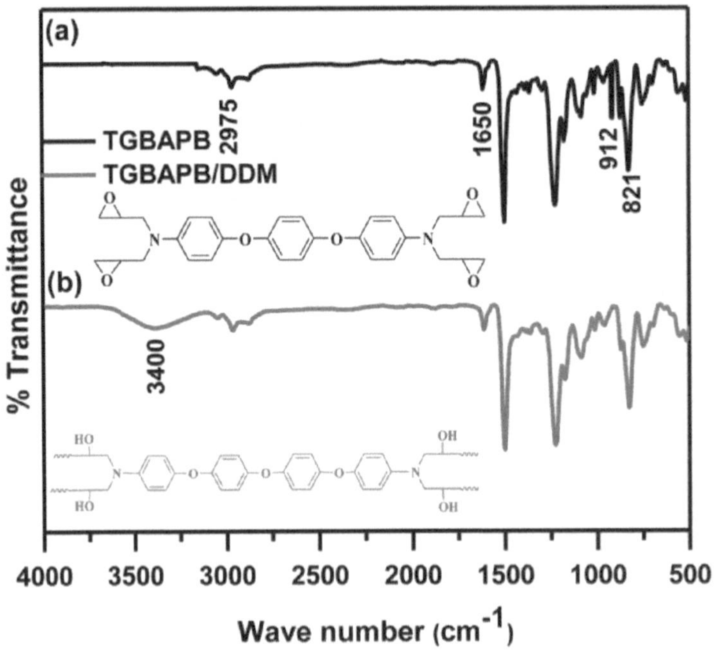

FIGURE 9.3 FT-IR spectra of TGBAPB epoxy resin cured with DDM.

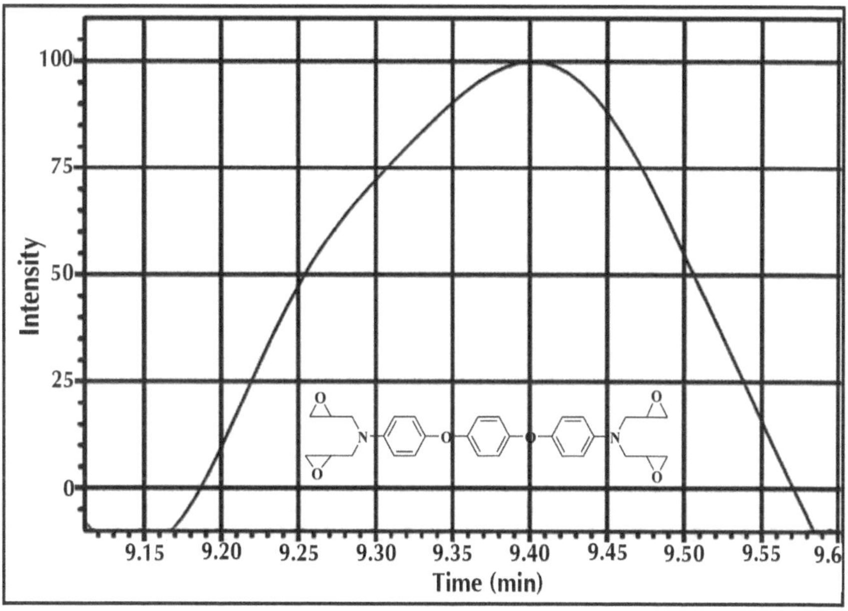

FIGURE 9.4 GPC spectra of TGBAPB epoxy resin.

9.3.2 MECHANICAL PROPERTIES

It was interesting to note that the 1 and 3 wt.% POSS-reinforced epoxy nanocomposites systems "b" and "c" exhibited the best values of mechanical properties in terms of tensile strength, flexural strength, and impact strength among all other systems and are shown in Table 9.2. and Figure 9.5. From Table 9.2 it can be clearly seen that the values of tensile strength, flexural strength, and impact strength of TGBAPB reinforced with 3 wt.% POSS (system "c") are 157.5 MPa, 211.5 MPa, and 203.4 J/m^{-1}, respectively when compared to those values of neat TGBAPB epoxy system "a". This may be due to the homogeneous dispersion of POSS molecules, which are covalently bonded with epoxy matrix, thereby providing additional cross-linking networks in the epoxy resins, besides the cross-linking provided by the DDM curing agent. Consequently, a significant enhancement in stiffness is observed in the case of POSS-reinforced systems ("b" and "c") than those of other systems of the present study.

As observed in previous case, no significant improvement in mechanical properties was observed beyond 3 wt.% of POSS loading into epoxy matrix. Similar observation was made by Jones et al.[22] for mechanical behavior of

POSS in epoxy nanocomposites. Upon increasing the POSS content to 5 wt.%, (system "d"), the mechanical properties were found to have decreased and are shown in Figure 9.5. This could be attributed to the aggregation and irregular dispersion of the POSS nanoparticles, which led to higher stress concentration and void formation. This could be one of the reasons for the increase in distance between interacting particles, which prevents particle–particle interaction. As a result of which, there exists a void of cross-linking, which increases the segmental mobility and concomitantly decreased the mechanical properties. In addition, the steric hindrance offered by the presence of unreacted amine-terminated POSS cages will raise the activation energy of the curing reaction and reduce the cross-link density. These results are in concurrence with those reported by Khine et al.[23] for aggregation of POSS at high percentage loading in epoxy system.

TABLE 9.2 Mechanical Properties of Neat (TGBAPB) Epoxy Matrix Reinforced with POSS Nanocomposites.

Matrix Systems	Tensile Strength (MPa)	Flexural Strength (MPa)	Impact Strength (J/m^{-1})
a	76.3 ± 3	142.5 ± 2	131.2 ± 2
b	123.3 ± 2	182.4 ± 3	173.5 ± 1
c	157.5 ± 1	211.5 ± 2	203.4 ± 5
d	120.4 ± 3	178.5 ± 1	172.4 ± 2

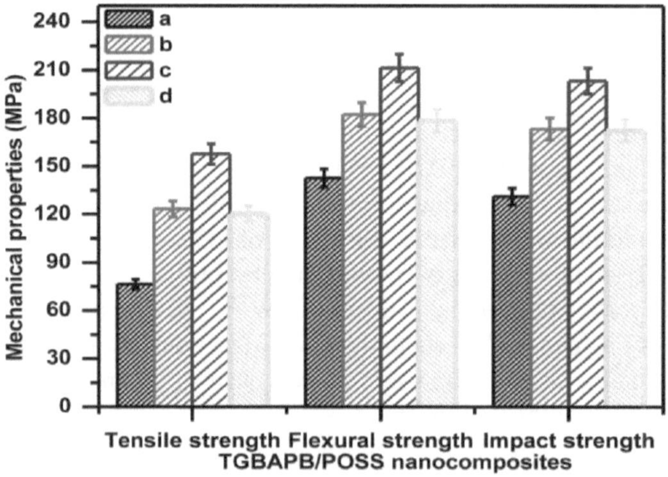

FIGURE 9.5 Mechanical properties of (a) Neat (TGBAPB), (b) TGBAPB/1 wt.% POSS, (c) TGBAPB/3 wt.% POSS, and (d) TGBAPB/5 wt.% POSS epoxy nanocomposites.

9.3.3 DMA ANALYSIS

It was interesting to observe that the storage modulus increases as the percentage content of nanoreinforcement increases. For example, an increase in storage modulus values was observed up to 3 wt.% loading. However, the values decreased beyond 3 wt.%. The increase in storage modulus observed up to 3 wt.% may be due to the effective dispersion of nanoreinforcement within the epoxy matrix, thereby producing an enhanced stiffness.[24] Table 9.3 shows the values of storage modulus of TGBAPB epoxy system after reinforcement with varying weight percentages of POSS. It can be seen from Figure 9.6 that the values are increased with increasing POSS content up to 3 wt.% (system "c"). This may be due to the organic–inorganic nature of the POSS caged tethers capable of enhancing the cross-linking density of the resultant nanocomposites owing to its unique molecular structure with multifunctional amine cross-linking sites. Similar observation was made by Zhang et al.[25] and Fu et al.[26] for thermal properties and toughening effect of POSS epoxy nanocomposites.

The restricted mobility of chains under loading provides good interfacial adhesion between POSS and matrices. Similar mechanism has been reported by Jones et al.[22] for thermal and mechanical behavior of POSS epoxy nanocomposites. At high concentration of POSS 5 wt.% (system "d"), the filler particles seem to have been dispersed irregularly into the epoxy matrix leading to the void of cross-linking, thereby increasing the segmental mobility of the polymer chain, which increases the distance between interacting particles. As a consequence, the particle–particle interaction is greatly reduced for the nanocomposites having 5 wt.% of nanoreinforcements. Moreover, the steric hindrance resulted by higher % of POSS (5 wt.%) may increase the activation energy subsequently reducing the cross-link density.

TABLE 9.3 DMA Results of Neat (TGBAPB) Epoxy Matrix and POSS Epoxy Nanocomposites.

Matrix Systems	Storage Modulus (GPa)	Glass transition Temperature, T_g (°C)
a	2.6	163
b	3.2	257
c	4.9	269
d	3.1	245

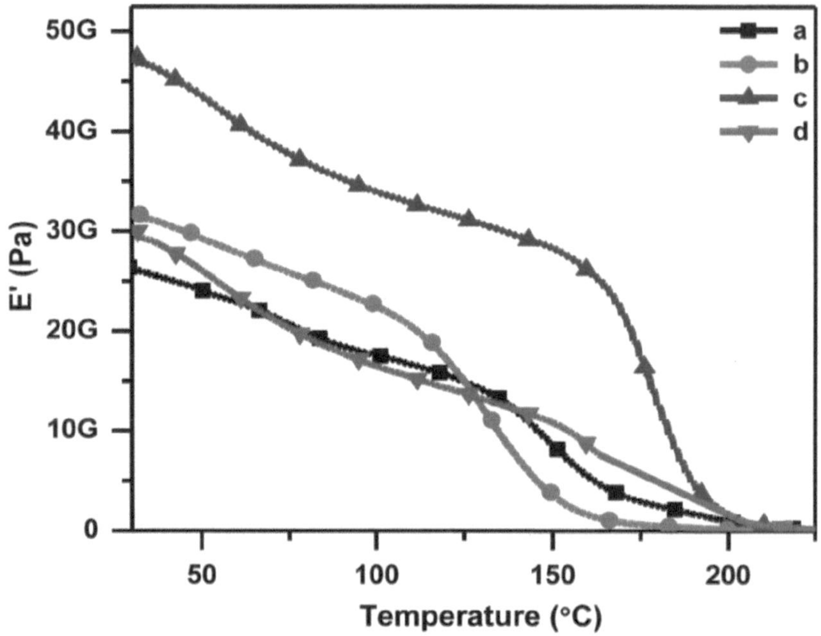

FIGURE 9.6 Storage modulus of (a) Neat (TGBAPB), (b) TGBAPB/1 wt.% POSS, (c) TGBAPB/3 wt.% POSS, and (d) TGBAPB/5 wt.% POSS epoxy nanocomposites.

9.3.4 *GLASS TRANSITION TEMPERATURE*

Incorporation of amino-functionalized POSS into the epoxy network prevents the movement of the polymer chains, and thereby influencing increased values of glass transition temperature. Glass transition temperature (T_g) was determined from the peak position of tan δ, which was increased from 163 to 269°C for POSS content up to 3 wt.% as shown in Figure 9.7. The values of T_g are given in Table 9.3. The values of T_g of TGBAPB epoxy systems "b" and "c" increased steadily up to 3 wt.% of POSS concentration. This may be due to the formation of covalent bond between the POSS cubes and organic molecular chain, which contributed to the enhancement of T_g.[22] At higher weight percentage of POSS 5 wt.% (system "d"), due to the increase in free volume content by the addition of these bulky cages, more flexibility is imparted, which leads to a decreased T_g[27] and is presented in Figure 9.7.

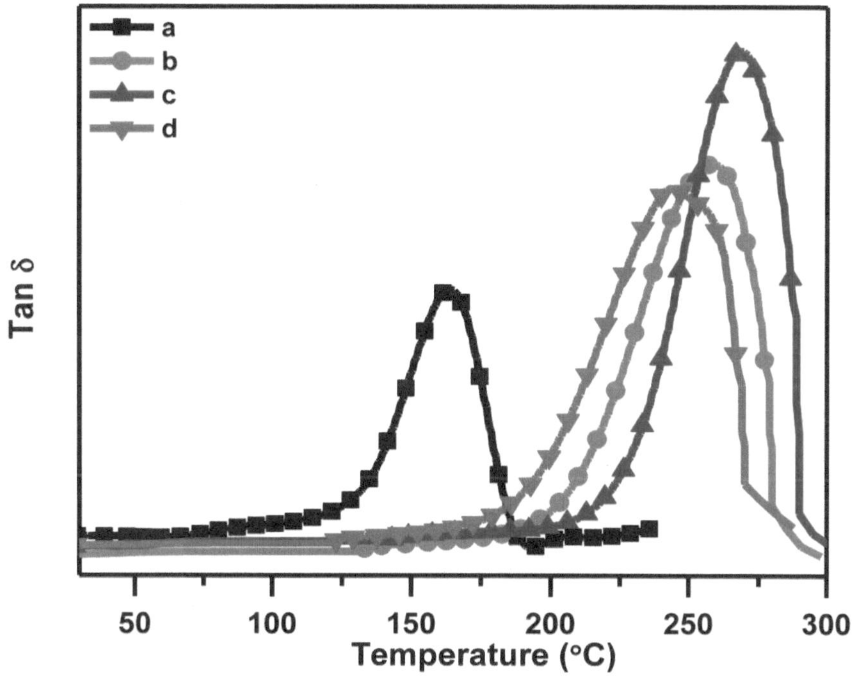

FIGURE 9.7 Tan δ of (a) Neat (TGBAPB), (b) TGBAPB/1 wt.% POSS, (c) TGBAPB/3 wt.% POSS, and (d) TGBAPB/5 wt.% POSS epoxy nanocomposites.

9.3.5 TGA ANALYSIS

The thermal stability of the matrix was studied by TGA and from the data the thermal degradation patterns of the matrix could be found. The data obtained from the TGA studies are presented in Table 9.4 and Figure 9.8. The TGA patterns of TGBAPB and POSS (1, 3, and 5%) show that the initial degradation temperature (IDT) was enhanced up to 3 wt.% POSS, which can be as observed from Figure 9.8. The significant increase in the values of thermal stability (Table 9.4) observed in systems "b" and "c" is due to the predominant properties of POSS reinforcement. It could be due to the presence of POSS moiety whose partial ionic nature along with high bond energy prevents the degradation of epoxy from heat and apparently high thermal energy is required to cause the degradation of POSS-reinforced tetrafunctional epoxy matrix.[28] While 5% POSS-reinforced TGBAPB epoxy system "d" exhibited decreased IDT and char yield value of 338°C and 43%. The reasons for the decreased thermal stability are due to porosity of POSS containing nanocomposites, which is composed of two portions: one portion comes from an external porosity as a result of the inclusion of bulky POSS, which can be interpreted as the increase in free volume of the nanocomposites caused by the presence of POSS cages between the polymer segments. The second portion of porosity can be attributed to the nanoporosity of the POSS core with a diameter of 0.53 nm. The decreasing cross-link density per unit volume in the hybrid systems leads to a decrease in the value of thermal properties at higher loading of POSS moieties into the epoxy networks.[29]

TABLE 9.4 Thermal Properties of Neat (TGBAPB) and POSS-Reinforced Epoxy Nanocomposites.

Matrix Systems	Initial Decomposition Temperature (°C)	Char Yield (%)
a	307	13
b	326	39
c	365	45
d	338	43

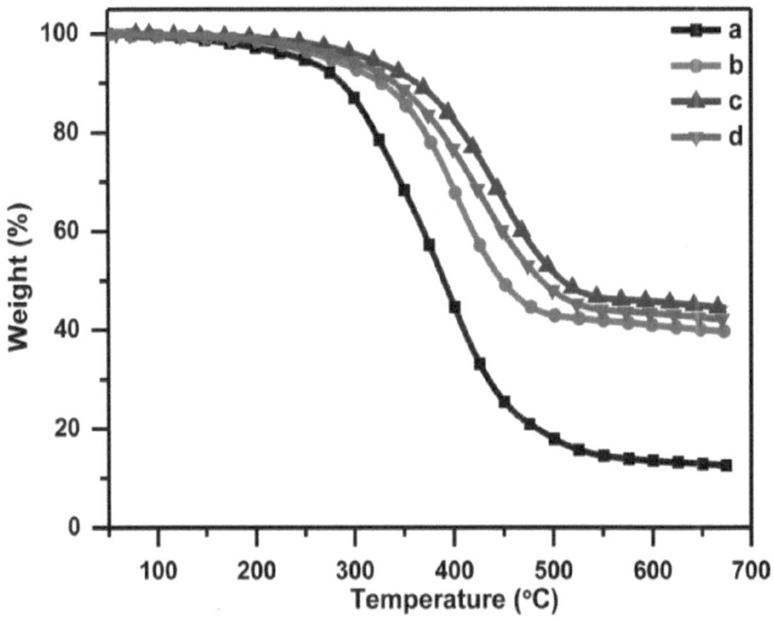

FIGURE 9.8 TGA of (a) Neat (TGBAPB), (b) TGBAPB/1 wt.% POSS, (c) TGBAPB/3 wt.% POSS, and (d) TGBAPB/5 wt.% POSS epoxy nanocomposites.

9.3.6 ELECTRICAL PROPERTIES

Incorporation of cage-like POSS molecules into the skeletally-modified ether-linked epoxy system TGBAPB resulted in a lower dielectric constant. The reduction in dielectric properties with increasing concentration of 5 wt.% of POSS (system "d") in the hybrid system is most likely due to the inherent porosity of POSS, when it is incorporated into epoxy systems "b", "c", and "d". It also imparts external porosity due to the development of voids in the hybrid systems. On the other hand, the thermally stable inorganic silica core of POSS molecules is simply less polar than the polymer segments, which in turn reduces the dielectric constant of the epoxy nanocomposites that are shown in Table 9.5. This proves that an amino-functionalized POSS-reinforced TGBAPB-modified epoxy could serve as a better candidate for electrical applications. From the data, it was observed that reinforcing, inorganic material into an organic system increases the insulating property. It is commonly known that the signal propagation delay time of the integrated circuits is proportional to the square root of the dielectric constant of the hybrid nanocomposites, and therefore the signal propagation is directly proportional to the dielectric constant of the hybrid composites. Thus, the low dielectric constant increases the speed of signal transmission between the chips in the packaging materials.[30] Hence, the hybrid nanomaterials developed in the present work may find better utility in insulation applications as well.

9.3.7 WATER ABSORPTION STUDIES

An incorporation of POSS (1, 3, and 5 wt.%) increased the hydrophobic character considerably according to their percentage composition. The % of water absorption can be seen to decrease with increasing nanoparticle content (Table 9.5). This could have been caused by two factors. Firstly, the volume of epoxy for water diffusion is reduced with increasing POSS content. Secondly, the presence of POSS can increase the path length for moisture diffusion. In the case of POSS-reinforced nanocomposites, the presence of Si—O—Si linkages favored negligible water absorption with a reduced free volume resulted from an increased cross-linking effect of POSS–amine.[31]

TABLE 9.5 Dielectric, Water Absorption Properties of Neat (TGBAPB) Epoxy and Nanocomposites.

Matrix systems	Dielectric Constant (ε′)	Water Absorption (%)
a	4.10	0.077
b	3.80	0.060
c	3.70	0.058
d	3.65	0.056

9.3.8 X-RAY DIFFRACTION STUDIES

XRD analysis was carried out on skeletally-modified epoxy system TGBAPB and its nanocomposites after reinforcement with 1–5 wt.% of POSS in order to understand its morphological behavior. The shift in peak toward lower 2θ values observed is shown in Figure 9.9. These XRD results matched well with 3 wt.% loading of nanoreinforcement POSS that showed a combination of intercalated and exfoliated nanocomposites platelets in epoxy systems. This could also be considered as a proof for increased cross-linking density and the reason for enhanced mechanical properties of the nanocomposite system.[32] The shift toward a lower angle indicates that the POSS reinforcement exhibited a combination of intercalation and exfoliation to a significant extent, when compared to the skeletally-modified epoxy system "a" without nanoreinforcements. In general, the improvement in storage modulus is attributed to the good dispersion of POSS and resulted in good interfacial adhesion between the POSS particles and the epoxy matrix so that the mobility of polymer chains is restricted under loading.[25] From the rising intensity of these peaks and the shift of these XRD peaks of systems, toward the lower angle with proportional increase in nano filler content, this indicates a good dispersion of the filler in the systems "b" and "c". The peak intensity of 5 wt.% POSS (system "d") loaded nanocomposites was significantly high and interplanar spacing was decreased indicating mixed exfoliated and intercalated behavior. The reason for the above mentioned behavior is the strong tendency of POSS to agglomerate and increased viscosity of resin when it is mixed with POSS.

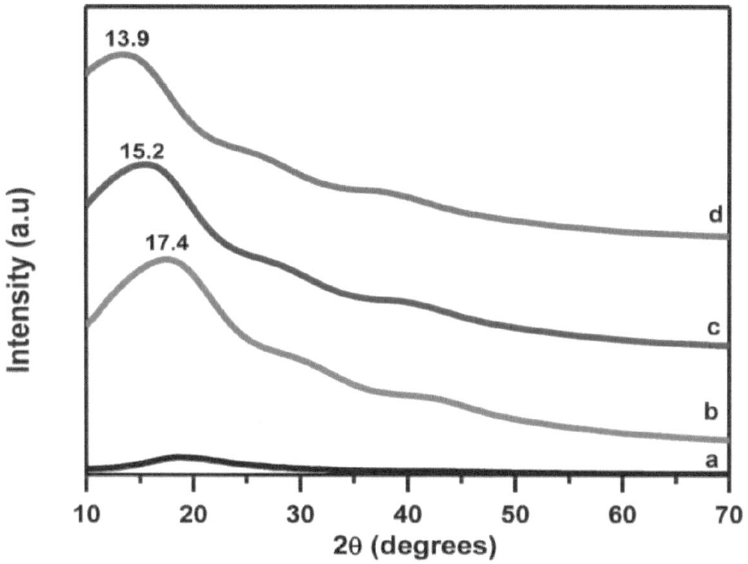

FIGURE 9.9 XRD spectra of (a) Neat (TGBAPB), (b) TGBAPB/1 wt.% POSS, (c) TGBAPB/3 wt.% POSS, and (d) TGBAPB/5 wt.% POSS epoxy nanocomposites.

9.3.9 MORPHOLOGY STUDIES

SEM and AFM of fractured surfaces of the nanocomposites showed considerably different fractographic features than that of TGBAPB epoxy matrix. Figure 9.10 shows the SEM micrographs of the TGBAPB epoxy matrix in which their surfaces were smooth (system "a"). However, upon incorporation of POSS particles, surface became rough. The roughness increases as the POSS content increases in the matrix systems "b", "c", and "d". For example, 3 wt.% (system "c") POSS (organic–inorganic hybrid) makes a better dispersion of POSS into ether-linked-modified epoxy matrix, which attributes an excellent compatibility between skeletally-modified epoxy matrix and amine-functionalized POSS. This may be due to the efficient adhesion and the influence of intermolecular interactions between skeletally-modified epoxy matrix and POSS as shown in Figure 9.10(b).

The increased surface area from the rough surface can be associated with higher energy necessary for crack propagation. Thus proper dispersion of nanocomposite particles would lead to form rough surfaces. Furthermore, the increased surface roughness implies that the path of the crack tip was distorted because of the presence of POSS, which make crack propagation more difficult.[32] On the other hand, it is proposed that intercalated nanocomposites promote toughness, whereas exfoliated nanocomposites platelets mainly improve stiffness of the polymer matrix, due to energy-absorbing shearing of intercalated nanocomposites layers.[33] It must be recalled that these materials do have a lateral dimension in the micron scale and it is possible that these may encourage crack stopping and pinning. The morphology studies have shown that the arrangement of the organic–inorganic layers cannot neatly be termed as either intercalated or exfoliated. It is likely that the overall structure of the nanocomposites of intercalated and exfoliated are in concurrence with XRD analysis. The fractured surfaces of POSS incorporated ether-linked-modified epoxy nanocomposites became more wrinkled and coarser, reflecting change of micro morphologies when compared to that of ether-linked-modified epoxy system "a" (TGBAPB) as shown in Figure 9.10(a).

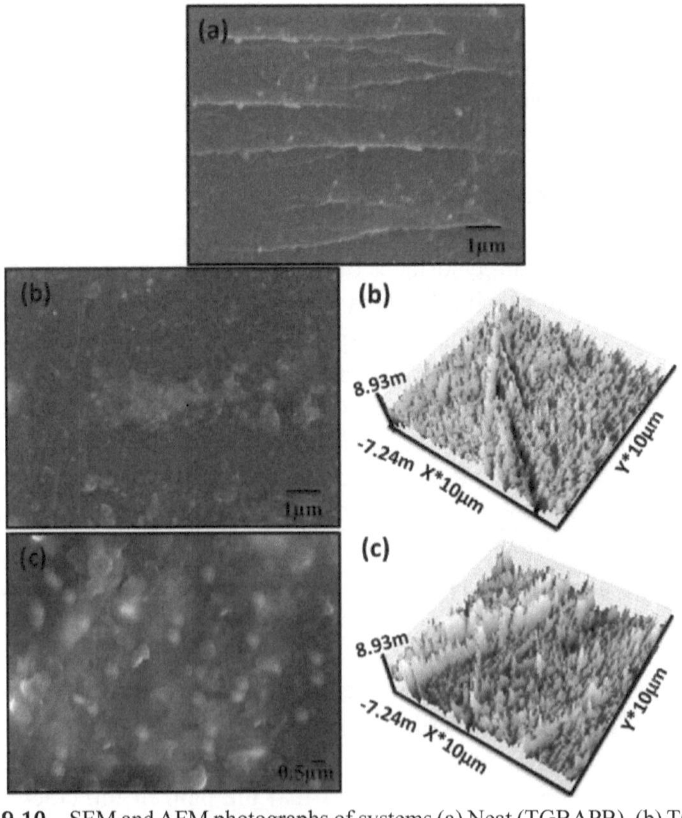

FIGURE 9.10 SEM and AFM photographs of systems (a) Neat (TGBAPB), (b) TGBAPB/3 wt.% POSS, and (c) TGBAPB/5 wt.% POSS epoxy nanocomposites.

A higher loading of POSS (5 wt.%) resulted in an increased surface roughness due to the pinning effect of incorporation of POSS initiated at a localized domain due to the presence of aggregates of POSS that act as stress concentrator. However, factors like inevitable aggregation of POSS particles at higher loading lead to adhesion problem at the POSS matrix interface at elevated temperatures, and apparent lack of orientation of the POSS particles might affect the values of nanocomposites. The results are depicted in Figure 9.10(c). The crack initiation was caused by the stress concentration developed by the agglomerated particle leading to failure. Particles would tend to agglomerate and prevent the intended particle to polymer interaction and eventually result in the reduction of storage modulus. Similar observations were made from DMA studies where 5 wt.% POSS reinforced nanocomposites exhibited inferior storage modulus due to the aggregation of excess nanocomposite particles.

9.3.10 TEM GRAPHS OF EPOXY NANOCOMPOSITES

Transmission electron microscopy (TEM) micrographs of the POSS epoxy nanocomposites showed considerably different fractographic features than that of TGBAPB epoxy matrix as shown in Figure 9.11. Figure 9.11(a) shows homogenous dispersion of POSS nanoparticles (up to 3 wt.% loading) within the epoxy system. The size of these reinforcements is consistent and there is no evidence of phase inversion (the particulate morphology is retained). Furthermore, there is no change in the morphology of the reinforcing particles present within epoxy nanocomposites at 3 wt.% loading. All these results point to retention of particle morphology in the skeletally-modified epoxy matrices. The skeletally-modified epoxy surfaces are essentially saturated with reinforcing POSS particles are relatively low concentrations of up to 3 wt.%.[34]

The lighter areas in the TEM images indicate the epoxy matrix and the black area depicts the POSS reinforcement. There is an excellent adhesion due to the formation of covalent bonding between the amino-functionalized POSS and TGBAPB epoxy matrix. Hence, there is a formation of a highly intercalated and partially exfoliated structure. This may be due to the parallel arrangement, which were considered for measurements of the interlayer distance. On the other hand, the higher weight percentage loading of 5 wt.% of POSS nanoparticles led to increased agglomeration as shown in Figure 9.11(b). The reinforcements are not broken but pulled out due to the weak interfacial bonding between the reinforcements and the epoxy matrix.

Apparently, the agglomerates reduce the reinforcing effects of reinforcement because they are acting as flaws in the epoxy matrix. This is why the poorly dispersed epoxy nanocomposites have lower mechanical properties than that of the well-dispersed ones.

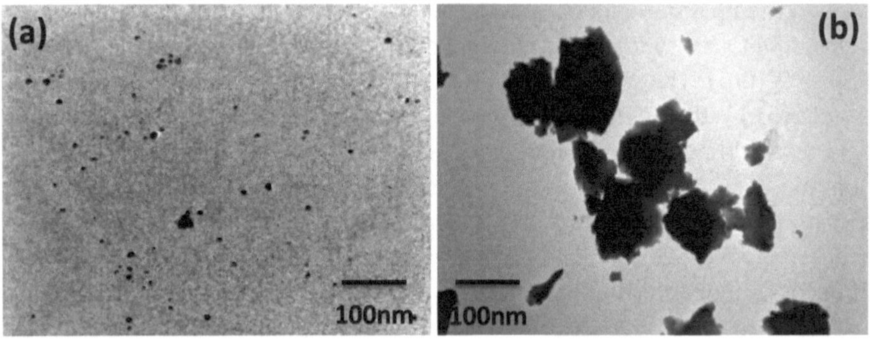

FIGURE 9.11 TEM image of systems (a) TGBAPB/3 wt.% POSS and (b) TGBAPB/5 wt.% POSS epoxy nanocomposites.

9.4 SUMMARY AND CONCLUSION

The scope of the present work is to synthesize TGBAPB epoxy matrix having high thermal stability, toughness, and mechanical properties by using precursor having skeletally-modified epoxy matrix to fabricate tetrafunctional epoxy nanocomposites by reinforcing amino-functionalized nanomaterials, namely, POSS, with different weight percentages and to find the suitability of the developed materials for possible use in high performance industrial applications. It was also observed that upon comparing the tensile strength, flexural strength, and impact properties of neat TGBAPB and POSS-reinforced systems, a beneficial increase in mechanical properties for the reinforced resin systems was observed. The addition of nanoreinforcement (POSS) produced a significant improvement on the thermal stability and char yield of the tetrafunctional epoxy resin system TGBAPB. The TGBAPB and POSS nanocomposites exhibited the least value of dielectric constant indicating the best insulation properties. Thus skeletally-modified tetraglycidyl epoxy TGBABP and its nanocomposites could be used for marine, maintenance, and tank lining applications. They could also be used as fluidized bed coating for electronic devices, corrosion protective coatings for metal surfaces, etc., providing high strength to weight ratio. They also find wide application in filament-wound epoxy composites used for high performance applications, namely, rocket motor casting, pressure vessels, and tanks. Their characteristic features like excellent adhesion, very good electrical insulation properties, mechanical properties, and chemical/heat resistance make them suitable for glass fiber-reinforced epoxy pipes used in the oil–gas, mining, chemical industries, concrete composites, and decorative flooring.

ACKNOWLEDGMENTS

Instrumentation facility provided under FIST-DST and DRS-UGC to Department of Chemistry, Anna University, Chennai is gratefully acknowledged.

KEYWORDS

- **TGBAPB epoxy resins**
- **hybrid polymers**
- **nanocomposites**
- **surface morphology**
- **nanoreinforcements**

BIBLIOGRAPHY

1. Nakamura, Y.; Yamaguchi, M.; Okubo, M.; Matsumoto. T. Effects of Particle Size on Mechanical and Impact Properties of Epoxy Resin Filled with Spherical Silica. *J. Appl. Polym. Sci.* **1992**, *45*, 1281–1289.
2. Shieh, J.Y.; Ho, T.H.; Wang, C.S. Aminosiloxane-Modified Epoxy Resins as Microelectronic Encapsulants. *Die Angew. Makromol. Chem.* **1995**, *224*, 21–32.
3. Weichang, L.; Russell, J.V.; George, P.S. Phosphorus-Containing Diamine for Flame Retardancy of High Functionality Epoxy Resins. Part II. The Thermal and Mechanical Properties of Mixed Amine Systems. *Polymer* **2006**, *47*, 2091–2098.
4. Ashokkumar, A.; Alagar, M.; Rao, R.M.V.G.K. Synthesis and Characterization of Siliconized Epoxy-1,3-bis(maleimido)benzene Intercross Linked Matrix Materials. *Polymer* **2002**, *43*, 693–702.
5. Anandakumar, S.; Denchev, Z.; Alagar, M. Synthesis and Thermal Characterization of Phosphorus Containing Siliconized Epoxy Resins. *Eur. Polym. J.* **2006**, *42*, 2419–2429.

6. Hsieh, K.H.; Han, J.L. Graft Interpenetrating Polymer Networks of Polyurethane and Epoxy. II. Toughening Mechanism. *J. Polym. Sci. Part. B: Polym. Phys.* **1990**, *28*, 783–794.
7. Mustata, F.; Bicu, I. Multifunctional Epoxy Resins: Synthesis and Characterization. *J. Appl. Polym. Sci.* 2000, *77*, 2430–2436.
8. Mustata, F.; Bicu, I.; Cascaval, C.N. Rheological and Thermal Behaviour of an Epoxy Resin Modified with Reactive Diluents. *J. Polym. Eng.* **1997**, *17*, 491–506.
9. Feher, F.J.; Wyndham, K.D.; Baldwin, R.K.; Soulivong, D.; Lichtenhan, J.D.; Ziller, J.W. Methods for Effecting Monofunctionalization of $(CH_2{=}CH)_8Si_8O_{12}$. *Chem. Commun.* **1999**, *14*, 1289–1290.
10. Feher, F.J.; Wyndham, K.D.; Soulivong, D.; Nguyen, F. Syntheses of Highly Functionalized Cube-Octameric Polyhedral Oligosilsesquioxanes ($R_8Si_8O_{12}$). *J. Chem. Soc. Dalton. Trans.* **1999**, *9*, 1491–1498.
11. Massam, J.; Pinnavaia, T.J. Clay Nanolayer Reinforcement of Glassy Epoxy Polymer. *MRS Proc.* **1998**, *520*, 223–232.
12. Brown, J.M.; Curliss, D.; Vaia, R.A. Thermoset-Layered Silicate Nanocomposites. Quaternary Ammonium Montmorillonite with Primary Diamine Cured Epoxies. *Chem. Mater.* **2000**, *12*, 3376–3384.
13. Kornmann, X.; Thomann, R.; Mülhaupt, R.; Finter, J.; Berglund, L.A. High Performance Epoxy-Layered Silicate Nanocomposites. *Polym. Eng. Sci.* **2002**, *42*, 1815–1826.
14. Tamaki, R.; Tanaka, Y.; Asuncion, M.Z.; Choi, J.; Laine, R.M. Octa(aminophenyl) silsesquioxane as a Nanoconstruction Site. *J. Am. Chem. Soc.* **2001**, *123*, 12416–12417.
15. Chou, C.H.; Hsu, S.L.; Dinakaran, K.; Chiu, M.Y.; Wei, K.H. Synthesis and Characterization of Luminescent Polyfluorenes Incorporating Side Chain Tethered Polyhedral Oligomeric Silsesquioxane Units. *Macromolecules* **2005**, *38*, 745–751.
16. Giannelis, E.P.; Krishnamoorti, R.; Manias, E. Polymer-Silicate Nanocomposites: Model Systems for Confined Polymers and Polymer Brushes. *Adv. Polym. Sci.* **1999**, *138*, 107–147.
17. Yong, N.; Sixun, Z.; Kangming, N. Morphology and Thermal Properties of Inorganic–Organic Hybrids Involving Epoxy Resin and Polyhedral Oligomeric Silsesquioxanes. *Polymer* **2004**, *45*, 5557–5568.
18. Duraibabu, D.; Ganeshbabu, T.; Manjumeena, R.; Anandakumar, S.; Dasan, P. Unique Coating Formulation for Corrosion and Microbial Prevention of Mild Steel. *Prog. Org. Coat.* **2014**, *77*, 657–664.
19. Huang, J.C.; He, C.B.; Xiao, Y.; Mya, K.Y.; Dai, J.; Siow, Y.P. Polyimide/POSS Nanocomposites: Interfacial Interaction, Thermal Properties and Mechanical Properties. *Polymer* **2002**, *44*, 4491–4499.
20. Nagendiran, S.; Hamerton, I.; Alagar, M. Octasilsesquioxane Reinforced DGEBA and TGDDM Epoxy Nanocomposites: Characterization of Thermal, Dielectric and Morphological Properties. *Acta. Mater.* **2010**, *58*, 3345–3356.
21. Laine, R.M. Nanobuilding Blocks Based on the $[OSiO_{1\text{-}5}]_x$ (x=6,8,10] Octasilsesquinoxanes. *J. Mater. Chem.* **2005**, *35*, 3725–3744.
22. Jones, I.K.; Zhou, Y.X.; Jeelani, S.; Mabry, J.M. Effect of Polyhedral-Oligomeric-Silsesquioxanes on Thermal and Mechanical Behavior of SC-15 Epoxy. *eXPRESS Polym. Lett.* **2008**, *2*, 494–501.
23. Khine, Y.M.; Chaobin, H.; Junchao, H.; Yang, X.; Jie, D.; Yeen, P.S. Preparation and Thermomechanical Properties of Epoxy Resins Modified by Octafunctional Cubic Silsesquioxane Epoxides. *J. Polym. Sci. Part. A: Pol. Chem.* **2004**, *42*, 3490–3503.

24. Duraibabu, D.; Ganeshbabu, T.; Saravanan, P.; Ananda Kumar, S. Development and Characterization of Novel Organic–Inorganic Hybrid Sol–Gel Films. *High. Perform. Polym.* DOI: 10.1177/0954008314528225. Published Online: April 8, 2014.
25. Zhang, Z.; Liang, G.; Wang, J.; Ren, P. Epoxy/POSS Organic–Inorganic Hybrids: Viscoelastic, Mechanical Properties and Micromorphologies. *Polym. Composite.* **2007**, *28*, 175–179.
26. Fu, H.K.; Huang, C.F.; Kuo, S.W.; Lin, H.C.; Yei, R.; Chang, F.C. Effect of an Organically Modified Nanoclay on Low-Surface-Energy Materials of Polybenzoxazine. *Macromol. Rapid. Comm.* **2008**, *29*, 1216–1220.
27. Chung, H.S.; Yi, P.C.; Chih, C.T.; Chin, L.C. Preparation, Characterization and Thermal Properties of Organic–Inorganic Composites Involving Epoxy and Polyhedraloligomeric Silsesquioxane (POSS). *J. Polym. Res.* **2010**, *17*, 673–681.
28. Lu, T.; Chen, T.; Liang, G. Synthesis, Thermal Properties, and Flame Retardance of the Epoxy Silsesquioxane Hybrid Resins. *Polym. Eng. Sci.* **2007**, *47*, 225–234.
29. Leu, C.M.; Chang, Y.T.; Wei, K.H. Synthesis and Dielectric Properties of Polyimide Tethered Polyhedral Oligomericsilsesquioxane (POSS) Nanocomposites via POSS Diamine. *Macromolecules* **2003**, *36*, 9122–9127.
30. Vengatesan, M.R.; Devaraju, S.; Ashok Kumar, A.; Alagar, M. Studies on Thermal and Dielectric Properties of Octa (maleimido phenyl) Silsesquioxane (OMPS) Polybenzoxazine (PBZ) Hybrid Nanocomposites. *High. Perform. Polym.* **2011**, *23*, 441–456.
31. Ananda Kumar, S.; Sasikumar, A. Studies on Novel Silicone/Phosphorus/Sulphur Containing Nano-Hybrid Epoxy Anticorrosive and Antifouling Coatings. *Prog. Org. Coat.* **2010**, *68*, 189–200.
32. Tianxi, L.; Wuiwui, C.T.; Yuejin, T.; Chaobin, H.; Sok, S.G.; Tai Shung, C. Morphology and Fracture Behavior of Intercalated Epoxy/Clay Nanocomposites. *J. Appl. Polym. Sci.* **2004**, *94*, 1236–1244.
33. Zilg, C.; Mulhaupt, R.; Finter. Morphology and Toughness/Stiffness Balance of Nanocomposites Based Upon Anhydride-Cured Epoxy Resins and Layered Silicates. *Macromol. Chem. Phys.* **1999**, *200*, 661–670.
34. Balakrishnan, S.; Start, P.R.; Raghavan, D.; Hudson, S.D. The Influence of Clay and Elastomer Concentration on the Morphology and Fracture Energy of Preformed Acrylic Rubber Dispersed Clay Filled Epoxy Nanocomposites. *Polymer* **2005**, *46*, 11255–11262.

CHAPTER 10

SYNERGISTICALLY C8 ETHER LINKED BISMALEIMIDE TOUGHENED AND ELECTRICALLY CONDUCTING CARBON BLACK EPOXY NANOCOMPOSITES

M. MANDHAKINI* and M. ALAGAR

Department of Chemical Engineering, Alagappa College of Technology, Anna University, Chennai 600025, India.
**E-mail: mandhakini7@gmail.com*

CONTENTS

ABSTRACT

The present chapter deals with the toughening of brittle epoxy matrix with C8 ether-linked bismaleimide (C8 e-BMI) and then study the reinforcing effect of carbon black (CB) in enhancing the conducting properties of insulating epoxy matrix. The Fourier transform infrared spectroscopy (FTIR) and Raman analysis indicate the formation of strong covalent bonds between CB and C8 e-BMI/epoxy matrix. The X-ray diffraction (XRD) and field emission scanning electron microscope (FESEM) analysis indicate the event of phase separation in 5 wt% CB loaded epoxy C8 e-BMI nanocomposites. The impact strength increased up to 5 wt% of CB loading with particle pull and crack deflection to be driving mechanism for enhancing the toughness of the nanocomposite and beyond 5 wt% the impact strength started to decrease due to aggregation of CB. The dynamic mechanical analysis (DMA) also indicates the toughness of the nanocomposites was improved with 5 wt% of CB loading due to the phase segregation between epoxy and C8 e-BMI in the presence of CB. The electrical conductivity was also increased with 5 wt% of CB due to classical conduction by Ohmic chain contact.

10.1 INTRODUCTION

Epoxy resin is the polymer matrix used most often with reinforcing fibers for advanced composite application. The epoxy resins have good stiffness, specific strength, dimensional stability, and chemical resistance, and show considerable adhesion to the embedded fiber. However, the main disadvantage of epoxy resin is its inherent brittle nature, which overshadows its other excellent properties. Over the years, many attempts have been made to modify epoxy by adding either rubber particles[1,2] or fillers to improve the matrix-dominated composite properties. In recent years, nanosized fillers, such as nanoparticles, carbon black, clay, and nanofibers have been considered as filler for epoxy to produce high performance composites with enhanced properties.[3] The inherent resistivity nature of the polymers is generally around 10^{-15} W m, and this enables them to serve the purpose of electrical insulators in the fields of electrical and electronics. The importance of electromagnetic interference (EMI) shielding relates to high demand on the reliability of electronics, and the rapid growth of radio frequency radiation sources. When compared with metallic conductor, conductive polymer composites show signs of superior properties in many aspects, such as low density, ease of shaping, and wide range of electrical conductivities as well as corrosion resistance.[4] Electrically conductive adhesives have attracted great attention because of their adhesive and electrical properties and numerous potential. A growing interest is observed in the potential of electrically conductive metal-loaded polymer adhesives for solder replacement in surface mount technology and other microelectronic application[5,6] in cases when soldering is difficult or impossible because of low thermal stability. When an adhesive is loaded with critical or higher concentration of electrically conductive filler, e.g., carbon black, graphite, or metallic powder, a transition from insulator to electrical conductor is observed.[7–10] Hence, an appropriate balance between the electrical conductivity and desirable mechanical behavior is one of the largest challenges for the application of filled polymer composites in various applications. With all these essential aspects in concern, the novelty of the work is encompassed in two steps: (a) to improve the toughness of the epoxy matrix with C8 e-BMI that having flexible ether groups and a long methylene chain within the matrix and (b) to exploit conductivity in the insulating epoxy matrix by reinforcing the C8 e-BMI-toughened matrix with conductive carbon black to achieve multifunctional properties in a cost-effective manner.

10.2 EXPERIMENTAL DETAILS

10.2.1 MATERIALS

The prepolymer used in this study Diglycidyl ether of bisphenol A, [DGEBA-LY556] was supplied by Ciba-Geigy Ltd., India with an epoxy equivalent of 180–190. The hardener 4,4 diaminophenylmethane (DDM) was supplied by Ciba-Geigy Ltd., India. The electroconductive CB [VULCAN XC 72], was provided by Cabot and dimethylformamide (DMF) was purchased from SRL, India. Maleic anhydride was purchased from SRL, India. C8 e-BMI was synthesized from maleic anhydride as per the literature reported.[11,12] The C8 e-BMI/DGEBA samples were prepared by blending C8 e-BMI in DGEBA at about 60°C through mechanical stirring with the ratio of C8 e-BMI/DGEBA ranging from 10:90 to 15:85 wt.%. The modified resin was mixed with a stoichiometric amount of DDM, stirred continuously until a homogeneous solution was obtained. The mixtures were cured in an aluminum mold after being degassed under vacuum for 10 min at room temperature. The liquid blends were cured at 120°C for 2 h and postcured at 180°C for 3 h to attain the complete curing reaction, and to allow the samples to cool gradually to room temperature.

10.2.3 MATERIAL PROCESSING OF CB/15 WT% C8 E-BMI-TOUGHENED EPOXY NANOCOMPOSITES

Initially, precalculated quantity of C8 e-BMI/CB in DMF was subjected to ultrasonication for 2 h to overcome the dispersion problem of CB in C8 e-BMI. This CB/C8 e-BMI mixture was added to calculated amount of epoxy and stirred using mechanical stirrer for 5 h at 60°C. In order to reduce the probability of voids, the liquid was preheated to 80°C to mitigate the viscosity and high vacuum system was used for about 30 min. After the bubbles were completely removed, precalculated amount of DDM was added and the mixture was transferred to plastic and Teflon-coated metal rectangular molds and cured for 2 h at 120°C and postcured at 180°C for 3 h. Finally, the cured material was trimmed and the test samples were machined for electrical and mechanical characterization.

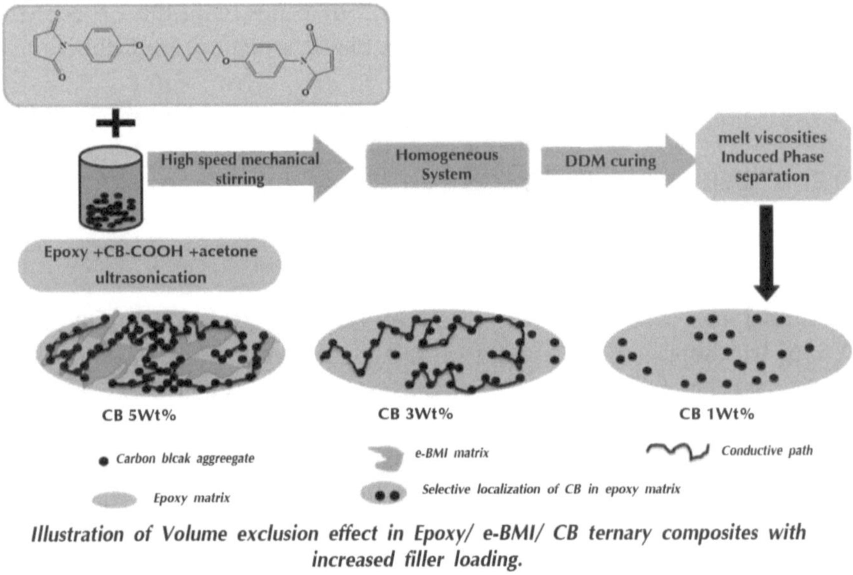

SCHEME 10.1 Schematic representation of C8 e-BMI/epoxy CB composite formation and phase separation as a function of CB concentration. (From Mandhakini, M., Mater. Design, 64, 706, 2014. With permission).

The agglomerates of CB are difficult to separate and infiltrate with the matrix so in order to overcome the problem of agglomeration and obtain better dispersion of nanosized fillers in a polymer matrix dispersion methods, such as ultrasonication and mechanical stirring is utilized in combination as they are simplest and most convenient methods. The schematic representation of C8 e-BMI/epoxy CB composite formation and phase separation as a function of CB concentration is given in Scheme 10.1.

10.3 RESULTS AND DISCUSSION

10.3.1 FTIR SPECTRAL STUDIES

Figure 10.1 shows the FTIR spectra of epoxy and C8 e-BMI-toughened epoxy matrices with varying wt% of CB content. Figure 10.1(a) corresponds to the neat epoxy and the FTIR spectrum of pure carbon black (Fig. 10.1b) shows bands at 1740 and 1600 cm^{-1} due to C=O and stretching vibrations of asymmetrically bonded C=C groups, respectively.

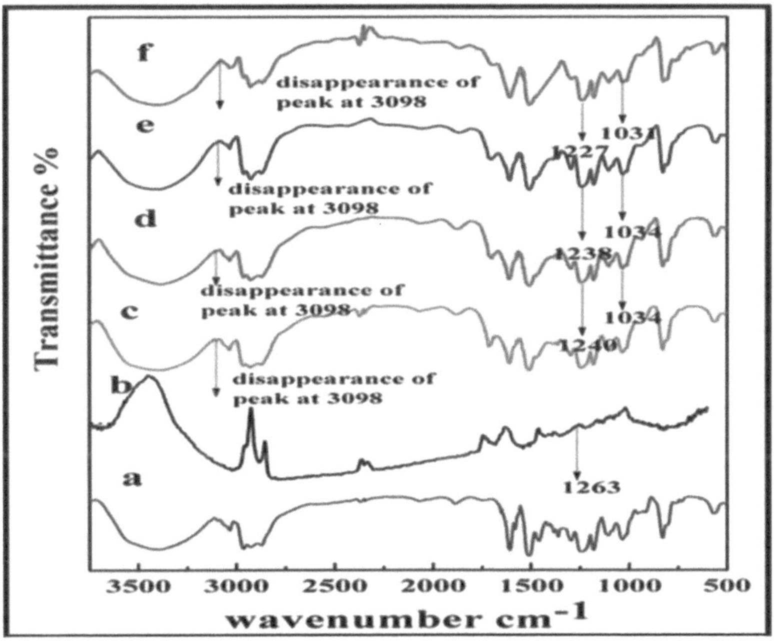

FIGURE 10.1 The FTIR spectra of C8 e-BMI/CB epoxy nanocomposites. (From Mandhakini, M., Mater. Design., 64,706, 2014. With permission).

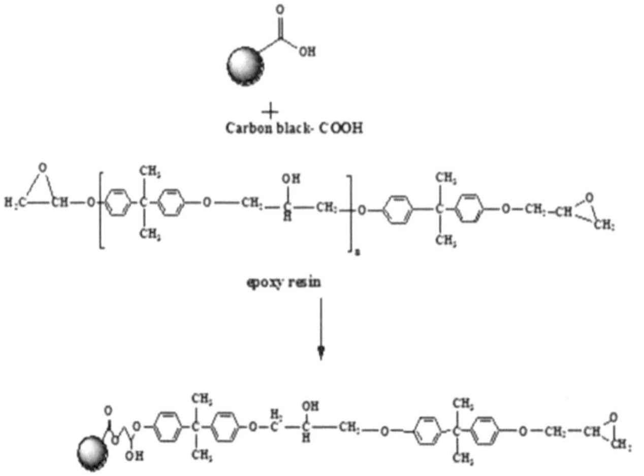

FIGURE 10.2 Chemical interaction between epoxy and CB in epoxy/C8 e-BMI/CB nanocomposites. (From Mandhakini, M., Mater. Design., 64,706, 2014. With permission).

A weak broad absorption peak between 1000 and 1400 cm^{-1} is attributed to the vibration of different functional groups or internal bonding like νC—O—C, νC—C. The appearance of strong bands between 3100 and 2800 cm^{-1} is attributed to the aliphatic and aromatic νC—H vibrations and this group of bands corresponds to the bonded hydrogen content in the samples. The vibration bands at 1460 and 1380 cm^{-1} are assigned to the deformation vibration of sp^3 C—CH_3 groups. The broad maximum band between 1000 and 1263 cm^{-1} is caused by different vibrations of functional groups like νC—O, νC—C, νH—C—O, νH—C—C.[13] Figure 10.1(c) shows the FTIR spectrum of C8 e-BMI-toughened epoxy matrix. The peak at 3099 cm^{-1} corresponding to H—C = vibration is absent for the C8 e-BMI-toughened epoxy systems and this confirms the reaction between epoxy and C8 e-BMI. The absence of oxirane peak at 915 cm^{-1} and the appearance of OH peak is due to the opening of the oxirane ring of the epoxy in the entire cured product and this confirms the occurrence of Michael addition reaction of DDM with epoxy and C8 e-BMI. When CB is incorporated to epoxy/C 8 e-BMI blend and cured, CB particles will tend to have greater affinity for epoxide, hydroxyl, and carboxyl functional groups, which are ornamented on the basal planes of CB and will have greater chemical interaction with oxirane ring of epoxy and hence, a strong bond is formed between epoxy/CB as shown in Figure 10.2.[14]

This is indicated from the shift observed in the symmetrical stretching of epoxy peak in CB from 1263 to 1234 ± 7 cm^{-1} and also the appearance of a new C—O stretching vibration corresponding to ester bond is observed at

1034 cm^{-1}.[15] The use of DDM as curative for DGEBA homopolymerization resulted in strong covalent bonds between CB and epoxy matrix through chain transfer reactions involving the secondary hydroxyl groups (Fig. 10.1d–f). The disappearance of the band at 917 cm^{-1} indicates that most of the epoxide groups of DGEBA were consumed in the reaction.[15]

10.3.2 RAMAN SPECTROSCOPY

Raman spectra were carried out to confirm the formation of covalent bonding between carbon black and C8 e-BMI/epoxy nanocomposites. The Raman spectrum of neat CB (Fig. 10.3a) displays two characteristic peaks: the first at 1286.5 cm^{-1} (D band) deriving from defect disordered zones of the graphitized walls; and the second at 1594 cm^{-1} deriving from in-plane vibration of graphite walls.[16] By contrasting the spectra of neat CB and CB reinforced C8 e-BMI toughed epoxy nanocomposites, it can be observed that the G band of neat carbon black is shifted from 1594 to 1591.6 cm^{-1} for 5 wt% incorporated C8 e-BMI/epoxy system (Fig. 10.3c), and this confirms the interaction between CB and epoxy matrix.

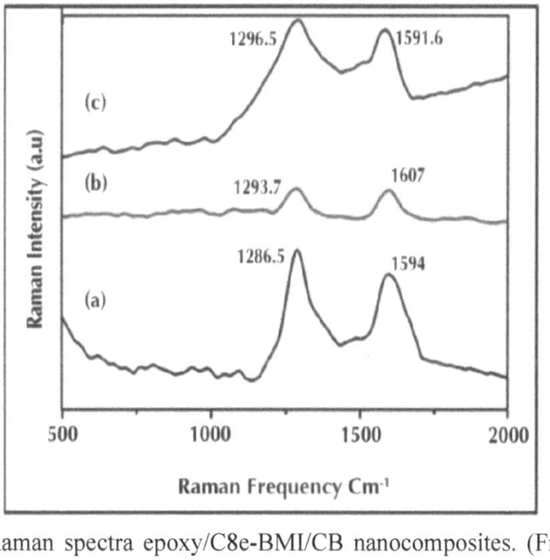

FIGURE 10.3 Raman spectra epoxy/C8e-BMI/CB nanocomposites. (From Mandhakini, M., Mater. Design., 64,706, 2014. With permission).

The R-value, which is the ratio of integrated intensity of D line to that of G line ($R = I_D/I_G$) is 1.32 for neat carbon black, and it shifts to 1.05 and

1.57 for 1 and 5 wt% CB epoxy nanocomposites, respectively. The increase in the disorder deformation is attributed to the increased interaction level between the epoxy matrix and CB due to[17] good dispersion levels of CB in the epoxy matrix.

10.3.3 XRD PATTERN

XRD was performed epoxy/C8 e-BMI composite samples with varying wt% of CB added. The XRD of neat CB shown in Figure 10.4(a) exhibit relatively broad and low intensity peak at 2 h = 24.7° due to the presence of large amounts of amorphous carbon in association with CB. The XRD of neat C8 e-BMI in Figure 10.4(c) shows strong and weak intensities in the region 2 h = 10°–40° that are assigned to the prominent crystal planes and the subordinate crystal planes. It is apparent that "charge transfer complexation" occurs between electron-rich diamine and the electron deficient anhydride regions in aromatic bismaleimide chains during the electronic polarization and due to this chain interactions between the stiff linear rigid-rod and segmented rigid-rod bismaleimide, the segments become aligned along their axes and this alignment contributes to the formation of short-range order and crystallinity in the neat bismaleimide resin system. The XRD pattern of DDM cured C8 e-BMI/epoxy matrix in Figure 10.4(d) shows a broad peak

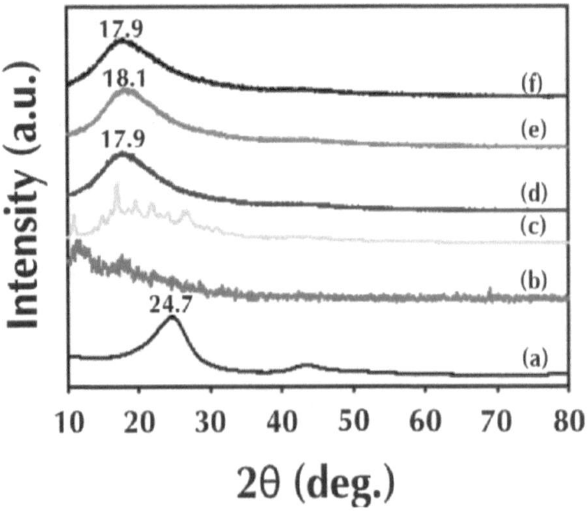

FIGURE 10.4 The XRD of epoxy/ C8 e -BMI/CB nanocomposites. (From Mandhakini, M., Mater. Design., 64,706, 2014. With permission).

due to the presence of noncoplanar conformation of the DDM unit and the bulky phenyl groups, which in turn decreases the intermolecular forces and packing between the polymer chains and hence, a decrease in the diffraction peaks was observed.[18] The XRD patterns of C8 e-BMI/CB/epoxy nanocomposites are shown in Figure 10.4(e) and (f) and the XRD pattern of 1 wt% CB/C8 e-BMI/epoxy system exhibits a peak at 17.9° and as the CB concentration is increased to 3 and 5 wt%, the intensity of the peaks are reduced due to the distortion of C8 e-BMI chains created by the physical dispersion of CB in the C8 e-BMI matrix, and this led to inferior orderliness of the final C8 e-BMI chains in the epoxy matrix.[19]

10.3.4 MICROSTRUCTURE ANALYSIS OF CB/C 8 e-BMI-TOUGHENED EPOXY NANOCOMPOSITES

The optical micrographs of epoxy/C8 e-BMI (85/15) matrices loaded with 1, 3, and 5 wt% of CB are displayed in Figure 10.5a–c, respectively. The microstructures of epoxy/C8 e-BMI (85/15) with 5 wt% CB shows phase separation and the CB particles are concentrated at the epoxy phase whereas in the case of 1 and 3 wt% the CB is distributed evenly without any phase separation. In order to analyze the microstructure of the prepared nanocomposites, FESEM was taken for 5 wt% CB nanocomposite at different magnifications and it is shown in Figure 10.6.

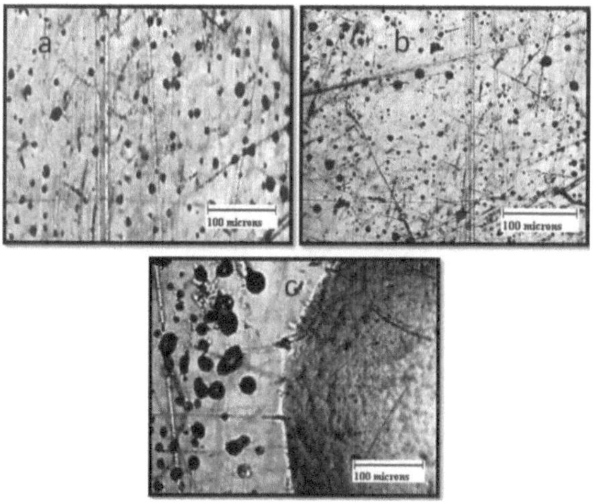

FIGURE 10.5 The optical micrographs of epoxy/ C8e -BMI/CB nanocomposites. (From Mandhakini, M., Mater. Design., 64,706, 2014. With permission).

FIGURE 10.6 FESEM of epoxy/ C8 e-BMI/5wt% CB nanocomposites at different magnifications. (From Mandhakini, M., Mater. Design., 64,706, 2014. With permission).

The FESEM reveals a heterogeneous morphology between the epoxy and C8 e-BMI in the presence of 5 wt% CB and this was in favor of the images obtained by optical microscope. The preferential localization of CB in epoxy is attributed to the good affinity of CB with epoxy phase when compared with that of C8 e-BMI phase.[20–22] In the case of 5 wt% CB-reinforced nanocomposite, the preferential localization of CB in epoxy phase is consistent with the results of optical micrographs and DMA results reported in Section 10.3.7. The melt viscosity of C8 e-BMI is higher than that of epoxy and hence, CB particles are localized at the C8 e-BMI/epoxy interfaces, and also in the lower viscous continuous epoxy-rich matrix thereby decreasing the volume of the epoxy matrix available for the CB particles, and hence, more CB interconnections are observed within the continuous epoxy-rich channels with respect to Figure 10.6(c). Thus, the aforementioned behavior suggests that a combination of the distinctly different melt viscosities synergistically achieved higher conductivity than that of the constituent polymers.[23]

10.3.5 IMPACT PROPERTIES

The impact strength of C8 e-BMI-toughened epoxy CB nanocomposites is shown in Figure 10.7. The impact strength of 15 wt% C8 e-BMI-toughened epoxy is 84 kJ/m^2 and it is higher than that of the neat epoxy matrix whose impact strength is 70 kJ/m^2. This is due to the flexible ether linkage and long aliphatic chain of C8 e-BMI. The impact strength of 15 wt% C8 e-BMI/5 wt% CB epoxy nanocomposite increases from 84 to 92 kJ/m^2 due to the presence of CB particles and the second phase toughening by the dispersed C8 e-BMI particles.

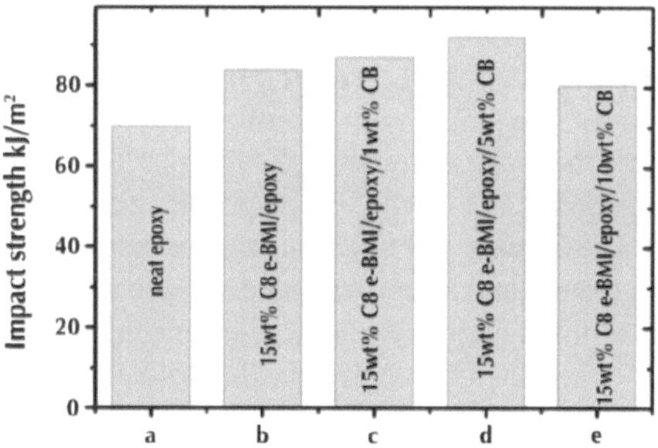

FIGURE 10.7 Impact strength of epoxy/ C8 e-BMI/CB nanocomposites. (From Mandhakini, M., Mater. Design., 64,706, 2014. With permission).

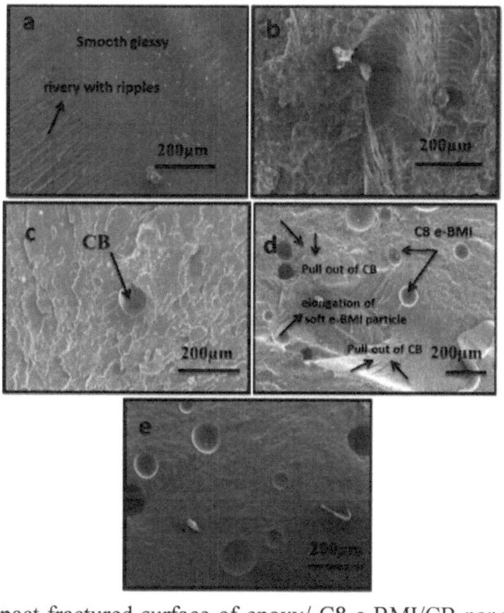

FIGURE 10.8 Impact fractured surface of epoxy/ C8 e-BMI/CB nanocomposites. (From Mandhakini, M., Mater. Design., 64,706, 2014. With permission).

When the sample is subjected to impact energy, the stress is transferred from one C8 e-BMI particle to another through the matrix ligament with shear yielding taking place throughout the deformation zone and contributing toughness to the nanocomposite. The impact strength attains a maximum value for 5 wt% CB content beyond, which a falling tendency was noted. The decreasing tendency of impact strength is attributed to the aggregation of CB and inefficient load transfer between CB/epoxy matrix.

10.3.6 MORPHOLOGY AND TOUGHENING BEHAVIOR

In order to analyze the mode of fracture, the morphology of the impact fractured surfaces of epoxy/C8 e-BMI/CB composites were recorded as shown in Figure 10.8. The impact fractured surface of neat epoxy (Fig. 10.8a) shows a smooth, glassy, and rivery fractured surface with ripples. The relative smoothness of the fractured surface indicates that no significant plastic deformation has occurred. The ripples are due to the brittle fracture of the epoxy network, which accounts for its poor impact strength, as there is no energy dissipation mechanism operating here. The fractured surface of 15 wt% C8 e-BMI/epoxy matrix showed in Figure 10.8(b) witness no vacant spaces left by removal of C8 e-BMI domains from the fractured surface confirming the existence of good interfacial adhesion and great interaction between the dispersed C8 e-BMI domains and epoxy matrix, for better stress transfer from the matrix to the C8 e-BMI domains. In the case of 1 wt% CB epoxy nanocomposite (Fig. 10.8c), the CB particles are mainly dispersed in the epoxy matrix and a small quantity of CB is found in the droplet. The fractured surface of 5 wt% CB loaded epoxy nanocomposite in Figure 10.8(d) clearly shows two distinct phases namely a continuous epoxy matrix and the dispersed C8 e-BMI phase. The heterogeneous morphology is developed by the CB particles, which hinder the complete dissolution of C8 e-BMI in the epoxy matrix, and this morphological development can be correlated with damping behavior obtained from DMA. When the crack front reaches the rigid CB particle, the crack growth is hindered by the rigid particle and the crack is forced to change its direction to a long irregular path paving way for secondary cracks in the course of crack propagation as a result of which more impact energy was consumed by the crack to dodge between the CB particles and hence, resulted in a much coarser and rougher fractured surface. The examined fractured surface also reveals circular cavities created by the "pullout" of the CB particles upon impact. Thus, both toughening and flexibilization effect is operative for improvement in impact strength of

CB filled epoxy/C8 e-BMI composites. At 10 wt% of CB loading, the effectiveness of C8 e-BMI particles to act as stress concentrator and dissipate the energy is mitigated resulting in decreased impact strength, created by massive phase separation taking place between epoxy/C8 e-BMI resulting in increase of C8 e-BMI domain size.[24,25]

10.4 DYNAMIC MECHANICAL ANALYSIS

10.4.1 STORAGE MODULUS

Figure 10.9 shows the temperature dependence of storage modulus of the neat epoxy and C8 e-BMI-toughened epoxy matrices with varying wt% of CB contents. From DMA it is evident that storage modulus follows the order; neat epoxy (18 GPa) > 15 wt% C8 e-BMI/epoxy/1 wt% CB (15 GPa) > 15 wt% C8 e-BMI/epoxy (14 GPa) > 15 wt% C8 e-BMI/epoxy 5 wt% CB (13 GPa). The long aliphatic chain and flexible ether linkages of C8 e-BMI impart flexibility to the stiff epoxy matrix and hence, the storage modulus of C8 e-BMI-toughened epoxy is lower than that of the neat epoxy matrix. The increase in the modulus with CB is due to good interfacial adhesion between CB/epoxy matrix and effective stress transfer from matrix to CB under loading. With the loading of 5 wt% CB, phase segregation occurs between epoxy and C8 e-BMI and improves the toughness of the composite with detrimental effect on storage modulus.

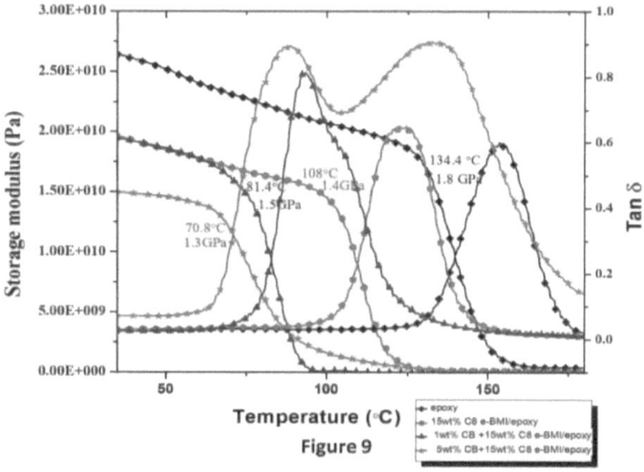

FIGURE 10.9 Dynamic mechanical property variation of epoxy/ C8 e-BMI/CB nanocomposites with temperature. (From Mandhakini, M., Mater. Design., 64,706, 2014. With permission).

10.4.2 DAMPING PARAMETERS

The variation of tan δ of neat epoxy and C8 e-BMI-toughened epoxy matrices with CB contents as a function of temperature is shown in Figure 10.9. From the figure, it is evident that the neat epoxy has single narrow well-defined relaxation tan δ peak at 154.9°C whereas the 15 wt% C8 e-BMI-toughened matrices shows a low T_g value of 120°C. The tan δ peak has become broader with CB wt% loading and 1 wt% CB reinforced 15 wt% C8 e-BMI-toughened epoxy matrices exhibits a relaxation peaks at 93.6°C. The 5 wt% CB composite is evident with two clear relaxation peaks at 88.3 and 132.3°C. As the filler content is increased to 5 wt%, a massive phase separation and a high immobilized short range layer of few nm thickness occurs near the surface of the filler [25–28] causing alteration in the polymer chain kinetics and conformational entropy, which has shifted the T_g to 132.3°C and on the other hand, the excess of CB particles, which are dispersed in the C8 e-BMI phase without chemical interaction has decreased the T_g to 88.3°C.

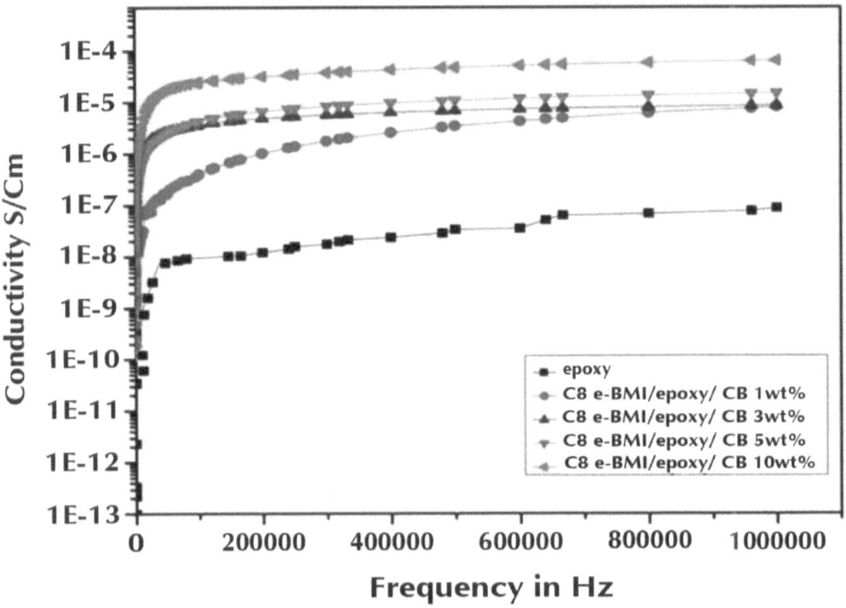

FIGURE 10.10 Graphical representation of conductivity as a function of the CB loading. (From Mandhakini, M., Mater. Design., 64,706, 2014. With permission).

10.4.3 ELECTRICAL CONDUCTIVITY

Figure 10.10 shows the graphical representation of conductivity as a function of the CB loading in epoxy/C8 e-BMI composites. At low loadings of CB (1 wt%), the gap between CB aggregates, through which the electrons are transmitted, is very large and hence, the resistivity of the composites was approximately same as that of the neat epoxy matrix. At 3 wt% CB loading, the gap between CB aggregates is close but they are not in actual contact with each other, as a result, the electron has to tunnel over a potential barrier to get out of the CB aggregate and cross the gap.[29,30] When the loading is increased to 5 wt%, the gap between the aggregates is further reduced, and thus, the classical conduction through Ohmic contact chain takes the leading mechanism for electrical conduction instead of tunneling.

10.5 CONCLUSIONS

This paper offers the insight into the toughening of epoxy matrix with long methylene chain C8 e-BMI and exploit conductivity by reinforcing CB at different weight percentage in order to achieve synergetic effect on both the mechanical properties and electrical conductivity of epoxy composites. Based on the thermo-mechanical, toughness, and conductivity results, the following conclusions can be drawn:

The combination of distinctly different melt-flow properties of the polymer synergistically achieved higher conductivity than with lower weight percentage of CB. The overall storage modulus or stiffness of the C8 e-BMI/epoxy nanocomposites with 5 wt% CB decreases and proves to be a very well toughened system. Toughening and flexibilization effect is operative for improvement in impact strength of CB filled epoxy/C8 e-BMI composites. The interfacial adhesion between the CB particles and the C8 e-BMI-toughened epoxy resin has enhanced the conductivity of the nanocomposite. Thus, the results of the experimental analysis clearly suggest that the combination of C8 e-BMI/epoxy with 5 wt% CB exhibits synergism between electrical and impact strength and hence, it is expected to perform very well as thermal pastes of high performance and solder as thermal interface material due to its conformability and spreadability.

KEYWORDS

- **nanocomposites**
- **conducting epoxy**
- **epoxy matrices**
- **Raman spectroscopy**
- **nanosized fillers**

BIBLIOGRAPHY

1. Xian, G.; Walter, R.; Haupert, F. Friction and Wear of Epoxy/TiO_2 Composites: Influence of Additional Short Carbon fibers, Aramid and PTFE Particles. *Compos. Sci. Technol.* **2006**, *66,* 3199–3209.
2. Dadfar, M.R.; Ghadami, F. Effect of Rubber Modification on Fracture Toughness Properties of Glass Reinforced Hot Cured Epoxy Composites. *Mater. Des.* **2013**, *47*, 16–20.
3. Vasconcelos, P.V.; Lino, F.J.; Magalhaes, A.; Neto, R.J.L. Impact Fracture Study of Epoxy Based Composites with Aluminium Particles and Milled fibres. *J. Mater. Process. Technol.* **2005**, *170*, 277–283.
4. Norman, R.H. Conductive Rubbers and Plastics: Their Production Application and Test Methods. Amsterdam, New York: Elsevier Pub. Co.; 1970.
5. Li-Ngee Hoab, Teng Fei Wuc Hiroshi Nishikawa,. Properties of Phenolic-Based Ag-filled Conductive Adhesive Affected by Different Coupling Agents. *J. Adhesion.* **2013**, *89*, 847–858.
6. Lu, D.; Wong, C.P. High Performance Conductive Adhesives. *IEEE Trans. Electron. Packag. Manufact.* **1999**, *22*, 324–330.
7. Lee, H.H.; Chou, K.S.; Shih, Z.W. Effect of Nano-Sized Particles on the Resistivity of Polymeric Conductive Adhesives. *Int. J. Adhes.* **2005**, *25*, 437–441.
8. Schueler, R.; Petermann, J.; Schulte, K.; Wentzel, H.P. Agglomeration and Electrical Percolation Behavior of Carbon Black Dispersed in Epoxy Resin. *J. Appl. Polym. Sci.* **1997**, *63*, 1741–1746.
9. Maminya, Y.P.; Davydenko, V.V.; Pissis, P.E.; Lebedev, V. Electrical and Thermal Conductivity of Polymers filled with Metal Powders. *Eur. Polym. J.* **2002**, *38,* 1887–1897.
10. Novak, I.; Krupa, I.; Chodak, I. Relation Between Electrical and Mechanical Properties in Polyurethane/Carbon Black Adhesives. *J. Mater. Sci. Lett.* **2002**, *21*, 1039–1041.
11. Li, P.H.; Wang, C.Y.; Li, G.; Jiang, J.M. Highly Organosoluble and Transparent Polyamides Containing Cyclohexane and Trifluoromethyl Moieties: Synthesis and Characterization. *Express. Polym. Lett.* **2009**, *3*, 703–712.
12. Ashok Kumar, A.; Alagar, M.; Rao, R.M.V.G.K. Synthesis Characterization of Siliconized Epoxy-1,3-Bis(Maleimido)Benzene Intercrosslinked Matrix Materials. *Polym. J.* **2002**, *43*, 693–702.

13. Jager, C.; Henning, T.; Schlogl, R.; Spillecke, O. Spectral Properties of Carbon Black. *J. Non. Cryst. Solids.* **1999**, *258*, 161–179.
14. Gao, W.; Alemany, L.B.; Ci, L.J.; Ajayan, P.M. New Insights into the Structure and Reduction of Graphite Oxide. *Nat. Chem.* **2009**, *1*, 403–408.
15. Bellamy, L.J. The Infra-Red Spectra of Complex Molecules. 3rd ed. London: Chapman and Hall; 1975.
16. Eklund, P.C.; Holden, J.M.; Jishi, R.A. Vibrational Modes of Carbon Nanotubes; Spectroscopy and Theory. *Carbon.* **1995**, *33*, 959–972.
17. Tishkova, V.; Raynal, P.I.; Puech, P.; Lonjon, A.; Marion, L.F.; Philippe, D. et al. Electrical Conductivity and Raman Imaging of Double Wall Carbon Nanotubes in a Polymer Matrix. *Compos. Sci. Technol.* **2011**, *71*, 1326–1330.
18. Satheesh Chandran, M.; Krishna, M.; Salini, K.; Rai, K.S. Preparation and Characterization of Chain-extended bismaleimide/carbon fibre composites. *Int. J. Polym. Sci.* 2010, Article ID 987357.
19. Xue P.; Bao Y.; Li Q.; Wu, C. Impact of Modification of Carbon Black on Morphology and Performance of Polyimide/Carbon Black Hybrid Composites. *Phys. Chem. Chem. Phys.* **2010**, *12*, 1342–11350.
20. Li Y.; Wang S.; Zhang Y.; Zhang Y. Carbon Black-filled Immiscible Polypropylene/ Epoxy Blends. *J. Appl. Polym. Sci.* **2006**, *99*, 461–471.
21. Zhongbin Xu.; Zhang Y.; Hong W.; Zheng S. Electric Conductivity and Thermorheology Properties of Polyacrylonitrile/Nylon6 Composites filled with Carbon Black. *Polym-Plast. Technol.* **2009**, *48*, 280–284.
22. Maxwell, J.C. *A Treatise on Electricity and Magnetism.* 3rd ed. Clarendon: Oxford; 1891; Wagner KW. Arch Elektrotech 2:371 [Berlin]; Sillars, R. *J. Inst. Electron. Eng.* **1914**; *1*(80), 378.
23. Gubbels, F.; Jerome, R.; Teyssie, P.H.; Vanlathem, E.; Deltour, R.; Calderone, A, et al. Selective Localization of Carbon Black in Immiscible Polymer Blends: A Useful Tool to Design Electrical Conductive Composites. *Macromolecules.* **1994**, *27*, 1972–1974.
24. Varaporn, T.; Sungthong, N.; Raksa, P. Rubber Toughening of Nylon 6 with Epoxidized Natural Rubber. *Polym. Test.* **2008**, *27*, 794–800.
25. Hergeth, W.D.; Steinau, U.J.; Bittrich, H.J.; Simon, G.; Schmutzler, K. Polymerization in the Presence of Seeds. Part IV: Emulsion Polymers Containing Inorganic filler Particles. *Polymer.* **1989**, *30*, 254–258.
26. Kotsilkova, R.; Fragiadakis, D.; Pissis, P. Reinforcement Effect of Carbon Nanofillers in an Epoxy Resin System: Rheology, Molecular Dynamics, and Mechanical Studies. *J. Polym. Sci. Part B: Polym. Phys.* **2005**, *43*, 522–533.
27. Sun, Y.; Zhang, Z.; Moon, K.S.; Wong, C.P. Glass Transition and Relaxation Behavior of Epoxy Composites. *J. Polym. Sci. Part B: Polym. Phys.* **2004**, *42*, 3849–3858.
28. Hemmati, M.; Rahimi, G.H.; Kaganj, A.B.; Sepehri, S.; Rashidi, A.M. Rheological and Mechanical Characterization of Multi-Walled Carbon Nanotubes/Polypropylene Composites. *J. Macromol. Sci. Phys.* **2008**; *47*, 1176–1187.
29. Medalia, A.I. Electrical Conduction in Carbon Black Composites. *Rubber Chem. Technol.* **1986**, *59*, 432–454.
30. Kimura T, Yoshimura N, Ogiso T, Maruyama K, Ikeda M. Effect of Elongation on Electric Resistance of Carbon-Polymer Systems. *Polymer.* **1999**, 40, :41–49.

CHAPTER 11

DEVELOPMENT AND CHARACTERIZATION OF POSS REINFORCED SILICONIZED EPOXY NANOCOMPOSITES

V. MADHUMITHA, D. DURAIBABU, and S. ANANDA KUMAR*

Department of Chemistry, Anna University, Chennai 600025, India.
**E-mail: sri_anand_72@yahoo.com*

CONTENTS

ABSTRACT

The research work describes the synthesis of siliconized epoxy resin using Di-phenyl silanediol and DGEBA. Its structural characterization was made using Fourier transform infrared spectroscopy. Gel permeation chromatography (GPC) ascertained its molecular weight. Siliconized epoxy nanocomposites were fabricated using siliconized epoxy resin with 2.5% and 5 wt. % of amine functionalized polyhedral oligomeric silsesquioxane (POSS) using di-amino diphenyl methane (DDM) as the curing agent. Thermal stability of the neat and nanocomposites was evaluated by dynamic mechanical analysis (DMA) and thermo gravimetric analysis (TGA). Scanning electron microscopy (SEM), transmission electron microscopy (TEM) and X-ray diffraction (XRD) techniques were used to investigate the morphology of the above said systems. Dielectric constant was also determined. The results obtained from the study are discussed in detail. The results reveal that the composites thus, synthesized in our study could be used for high-performance applications like aerospace sector.

11.1 INTRODUCTION

The pliancy of basic properties of epoxy tailored by appropriate choice of monomers, hardeners, and chemical reactions, impart versatility to these resins.[1] This is one of the prime reasons for the cornucopia of applications, ranging from paints and adhesives to aerospace, electrical, and industrial composites.[2] Its characteristic features include excellent adhesion, mechanical properties, and chemical/heat resistance. In addition, it also has very good electrical insulation properties.[3] Concomitantly a few drawbacks associated with these resins are poor oxidative stability, moisture sensitivity, and limited thermal stability.[4] The high-performance resins such as tetraglycidyl epoxies are more expensive as well. Thermal stability, weathering resistance, low surface tension, maintaining elasticity at low temperatures, hydrophobicity, and lubricating properties make silicone-containing resins widely accepted in areas, such as high temperature resistant coatings, weather resistant exterior coatings, and impregnation of concrete. Siliconized epoxy hybrids are used, specifically in cases requiring better corrosion protection and high thermal and chemical resistance.[4] The presence of organic radicals such as phenyls attached with the silicon leads to an increased solubility, flexibility, and toughness.[1] Reinforcing polymers with inorganic nanoclusters having organic tethers leads to enhancement in properties due to the better interaction between the matrix and the reinforcement. Polyoligomeric silsesquioxanes (POSS) are cubic-caged structural hybrids of inorganic–organic composition that helps to control the nanoarchitecture when they are brought inside a polymer network. They are derived by hydrolysis and condensation of trifunctional organosilanes and can be incorporated into polymers by either copolymerization or physical blending.[5,6] Several studies on reinforcement of POSS with epoxy resins have shown amelioration in thermomechanical, dielectric, and flameretardancy properties.[7–10] Thus, the scope of this research is to synthesize POSS reinforced siliconized epoxy and studies its thermal and mechanical properties. The composites thus, synthesized could be used for high-performance applications like aerospace sector.

11.2 POLYHEDRAL OLIGOMERIC SILSESQUIOXANE (POSS)-BASED POLYMER

Silsesquioxanes are nanostructured materials having the empirical formula $RSiO_{1.5}$, where *R* is a hydrogen atom or an organic functional group, such as an alkyl, alkene, acrylate, hydroxyl, or epoxide unit. Scott et al. [1946][11] discovered the first oligomeric organosilsesquioxane, $(CH_3SiO_{1.5})_n$, along with other volatile compounds through the thermolysis of polymeric products prepared from cohydrolysis of methyl trichlorosilane and dimethyl chlorosilane. Although silsesquioxane chemistry has been studied for more than half a century, interest in this field has increased dramatically in recent years. Baney et al. [1995][12] reviewed the preparation, properties, structures, and applications of silsequioxanes, especially those of ladder-like polysilsesquioxanes.

More recently, attention has been concentrated on silsesquioxane possessing the specific cage structures Figure 11.1. These polyhedral oligomericsilsesquioxanes are commonly referred to as by the acronym "POSS". Derivatives of POSS are true hybrid inorganic/organic chemical composites that possess an inner inorganic silicon and oxygen core $(SiO_{1.5})_n$ and external organic substituents that can feature a range of polar or nanpolar functional groups. POSS nanostructures having diameters ranging from 1 to 3 nm can be considered as the smallest possible particles of silica, that is, molecular silica. Unlike most silicones or fillers, POSS molecules contain organic substituents on their outer surfaces, making them compatible or miscible with most polymers (Laine 2005).[13]

POSS may also be used as a nanoprecursor in polymeric systems. This is accomplished through the introduction of polymerizable moieties at the corner silicon atoms in the POSS molecules that may be reacted to form highly cross-linked polymers. Functional groups that have been employed in the preparation of cross-linked hybrid POSS materials are epoxy, amine, methacryloyl, vinyl, alkyl halide, hydriodo, alcohols, isocynates, and maleimide (Laine 2005, Jothibasu et al. 2008, Nagendiran et al. 2010).[13–15]

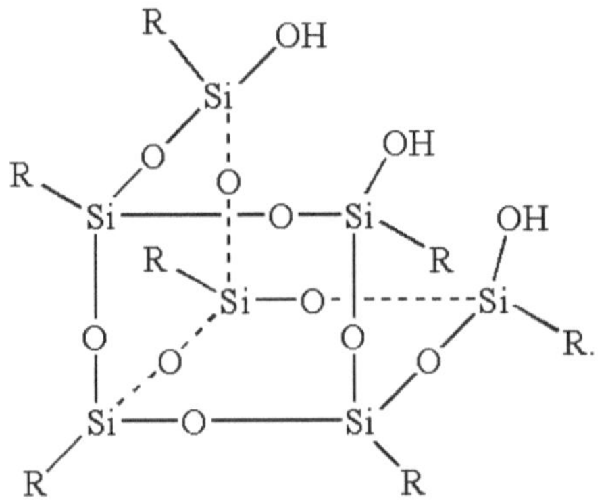

FIGURE 11.1 Structure of POSS.

11.2.1 ADVANTAGES OF POSS INCORPORATION IN POLYMERIC MATERIALS

- Use of POSS segments in polymeric materials result in enhancement of the physical properties of the compositions.
- Compared to common fire retardant polymers, POSS-containing polymers show delayed combustion and major reductions in heat evolution.
- POSS possess the ability to control the chain motion results in usage temperature enhancement of nearly all types of thermoplastic and thermosetting polymers. In many cases, the glass transition can be increased even up to the decomposition temperature of the polymer.
- Use of POSS additives often eliminates the need to use common (dense) fillers such as silica.
- Depending on loading level, bulk density reductions of up to 10% have been observed with viscosity reductions of up to 24% relative to silica.
- Incorporation of POSS increases the modulus and hardness, while maintaining the stress and strain characteristics of the base resin.
- Organo-functionalized POSS technology can be tailored to meet resin and consumer compatibility needs.

11.3 EXPERIMENTAL

11.3.1 SYNTHESIS OF OCTAPHENYLSILSESQUIOXANE (OPS)

The OPS was synthesized in a manner reported by Huang et al.[16] Total, 12.4 mL (16.35 g, 0.077 mol.) of phenyl trichlorosilane was dissolved in 75 mL of benzene and 150 mL of water, the mixture was stirred for 5 h at 25°C. The acid layer formed was removed and the benzene layer was washed with water until pH reached neutral. It was then dried with anhydrous magnesium sulfate. Total 1.8 mL (4.5 mmol) subsequently 4% benzyl trimethyl ammonium hydroxide/methanol solution was added to this organic layer and the whole mixture was refluxed for 4 h, and then allowed to stand for 4 days. The mixture was refluxed again for 24 h, cooled, and then filtered, which resulted in a white microcrystalline powder (Scheme 11.1). The product thus, obtained was extracted, using dry benzene in a Soxhlet extractor to remove the soluble resin. It was further dried overnight in vacuum at 70°C. The yield was 8.9 g (92.2%).[17]

11.3.2 SYNTHESIS OF OCTA (NITROPHENYL) SILSESQUIOXANE (ONPS)

The ONPS was synthesized by following Laine's method.[18] Total 90 mL of fuming nitric acid was added in small portions to 10 g (9.7 mmol) of OPS and placed under stirring in an ice bath (0°C) for half an hour. It was then allowed to stir at ambient temperature for 20 h (Scheme 11.1). After filtration through glass wool, the solution was then poured over 100 g of ice to precipitate the product. It was then filtered to obtain a light yellow color precipitate, which was washed with water until the pH reached 6 and was finally washed with ethanol. The residual solvent was removed by drying the powder under vacuum at room temperature. The yield was 11.5 g (94.8%) of ONPS.[6]

11.3.3 SYNTHESIS OF OCTA (AMINOPHENYL) SILSESQUIOXANE (OAPS)

Total 10.0 g (7.16 mmol) of ONPS and 5 wt % Pd/C (1.22 g) (0.574 mmol) was taken in a 500 mL RB flask equipped with a condenser under nitrogen, distilled THF (200 mL) was poured into it, and stirred at room temperature.

Through an addition funnel, 16 mL of formic acid (80%) was added in drops. Then the mixture was heated to 60°C (Scheme 11.1). Bilayer formation was observed after an hour. It contained a top layer and a bottom black suspension. After cooling, the mixture to room temperature 30 mL of THF was added to the flask to dissolve the remaining slurry. The suspension was filtered again. The mixture was then subjected to ethyl acetate and water wash consecutively. Total 15 g of anhydrous magnesium sulfate was added to the organic layer to dry it. The product was precipitated by adding 2 L of hexane. It was filtered and dissolved in 30:50 THF/ethyl acetate and reprecipitated into 1 L hexane. The powder was dried under vacuum to yield 6 g of OAPS (70.5%).[18]

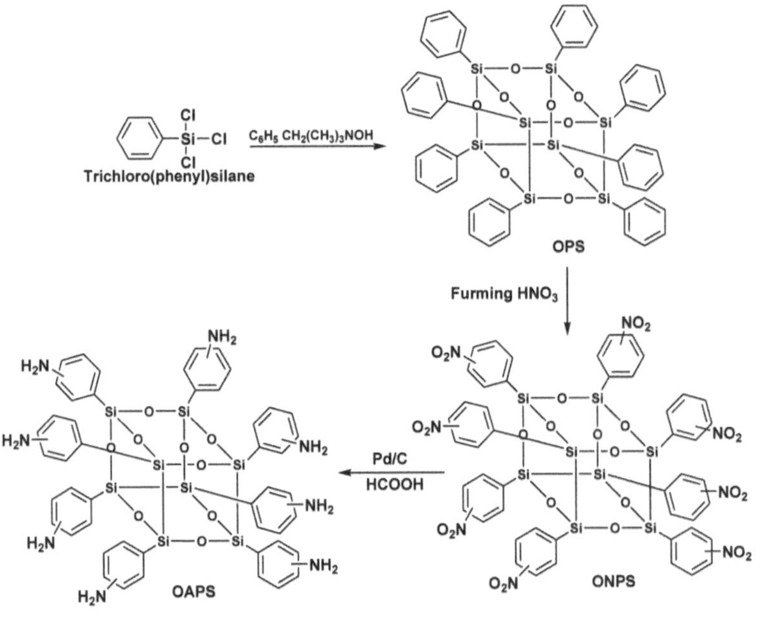

SCHEME 11.1 Synthesis of octaaminophenylsilsesquioxane.

11.3.4 PREPARATION OF SILICONIZED EPOXY RESIN MATRIX

High-molecular weight siliconized epoxy was obtained by reacting 7.9 g of diphenylsilanediol with 19 g of DGEBA using 0.012 g of tin (II) chloride as the catalyst. The mixture was kept under continuous stirring for 2 h at 140°C are shown in Scheme 11.2:[19]

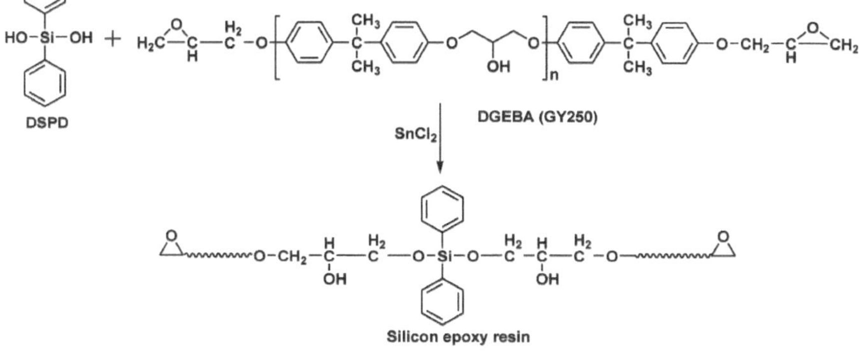

SCHEME 11.2 Synthesis of siliconized epoxy resin.

11.3.5 SAMPLE PREPARATION OF SILICONIZED EPOXY/POSS NANOCOMPOSITES

Total, 0.125 g (2.5%) and 0.25 g (5%) of OAPS was dissolved and dried in THF separately. It was then mixed with 5 g of siliconized epoxy resin and THF, heated until the solvent evaporated and degassed under vacuum at 60°C to remove the residual solvent. Subsequently, the stoichiometric amount of curing agent (DDM) was added and poured into rectangular steel mold and cured for 2 h at 120°C and postcured at 180°C for 3 h (Fig. 11.2 and Scheme 11.3).

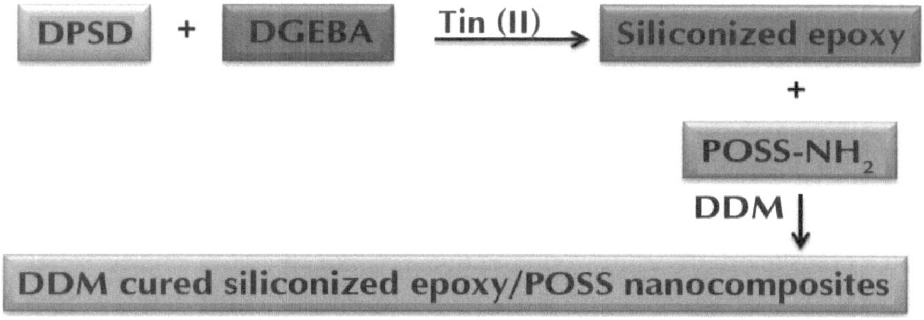

FIGURE 11.2 Flowchart representation for the preparation of siliconized epoxy/POSS-reinforced siliconized epoxy nanocomposites.

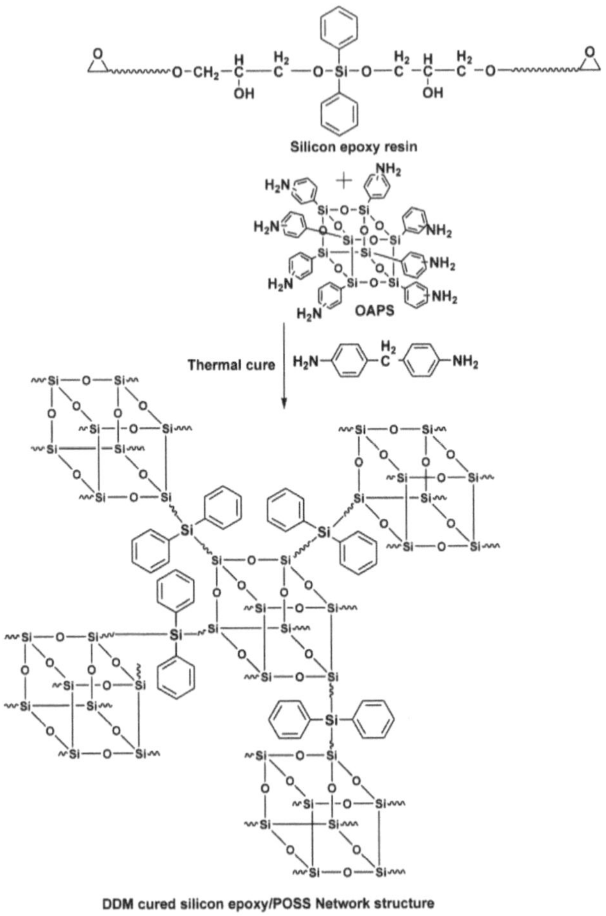

SCHEME 11.3 DDM cured epoxy/POSS network structure.

11.4 RESULTS AND DISCUSSIONS

11.5 FTIR SPECTROSCOPY

The FTIR spectra of OPS, ONPS, and OAPS are shown in Figure 11.3(a), (b), and (c), respectively. The peaks obtained at 3065, 1597, and 1115 cm^{-1} correspond to the Si—O—Si linkages of OPS. The FTIR spectra of ONPS and OAPS further indicated a complete conversion of nitro groups to amino groups. The bands those appeared at 1350 and1527 cm^{-1} in the IR spectrum of ONPS correspond to the nitro group introduced into the OPS molecule by nitration (N=O). Furthermore, the disappearance of these peaks and the appearance of new peak at 3360 cm^{-1} correspond to the amine (N—H) functional group.[20]

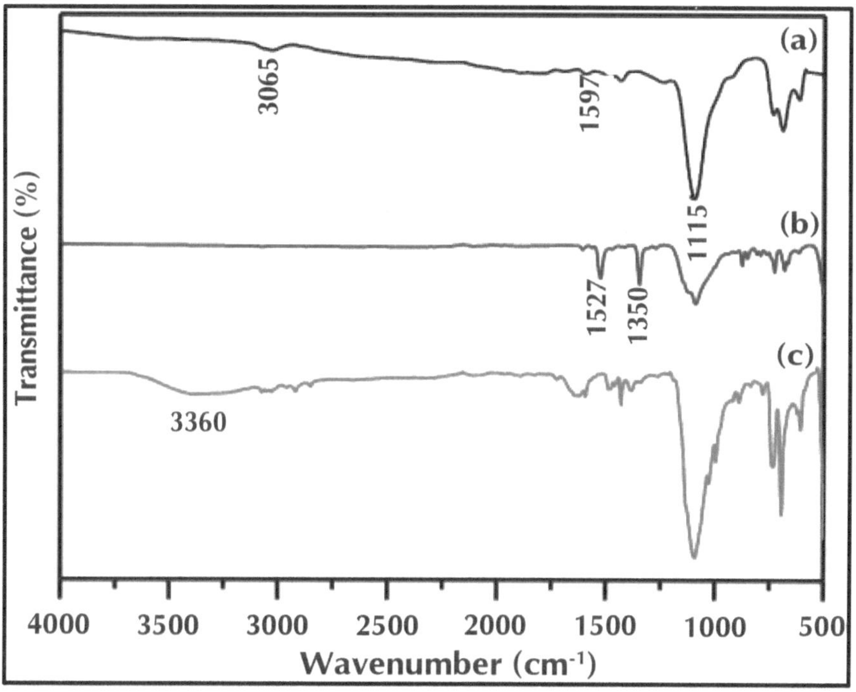

FIGURE 11.3 (a) FTIR spectrum of OPS, (b) FTIR spectrum of ONPS, and (c) FTIR spectrum of OAPS.

The FTIR spectra of neat siliconized epoxy and DDM cured siliconized epoxy are shown in Figure 11.3(d) and (e). The peak of Figure 11.3d at 1527 cm^{-1} indicates the Si—phenyl bond of DPSD. The absence of Si—OH peak at 2900 cm^{-1} of DPSD and the presence of Si—O—C peak at 1323 cm^{-1} indicate the completion of reaction between the hydroxyl group of DPSD and the glycidyl ring of DGEBA. The glycidyl ring undergoes ring opening and forms OH groups upon the addition of DDM curing agent. The disappearance of stretching frequency of the glycidyl ring (913 cm^{-1}) and appearance of peak at 3400 cm^{-1}, which corresponds to the secondary OH group formed during the curing reaction of siliconized epoxy with DDM, indicate the cross-linking reaction between epoxy and curing agent.[19] The GPC analysis of siliconized epoxy (Fig. 11.4), determined its number average molecular weight as 752, weight average molecular weight as 782 and polydispersity index to be 1.04.

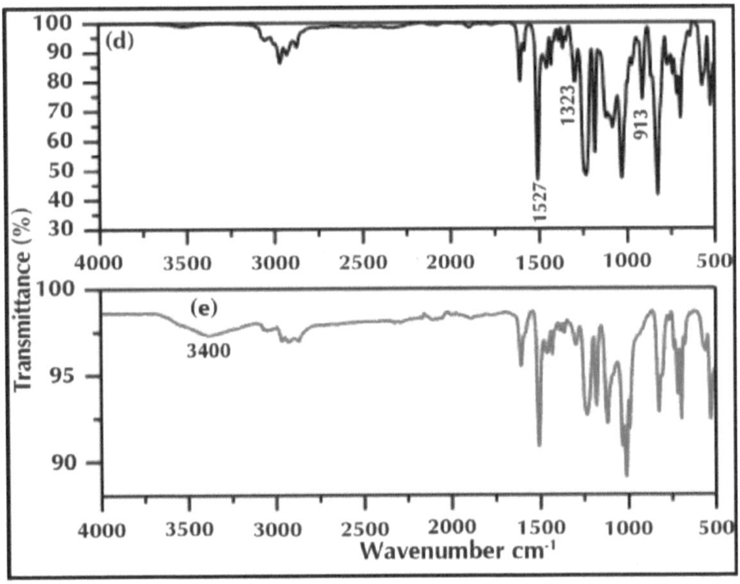

FIGURE 11.3 (d) FTIR spectrum of siliconized epoxy, (e) FTIR spectrum of siliconized epoxy cured with DDM.

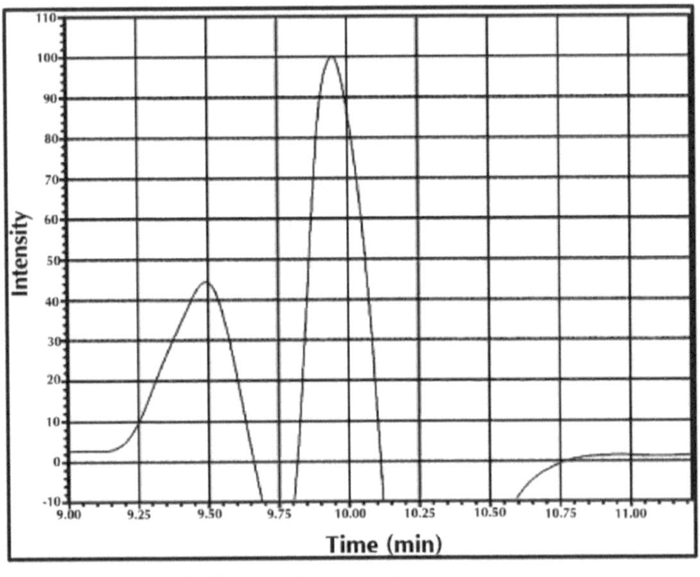

FIGURE 11.4 GPC data of siliconized epoxy resin.

11.6 DYNAMIC MECHANICAL ANALYSIS

The DMA studies were carried out to determine the storage modulus (E′) and glass transition temperature (tan δ) of the specimen, which explained the viscoelastic behavior of the polymers. Figures 11.5 and 11.6 illustrate the storage modulus and tan δ data pertaining to the neat siliconized epoxy (system "a"), 2.5% (system "b"), and 5% (system "c") POSS-reinforced siliconized epoxy nanocomposites, respectively. It can be seen from the DMA analysis that the system "b" (3.2 GPa) exhibits the best storage modulus (E′) whereas system "c" (1.0 GPa) exhibits the least storage modulus (E′).

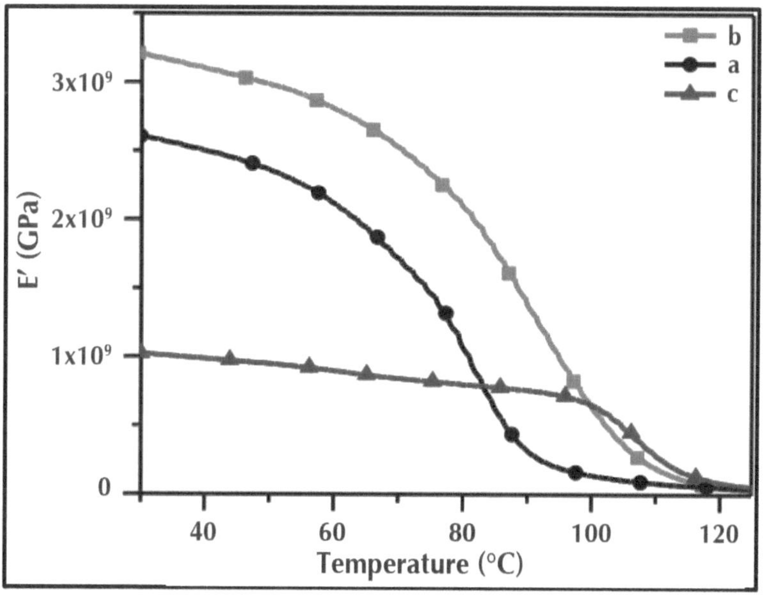

FIGURE 11.5 Storage modulus profile of siliconized epoxy and its nanocomposites cured with DDM (a) neat siliconized epoxy, (b) 2.5% POSS-reinforced siliconized epoxy nanocomposite, and (c) 5% POSS-reinforced siliconized epoxy nanocomposite.

The neat siliconized epoxy resin falls between the two reinforced composition namely systems "b" and "c", respectively. At low 2.5 wt% of POSS (system "b") content, an increased level of compatibility between siliconized epoxy resin and POSS was observed, which filled the defects present in the polymer. The increase in storage modulus may be due to the increased interfacial reaction between amine functionalized POSS and the siliconized epoxy resin. This interfacial interaction enhanced the interfacial thickness thereby facilitating an even stress transfer. Due to the increased cross-linking density, the storage modulus also increased.[21]

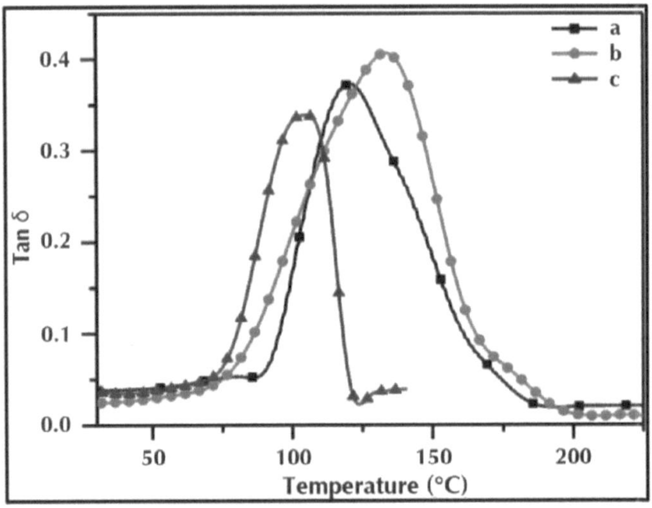

FIGURE 11.6 Glass transition profile of siliconized epoxy and its nanocomposites cured with DDM (a) neat siliconized epoxy, (b) 2.5% POSS-reinforced siliconized epoxy nanocomposite, and (c) 5% POSS-reinforced siliconized epoxy nanocomposite.

In system "c", due to the aggregation resulted from loading of higher percentages of POSS, higher stress concentration and void formation occurs and the inter particle distance will be large thereby preventing particle–particle interaction in the case of system "c". Hence, the reinforcing effect decreases, which is indicated by the lower storage modulus value imparted by system "c". Moreover, due to its irregular dispersion, there exists a void of cross-linking, which increases the segmental mobility and concomitantly decreased the storage modulus. In addition, the steric hindrance from the presence of unreacted amine terminated POSS will raise the activation energy of the curing reaction and reduce the cross-link density. The reduced cross-link density can be evidenced by the narrow tan δ peak.[22–24]

The Si—O—Si linkages introduce flexibility to the epoxy matrix. The presence of single T_g is indicative of good compatibility between the siliconized epoxy resin and POSS. The increase in T_g at low POSS concentration can be attributed to the increased cross-linking, which restricts the chain flexibility and mobility. However, at higher weight percentages, due to the increase in free volume content by the addition of these bulky cages, more flexibility is imparted, which leads to a decreased T_g[25] as indicated in Figure 11.6.

TABLE 11.1 DMA Results of Neat Siliconized and POSS-Reinforced Siliconized Epoxy Nanocomposites.

Systems	Resin systems	Storage modulus (GPa)	Glass transition temperature T_g (°C)
a	Neat siliconized epoxy	2.61	120
b	2.5% POSS-reinforced siliconized epoxy	3.22	134
c	5% POSS-reinforced siliconized epoxy	1.01	105

11.7 THERMOGRAVIMETRIC ANALYSIS

The data obtained from the thermogravimetric analysis can be observed from Figure 11.7 that system "c" exhibits the best thermal stability. System "b" exhibits better thermal properties than siliconized epoxy are shown in Table 11.2. The partial ionic nature and high bond energy of Si—O—Si is the reason for the increased thermal stability.[26–28] This nature prevented the epoxy degradation from heat and as a consequence, high thermal energy was needed to cause the degradation in systems "b" and "c." Higher silicon content led to increased inorganic components in the cured materials and thereby greater char content at higher temperatures was observed for system "c". The degradation temperature of POSS-reinforced nanocomposite systems was higher than those of neat siliconized epoxy system. This may be attributed to the prevention of oxidation of the inner part of the epoxy matrix at higher silicon concentrations by the formation of an inert silica layer over the surface of epoxy during when decomposition takes place.[29] The low surface energy of silica, makes it to migrate to the surface of the siliconized epoxy matrix, which results in the formation of a self-healing layer, which may also be accounted for a delayed degradation observed in system "b" and "c," respectively.

TABLE 11.2 Properties of Neat Siliconized and POSS-Reinforced Siliconized Epoxy Nanocomposites.

Systems	Initial degradation temperature (IDT)	Char yield at 800°C	Dielectric constant (MH_Z)	Water absorption (%)
a	267	7.1	3.3	11
b	287	15.1	3.15	8.6
c	334	24.46	3.09	7

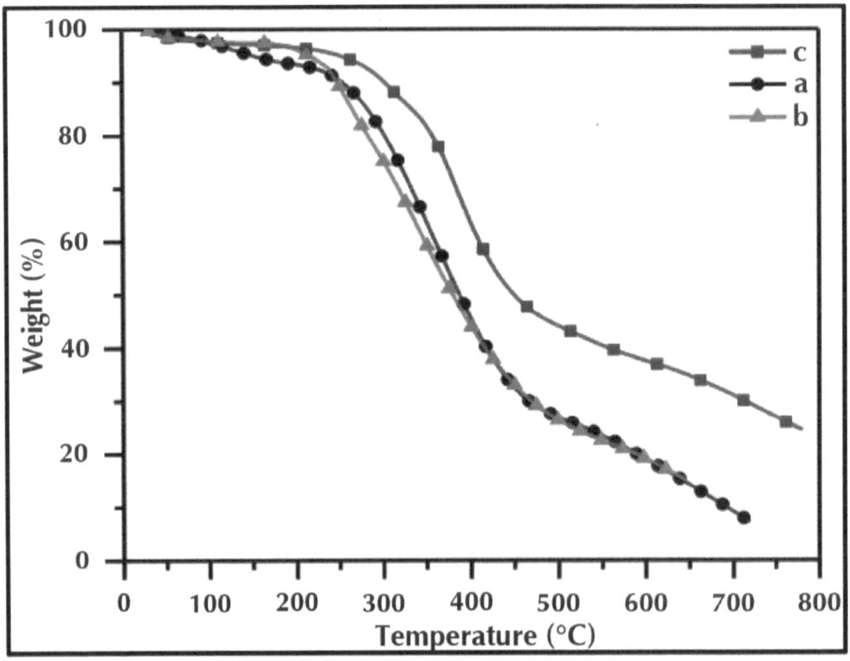

FIGURE 11.7 Thermal analysis of siliconized epoxy and its nanocomposites cured with DDM (a) neat siliconized epoxy, (b) 2.5% POSS-reinforced siliconized epoxy nanocomposite, and (c) 5% POSS-reinforced siliconized epoxy nanocomposite.

11.8 XRD (X-RAY DIFFRACTION) ANALYSIS

The presence and absence of peaks in the XRD pattern is indicative of intercalated and exfoliated nanocomposites, respectively. The XRD patterns of system "a" and "b" are similar to one another without the appearance of any significant peaks as shown in Figure 11.8. This confirms the formation of completely exfoliated structure in system "b".[30] This is an indication of good dispersion of POSS in system "b." However, system "c" exhibited an intercalated pattern indicating its improper dispersion. This aggregation led to an increase of the inter-particle distance, which prevented particle–particle interaction and also a void of cross-linking that led to decreased storage modulus in system "c."[23–25]

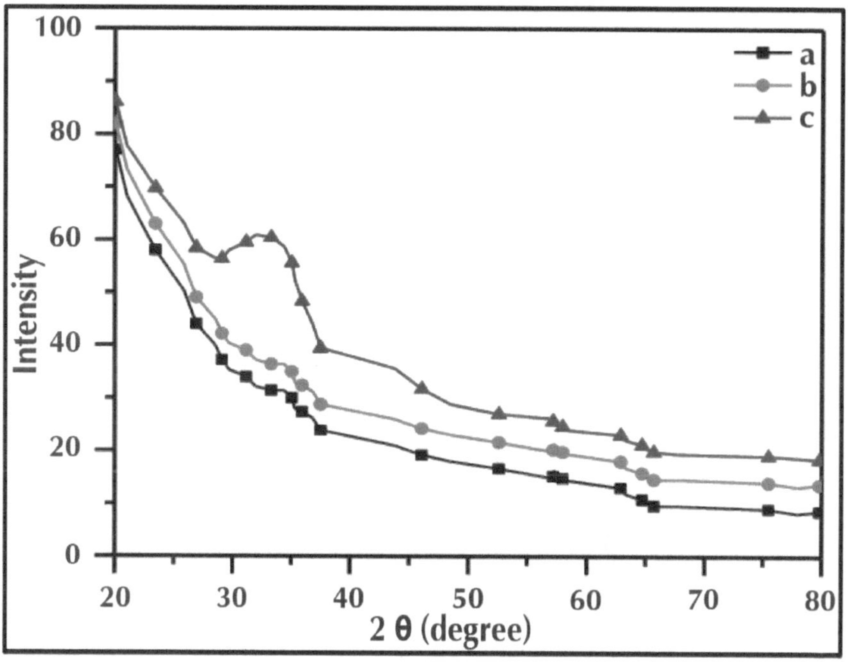

FIGURE 11.8 XRD pattern of siliconized epoxy and its nanocomposites cured with DDM (a) neat siliconized epoxy, (b) 2.5% POSS-reinforced siliconized epoxy nanocomposite, and (c) 5% POSS-reinforced siliconized epoxy nanocomposite.

11.9 SEM ANALYSIS

Figure 11.9 shows the SEM images of the siliconized epoxy and POSS-reinforced siliconized epoxy nanocomposite systems. The siliconized epoxies yield smooth and homogenous structures as shown in Figure 11.9(a).[28] Figure 11.9(b) and (c) show the dispersion of 2.5 and 5% of POSS fillers into the modified epoxy system. They exhibit heterogeneous morphology; these systems also exhibited increased surface roughness. The increased surface roughness may be attributed to the pinning effect of POSS incorporation and this may possibly prevent crack opening and might be the reason for the increased strength observed for the siliconized epoxy nanocomposites.[10]

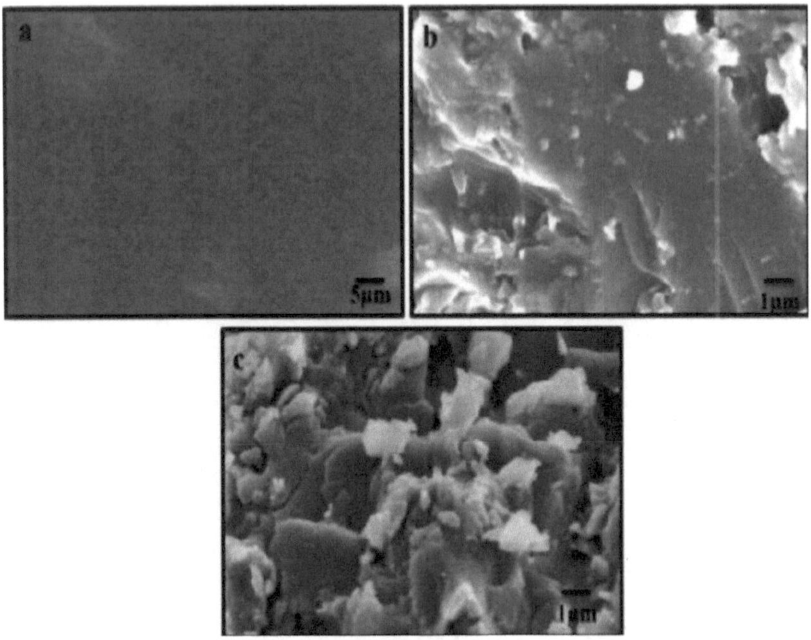

FIGURE 11.9 SEM images of epoxy nanocomposites (a) neat siliconized epoxy, (b) 2.5% POSS-reinforced siliconized epoxy nanocomposite, and (c) 5% POSS-reinforced siliconized epoxy nanocomposite.

11.10 TEM (TRANSMISSION ELECTRON MICROSCOPE) ANALYSIS

The TEM images of systems "a" and "b" are represented in Figure 11.10(a) and (b), respectively. The light grey area observed in TEM indicates the presence of siliconized epoxy matrix. The dark grey portions indicate the presence of POSS nanoreinforcements. From the TEM micrographs it is clearly seen that the system "a" (Fig. 11.10a) exhibits an uniform distribution of POSS within the siliconized epoxy matrix indicating better compatibility between the matrix and reinforcement.[31] In contrast to this observation, system "b" (Fig. 11.10b) exhibits aggregation of POSS within the siliconized epoxy matrix indicating its proper distribution within the matrix. This in proper aggregation of POSS within the matrix is caused mainly due to higher loading of (5 wt%) POSS, which results in increased inter-particle distance thereby preventing the particle-particle interaction. Apparently, reduce interfacial thickness and increase free volume.[32] This observation further supports the data obtained from DMA and XRD accounting for a decreased storage modulus observed in the case of system "c".

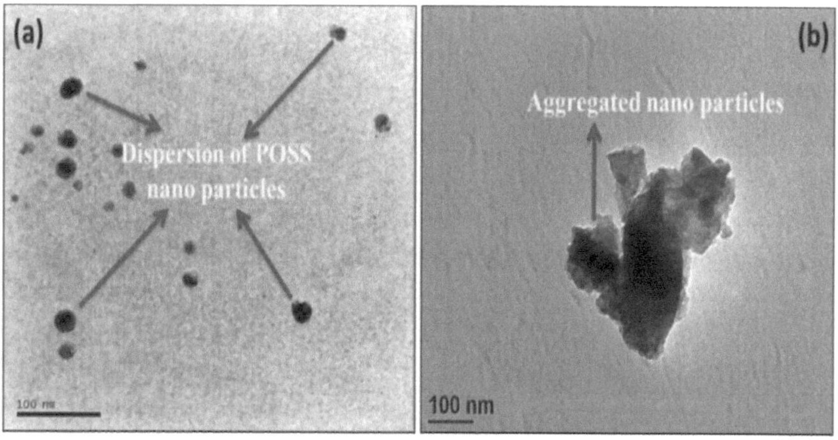

FIGURE 11.10 TEM images of epoxy nanocomposites (a) neat siliconized epoxy, (b) 2.5% POSS-reinforced siliconized epoxy nanocomposite, and (c) 5% POSS-reinforced siliconized epoxy nanocomposite.

11.11 DIELECTRIC CONSTANT

The values of the dielectric constant of systems "a", "b", and "c" are given in Table 11.2 and Figure 11.10(a)–(c), respectively. As shown in Figure 11.10(a)–(c) the POSS reinforced siliconized epoxy systems showed a lower dielectric constant value than that of the neat siliconized epoxy system. The dielectric constant values can be lowered significantly with the incorporation of less polarizable components. The introduction of POSS into the siliconized epoxy resin reduces its polarizability and increases the free volume in the resulting hybrid nanocomposites.[6,10] As expected, with the increase in POSS content (2.5 and 5 wt%), the dielectric constant value decreased indicating its insulating nature.

11.12 WATER ABSORPTION

Table 11.2 shows the percentage of water absorption by the siliconized epoxy and POSS-reinforced siliconized epoxy nanocomposites systems. Free rotation and ionic polarization of these Si—O—Si linkages allowed the siloxane chains to align accordingly, thereby resulting in improved hydrophobic properties.[33] In addition, the Si—O—Si linkage of the POSS-amine filler further increases the water resisting property due to its inherent hydrophobicity.[4] Therefore, as the content of POSS increased, the resistance to water absorption also increased. This is clearly seen from the water absorption values imparted by siliconized epoxy nanocomposites "b" and "c", respectively.

11.13 CONCLUSIONS

Neat siliconized epoxy and POSS-reinforced siliconized epoxy nanocomposites were successfully synthesized and the intermediates were characterized by FTIR spectroscopy to confirm their structure. The molecular weight of the resin was determined by GPC. The morphology of the systems was studied using XRD, TEM, and SEM. Its water absorption and dielectric constant were also measured. From the results of mechanical, thermal and dielectric studies, it was inferred that the system reinforced with 2.5 % of POSS appears to be the best formulation, as it exhibited an ideal set of thermomechanical properties required for high-performance applications than those of neat siliconized epoxy and 5% compositions. Based on the data resulted from the above study, it is concluded that the 2.5 % system may be used as an effective composite material for a possible use in advanced composites for enhanced performance in comparison with the materials that are currently being used for the same.

ACKNOWLEDGMENTS

Instrumentation facility provided under FIST-DST and DRS-UGC to Department of chemistry, Anna University, Chennai is gratefully acknowledged.

KEYWORDS

- POSS
- DGEBA
- DMA
- DDM
- SEM
- TEM

BIBLIOGRAPHY

1. Maureen, A.; Boyle, Cary, J.; Martin, John, D. Neuner Hexcel Corporation. Epoxy Resins. D.B. Miracle and S.L. Donaldson, editors. ASM Handbook, Composites. 78–89.
2. B. Ellis (Ed), Chemistry and Technology of Epoxy resins, Chapman & Hall 1993.
3. Alagar, M.; Ashok Kumar, A.; Mahesh, K.P.O.; Dinakaran, K. Studies on Thermal and Morphological Characteristics of E-Glass/Kevlar 49 Reinforced Siliconized Epoxy Composites. *Eur. Polym. J.* **2000**, *36*, 2449–2454.
4. Ananda Kumar, S.; Denchev, Z. Development and Characterization of Phosphorus Containing Siliconized Epoxy Resin Coatings. *Prog. Org. Coat.* **2009**, *66*, 1–7.
5. Mateˇjka, L.; Murias, P. Pleštil, J. Effect of POSS on Thermomechanical Properties of Epoxy—POSS Nanocomposites. *Eur. Polym. J.* **2012**, *48*, 260–274.
6. Nagendiran, S.; Alagar, M.; Ian, Hamerton. Octasilsesquioxane Reinforced DGEBA and TGDDM Epoxy Nanocomposites: Characterization of Thermal, Dielectric and Morphological Properties. *Acta Materialia.* **2010**, *58*, 3345–3356.
7. Li, L.; Li, X.; Yang, R. Mechanical, Thermal Properties, and Flame Retardancy of PC/ Ultrafine Octaphenyl-POSS Composites. *J. Appl. Polym. Sci.* **2012**, *124*, 3807–3814.
8. Shiao-Wei, K.; Feng-Chih, C. POSS Related Polymer Nanocomposites. *Prog. Polym. Sci.* **2011**, *36*, 1649–96.
9. Weian, Z.; Axel, H.E.; Müllerb. Architecture, Self-Assembly and Properties of Well-Defined Hybrid Polymers Based on Polyhedral Oligomericsilsequioxane (POSS). *Prog. Polym. Sci.* **2013**, *38*, 1121–1162.
10. Zengping, Z.; Guozheng, L.; Jieliang, W.; Penggang Ren. Epoxy/POSS Organic–Inorganic Hybrids: Viscoelastic, Mechanical Properties and Micromorphologies. *Polym. Compos.* **2007**, *28*, 175–179.
11. Scott, D.W. Thermal Rearrangement of Branched-Chain Methylpolysiloxanes. *J. Am. Chem. Soc.* **1946**, *68*, 356–358.
12. Baney, R.H.; Itoh, M.; Sakakibara, A.; Suzuki, T. Silsesquioxanes. *Chem. Rev.* **1995**, *95*, 1409–1430.
13. Laine, R.M. Nanobuilding Blocks Based on the [OSiO1-5] x (x = 6, 8, 10) Octasilsesquinoxanes. *J. Mater. Chem.* **2005**, *35*, 3725–3744.
14. Jothibasu, S.; Premkumar, S.; Alagar, M.; Hamerton, I. Synthesis and Characterization of a POSS-Maleimide Precursor for Hybrid Nanocomposites. *High Perform. Polym.* **2008**, *20*, 67–85.
15. Nagendiran, S.; Hamerton, I.; Alagar, M. Octasilsesquioxane-Reinforced DGEBA and TGDDM Epoxy Nanocomposites: Characterization of Thermal, Dielectric and Morphological Properties. *Acta Mater.* **2010**, *58*, 3345–3356.
16. Huang, J.-C.; He, C.-B.; Xiao, Y.; Khine Yi Mya, Dai, J.; Ping, Y. Thermomechanical Properties of Polyimide-Epoxy Nanocomposites from Cubic Silsesquioxane Epoxides. *J. Mater. Chem.* **2004,** *14*, 2858–2863.
17. Huang, J.C.; He, C.B.; Xiao, Y.; Mya, K.Y.; Dai, J.; Siow, Y.P. Polyimide/POSS Nanocomposites: Interfacial Interaction, Thermal Properties and Mechanical Properties. *Polymer.* **2002**, *44*, 4491–4499.
18. Tamaki, R.; Tanaka, Y.; Michael, Z.; Choi, A.J.; Richard, M.; Laine. Octa(Aminophenyl) Silsesquioxane as a Nanoconstruction Site. *J. Am. Chem. Soc.* **2001**, *123*, 12416–12417.
19. Wu, C.S.; Liu, Y.L.; Chiu, Y.S. Epoxy Resins Possessing Flame Retardant Elements from Silicon Incorporated Epoxy Compounds Cured with Phosphorous or Nitrogen Containing Curing Agents. *Polymer.* **2002**, *43*, 4277–4284.

20. Erfan Suryani Abd Rashid, Ariffin, K.; Kooi, C.C.; Akil, H.M. Preparation and Properties of POSS/Epoxy Composites for Electronic Packaging Applications. *Mater. Des.* **2009,** *30*, 1–8.
21. Jones, I.K.; Zhou1, Y.X.; Jeelani, S.; Mabry, J.M. Effect of Polyhedraloligomericsil-Sesquioxanes on Thermal and Mechanical Behavior of SC-15 Epoxy. *Express. Polym. Let.* **2008**, 2, 494–501.
22. Liu, T.; Tjiu, W.C.; Tong, Y.; He C.; Goh, S.S.; Chung, T.-S. Morphology and Fracture Behavior of Intercalated Epoxy/Clay Nanocomposites *J. Appl. Polym. Sci.* **2004**, *94*, 1236–1244.
23. Mya, K.Y.; He, C.; Huang, J.; Xiao, Y.; Dai, J.; Siow, Y.P. Preparation and Thermomechanical Properties of Epoxy Resins Modified by Octafunctional Cubic Silsesquioxane Epoxides. *J. Polym. Sci. Part. A: Polym. Chem.* **2004**, 42, 3490–3403.
24. Zhang, Z.; Liang, G.; Wang, X. The Effect of POSS on the Thermal Properties of Epoxy *Polym. Bull.* **2007**, *58*, 1013–1020.
25. Su, C.-H.; Chiu, Yi.-P.; Teng, C.-C.; Chian, C.-L. Preparation, Characterization and Thermal Properties of Organic–Inorganic Composites Involving Epoxy and Polyhedral Oligomericsilsesquioxane (POSS). *J. Polym. Res.* **2010**, *17*, 673–681.
26. Shyue-tzoolin, Steve K.; Huang. Thermal Degradation Study of Siloxane—DGEBA Epoxy Copolymers. *Eur. Polym. J.* **1997**, *33*, 365–373.
27. Ananda Kumar, S.; Sankara Narayanan, T.S.N. Thermal Properties of Siliconized Epoxy Interpenetrating Coatings. *Prog. Org. Coat.* **2002**, *45*, 323–330.
28. Ananda Kumar, S.; Denchev, Z.; Alagar, M. Synthesis and thermal characterization of Phosphorus Containing Siliconized Epoxy Resins. *Eur. Polym. J.* **2006**, *42*, 2419–2429.
29. Guizhi, Li.; Lichang, W.; Hanli, Ni.; Charles, U. Pittman. Polyhedral Oligomeric Silsesquioxane (POSS) Polymers and Copolymers: A Review. *J. Inorg. Orgmet. Polym.* **2001**, *11*, 123–154.
30. Ara´ujo, E.M.; Barbosa, R.; Rodrigues, A.W.B.; Melo, T.J.A.; Ito, E.N. Processing and Characterization of Polyethylene/Brazilian Clay Nanocomposites. *Mater. Sci. Eng. A.* **2007**, *445–46*, 141–147.
31. Bocek, J.; Matejka, L.; Mentlik, V.; Trnka, P.; Slouf, M. Electrical and Thermomechanical Properties of Epoxy-POSS Nanocomposites. *Eur. Polym. J.* **2011**, *47*, 861–872.
32. Libor, M.; Piotr, M.; Josef, P. Effect of POSS on Thermomechanical Properties of Epoxy POSS Nanocomposites. *Eur. Polym. J.* **2012**, *48*, 260–274.
33. Hans, R.; Kricheldorf, Herausgeber, Hans R.; Kricheldorf. Silicon in Polymer Synthesis. Verlag: Springer, Berlin, **1996**, 469–491.

CHAPTER 12

CYANATE ESTERS BASED ORGANIC–INORGANIC HYBRID NANOCOMPOSITES FOR LOW-K DIELECTRIC APPLICATIONS

M. ALAGAR and S. DEVARAJU

Department of Chemical Engineering, Alagappa College of Technology, Anna University, Chennai 600025, India, mkalagar@yahoo.com

CONTENTS

ABSTRACT

Cyanate esters (CEs) are currently in widespread use because of their high thermal stability, excellent mechanical properties, good flame resistance, low outgassing, and good radiation resistance. Applications of CE resins include structural aerospace, electronic, microwave-transparent composites, encapsulant, and as an adhesives. High-speed electronic device circuits require low-k material to realize the faster signal transmission without cross-talk. To achieve this, it is important to develop the low-k dielectric materials that are needed for the efficient integrated circuits. It is well established that the low-k silica and other related materials can prevent the signal crossover with low-power consumption. In this view, many efforts have been taken to reduce the dielectric constant (<1.8) with different kind of materials. Especially, nanolevel porous inorganic materials, variety of polymeric materials and their combinations are investigated for low-k dielectrics. Hence, in the present chapter, development of organic–inorganic hybrid polymer nanocomposites based on CEs hybridizing with porous nano reinforcement like POSS, Mesoporous SBA-15, and silica under appropriate conditions to yield low dielectric constant is discussed. In addition, the thermal and morphological properties of the nanocomposites are discussed to ascertain their high-performance characteristics.

12.1 INTRODUCTION

Organic–inorganic hybrid nanocomposites can be defined as materials made up of organic and inorganic components combined over length scales ranging from a few angstroms to tens of nanometers. The development of organic–inorganic hybrid materials is mainly due to the development of soft inorganic chemistry processes. Homogeneously dispersed organic–inorganic hybrid nanocomposites can be obtained by increasing interactions between both components through the formation of hydrogen bonds or covalent bonds, by mixing different organic oligomers or via the adequate choice of the inorganic precursors. Such nanocomposite materials have attracted a considerable interest during the past few decades, as they exhibit unusual combinations of properties originating from the synergism of organic and inorganic components. A great deal of interest for hybrid nanocomposites is budding with regard to the fundamental research and from an applications point of view due to their potential as candidate materials for bridging the gap between organic polymers and inorganic ceramics.

The polymer nanocomposites have attracted much interest because of their high thermal stability, good flame retardancy, good mechanical properties, and light weight as well as high strength of such nanocomposite materials, the weight of automotives, satellite, and space craft components can be drastically reduced, which is most important in both the economical and efficiency aspects. Therefore, the applications of polymer nanocomposites have been also extended for space research and part of spacecraft materials. However, these nanocomposites, which can replace the metallic exterior and infrastructure of satellites and space vehicles, cannot survive the aggressive environment of space unless protected.

Cyanate esters (CEs) are currently in widespread use because of their high thermal stability, excellent mechanical properties, good flame resistance, low outgassing, and good radiation resistance. Applications of CE resins include structural aerospace, electronics, microwave-transparent composites, encapsulants, and adhesives. CE resins are superior in several ways to conventional epoxy, bismaleimide (BMI) resins, which either lack thermal stability in high-temperature applications or require narrow processing method for application. For example, the moisture absorption rate of CEs is lower than that of epoxy and BMI resins. The high glass transition temperature (T_g) (>250°C) of CE fills a temperature regime intermediate between that of epoxy resins and BMI resins.

12.2 IMPORTANCE OF LOW-K MATERIALS

Design of smart and miniaturized microelectronic devices, such as integrated circuits, memory devices, etc., have received enormous research interest owing to their negligible time delay, which are necessary for ultra-fast electronic devices [1]. In general, the square of the dielectric constant is inversely proportional to the propagation velocity of the signal. Consequently, the high speed electronic device circuits require low-k material to realize the faster signal transmission without cross talk [2, 3]. To achieve this, it is important to develop the low-k dielectric materials that are needed for the efficient integrated circuits. It is well established that the low-k silica and other related materials can prevent the signal cross over with low-power consumption [4]. In this view, many efforts have been taken to reduce the dielectric constant (<1.8) with different kind of materials [5–7]. Especially, nanolevel porous inorganic materials, variety of polymeric materials and their combinations are investigated for low-k dielectrics. At the same time, the low-k composite materials should be compatible and have a strong adhesion with the improved thermal and mechanical stability. Based on these aspects, the composite materials of interconnected structures with air gaps and nanoporous have been studied extensively.

Hence, in the present chapter, development of organic–inorganic hybrid polymer nanocomposites based on CEs hybridizing with porous nanoreinforcement like POSS, Mesoporous SBA-15, and silica under appropriate conditions to yield low dielectric constant is discussed. In addition, the thermal and morphological properties of the nanocomposites are also discussed to ascertain their high-performance characteristics.

12.3 EXPERIMENTAL

12.3.1 SYNTHESIS CYANATE ESTER MONOMERS

12.3.3.1 SYNTHESIS OF 4,4′-[ETHANE-1,2-DIYLBIS (OXY-1,1-PHENYLENE)] DICYANATE ($AECE_1$)

A three-necked round-bottomed flask, equipped with a magnetic stirrer bar, a dropping funnel, and a drying tube, was charged with 4,4′-[Ethane-1,2-diylbis(oxy)]diphenol (5 g, 0.0203 mol) and cyanogen bromide (4.7 g, 0.0447 mol). The reactants were dissolved in acetone (75 mL), and the solution was cooled to below −5°C in a salt-ice bath. Triethylamine (5.1 g, 0.0507 mol) was added dropwise via the dropping funnel to the reaction mixture and was maintained below −5°C. After adding triethylamine, the reaction mixture was left stirring for 1 h and then allowed to warm to room temperature. The mixture was washed by fast stirring with water to remove triethylamine hydrobromide (by-product) and the remaining precipitate was filtered and washed with water. A light white solid product was obtained by recrystallization from a 1:1 mixture of ethanol and water, to yield 3.9 g (65%). (Scheme 12.1).

Similarly, other monomers like 4,4′-[butane-1,4-diylbis(oxy-1,1-phenylene)] dicyanate ($AECE_2$), 4,4′-[hexane-1,6-diylbis(oxy-1,1-phenylene)] dicyanate ($AECE_3$), 4,4′-[octane-1,8-diylbis(oxy-1,1-phenylene)] dicyanate ($AECE_4$) also prepared with above procedures with respect to their corresponding diols (Scheme 12.1). All the CE monomers structure was confirmed by FT-IR (Figure 12.1), ^{1}H NMR (Figures 12.2–12.5) and ^{13}C NMR spectra (Figures 12.6–12.9).

Figure 12.1 Illustrate the Fourier transform infrared spectra (FTIR) of $AECE_{1-4}$.

$AECE_1$: (KBr, cm^{-1}): 2937(symmetric stretching), 2862(asymmetric stretching), 2277, 2229 (—OCN stretching), 1243(Ar—O—CH_2).
$AECE_2$: (KBr, cm^{-1}): 2940 (symmetric stretching), 2858(asymmetric stretching), 2278, 2233 (—OCN stretching), 1249 (Ar—O—CH_2).
$AECE_3$: (KBr, cm^{-1}): 2941 (symmetric stretching), 2862 (asymmetric stretching), 2272, 2230 (—OCN stretching), 1245(Ar—O—CH_2).
$AECE_4$: (KBr, cm^{-1}): 2937(symmetric stretching), 2856 (asymmetric stretching), 2278, 2230 (—OCN stretching), 1242 (Ar—O—CH_2).

Figures 12.2–12.5 Illustrate proton nuclear magnetic resonance (^1H NMR) spectra of $AECE_{1-4}$

$AECE_1$: (300 MHz, $CDCl_3$) δ (ppm): 7.2 (d, 4H, ArH), 7.0 (d, 4H, ArH), 4.3 (t, 4H, Ar—O—CH_2).
$AECE_2$: (300 MHz, $CDCl_3$) δ (ppm): 7.2 (d, 4H, ArH), 6.9 (d, 4H, ArH), 4.0 (t, 4H, Ar—O—CH_2), 2.0 (m, 4H, —CH_2–).
$AECE_3$: (300 MHz, $CDCl_3$) δ (ppm): 7.2 (d, 4H, ArH), 6.9 (d, 4H, ArH), 4.0 (t, 4H, Ar—O—CH_2), 1.8 (m, 4H, —CH_2–), 1.5(m, 4H, —CH_2–).
$AECE_4$: (300 MHz, $CDCl_3$) δ (ppm: 7.2 (d, 4H, ArH), 6.8 (d, 4H, ArH), 4.0 (t, 4H, Ar—O—CH_2), 1.9–1.7 (m, 8H, —CH_2–), 1.4 (m, 4H, —CH_2–).

Figures 12.6–12.9 Illustrate carbon nuclear magnetic resonance (^{13}C NMR) spectra of $AECE_{1-4}$

$AECE_1$: δ (ppm): 157, 146, 116, 115, 109 (aromatic carbon), 68(aliphatic carbon).
$AECE_2$: δ (ppm): 157, 146, 116, 115, 109 (aromatic carbon), 68, 25 (aliphatic carbon).
$AECE_3$: δ (ppm): 157, 146, 116, 115, 109 (aromatic carbon), 68, 29, 25, (aliphatic carbon).
$AECE_4$: δ (ppm): 156, 145, 115, 114, 107 (aromatic carbon), 67, 27, 21 (aliphatic carbon).

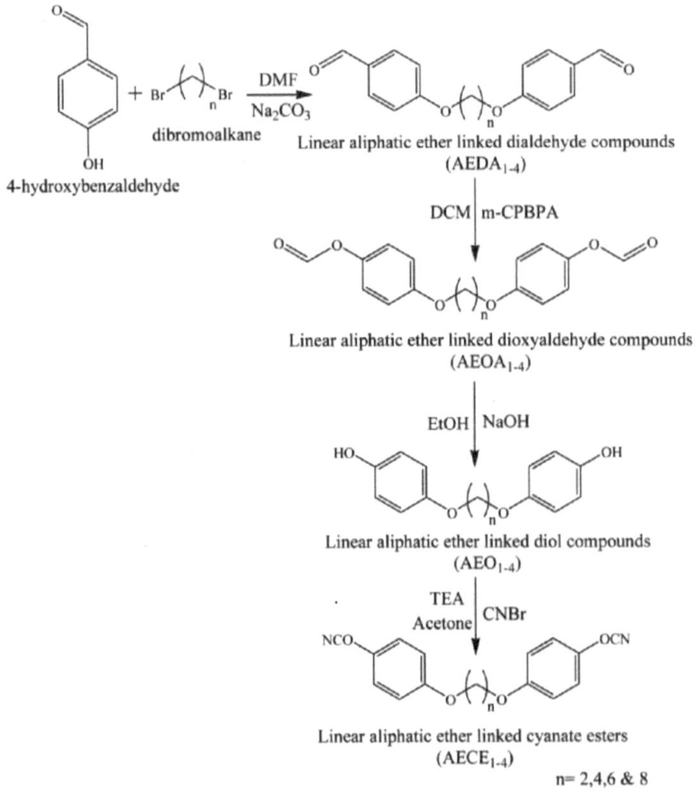

SCHEME 12.1 Synthesis of linear aliphatic ether linked cyanate ester monomers ($AECE_{1-4}$).

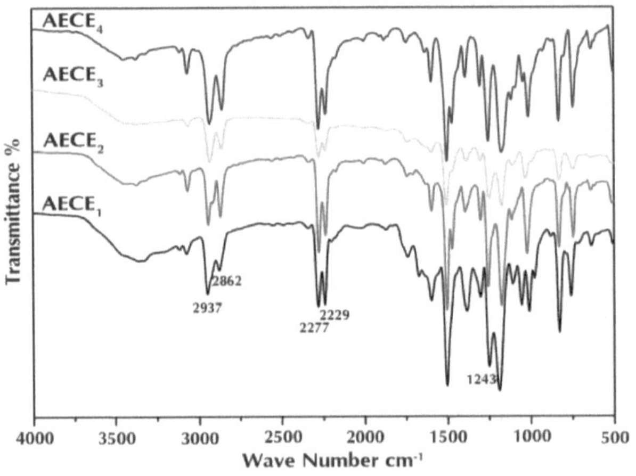

FIGURE 12.1 Fourier transform infrared spectra (FTIR) of $AECE_{1-4}$.

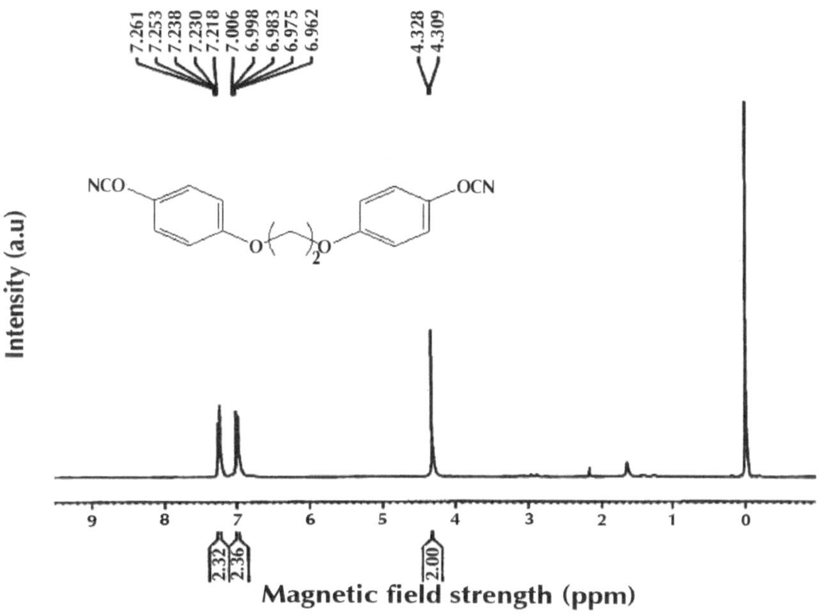

FIGURE 12.2 ¹H NMR spectrum of $AECE_1$.

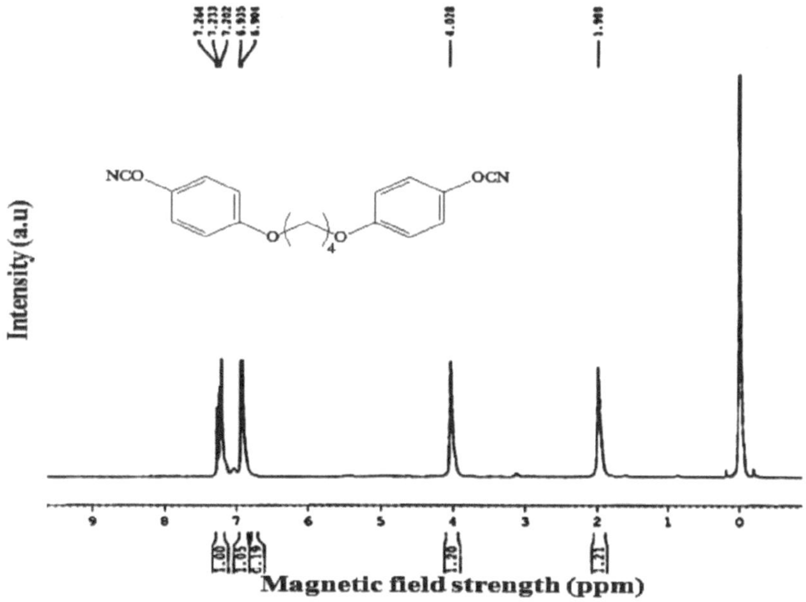

FIGURE 12.3 ¹HNMR spectrum of $AECE_2$.

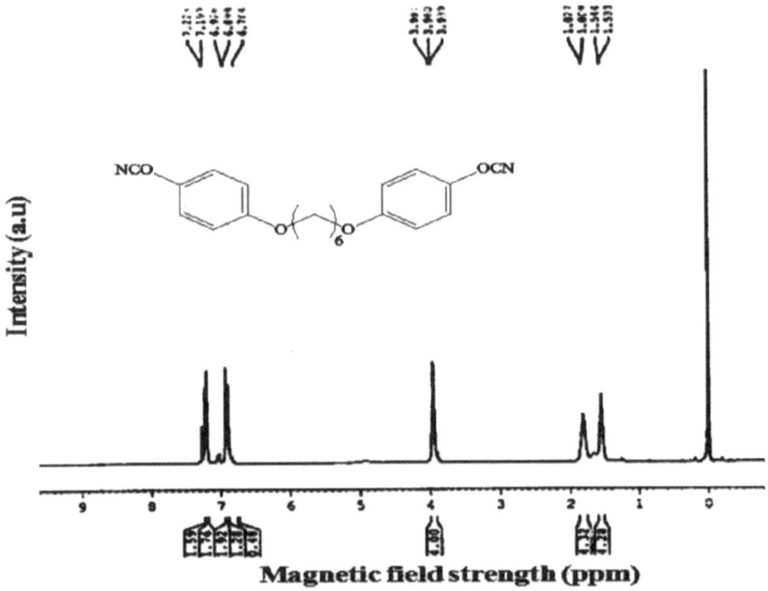

FIGURE 12.4 ^{1}H NMR spectrum of $AECE_3$.

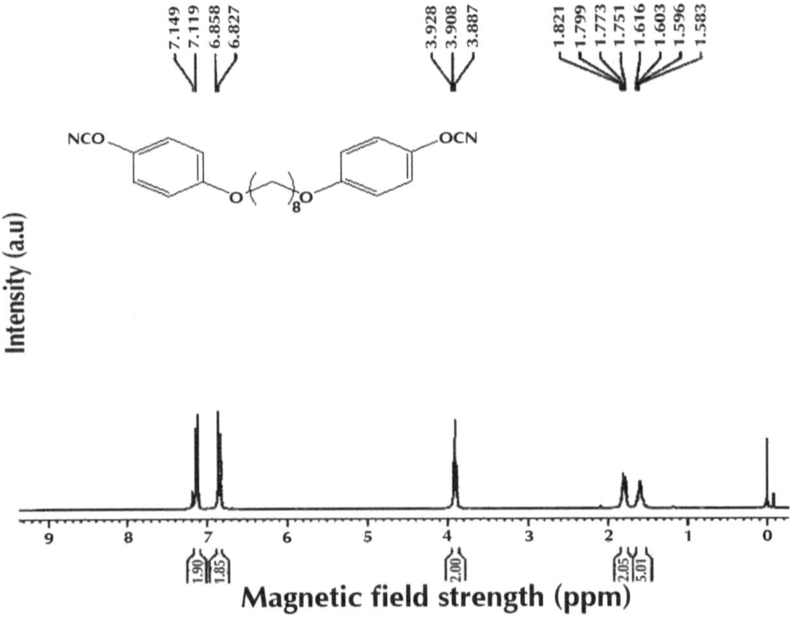

FIGURE 12.5 ^{1}H NMR spectrum of $AECE_4$.

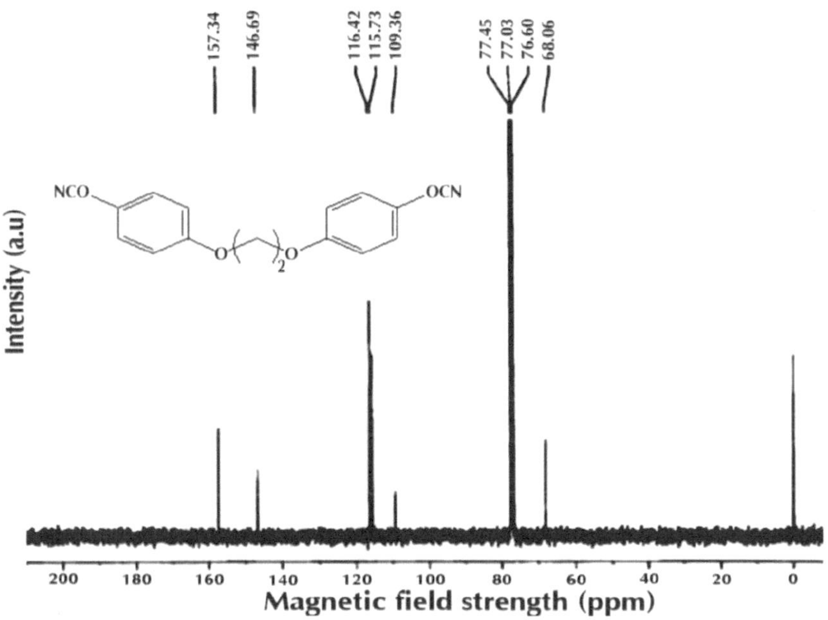

FIGURE 12.6 ^{13}C NMR spectrum of $AECE_1$.

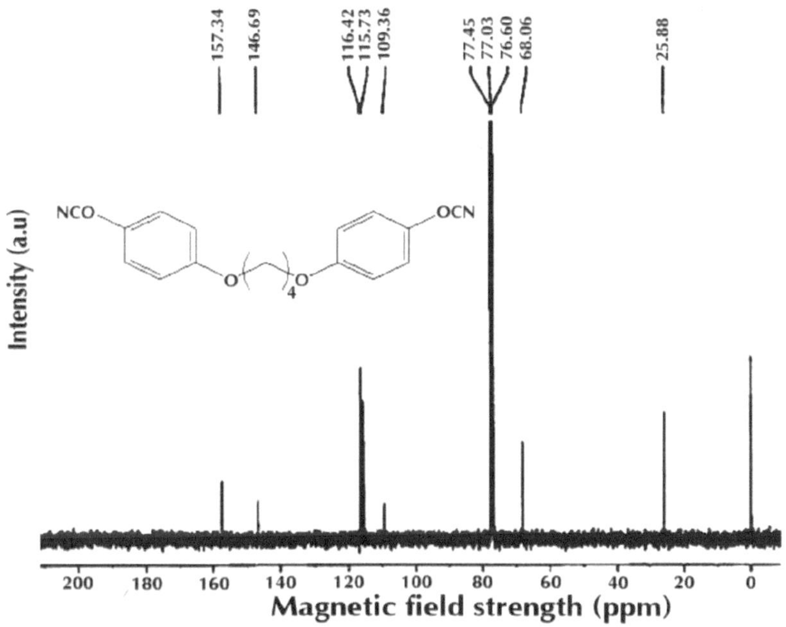

FIGURE 12.7 ^{13}C NMR spectrum of $AECE_2$.

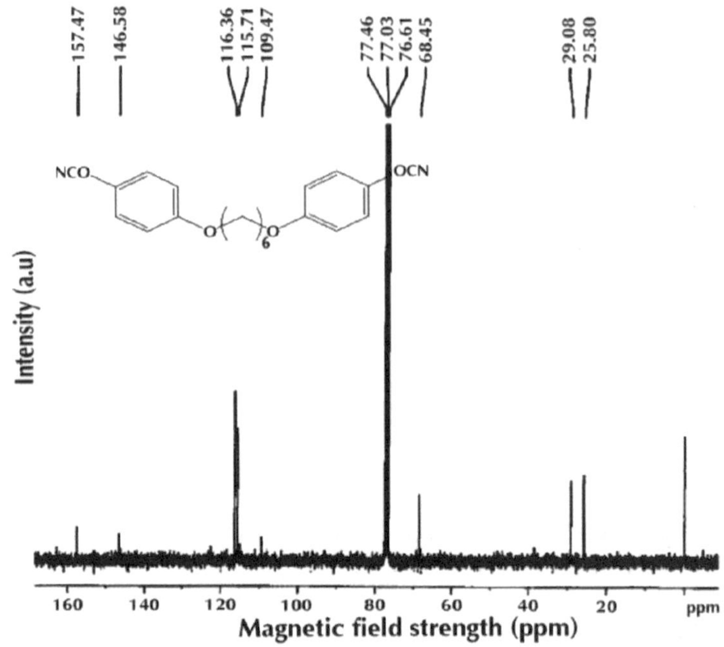

FIGURE 12.8 ^{13}C NMR spectrum of $AECE_3$.

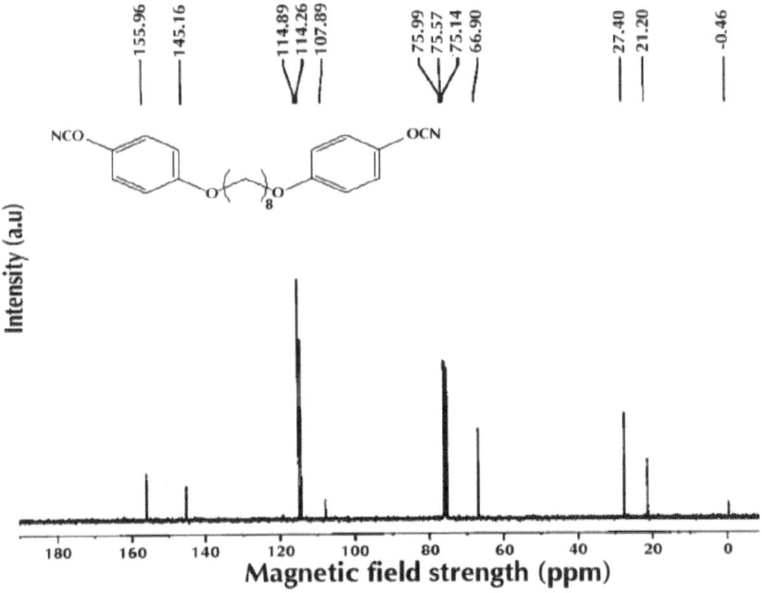

FIGURE 12.9 ^{13}C NMR spectrum of $AECE_4$.

12.3.3.2 SYNTHESIS OF 1,4-BIS (2-(4-CYANATOPHENYL)-2-PROPYL) BENZENE (BCE)

A three-necked round-bottomed flask, equipped with a magnetic stirrer, a dropping funnel, and a drying tube, was charged with 1,4-Bis(2-(4-hydroxyphenyl)-2-propyl)benzene (10 g, 0.0289 mol) and cyanogen bromide (6.7 g, 0.0635 mol). The reactants were dissolved in acetone (75 mL), and the solution was cooled to below −5°C in a salt-ice bath. Triethylamine (7.3 g, 0.0722 mol) was added dropwise via the dropping funnel to the reaction mixture. After adding triethylamine, the reaction mixture was stirred for 1 h at −5°C for 1 h at room temperature. The precipitate was washed with water to remove triethylamine hydrobromide (by-product) and filtered. A white solid was obtained by recrystallization from a 50/50 mixture of ethanol and water, to yield 7.2 g (63%) (Scheme 12.2).

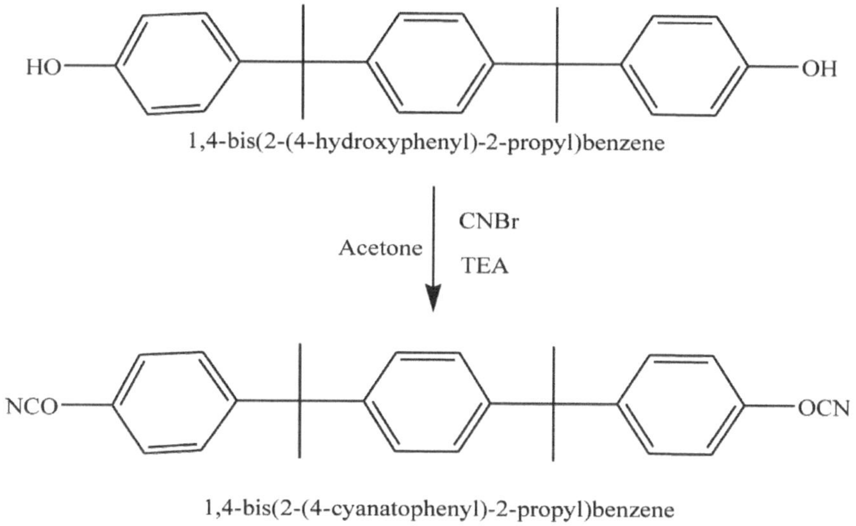

SCHEME 12.2 Synthesis of 1,4-bis(2-(4-cyanatophenyl)-2-propyl) benzene (BCE).

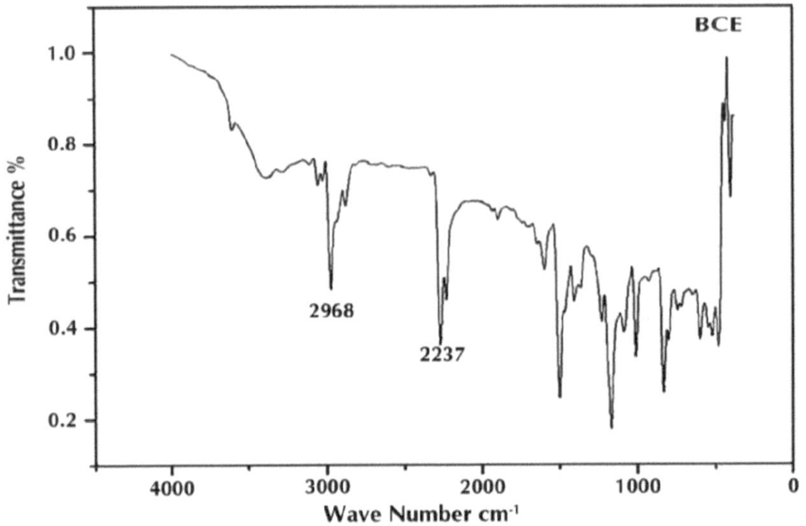

FIGURE 12.10 Fourier transform infrared spectrum of BCE.

Figure 12.10 Illustrates the FTIR spectrum of BCE: (KBr, cm^{-1}): 2968 (Symmetric stretching), 2862 (asymmetric stretching), 2237 (—OCN), 1505, 1227.

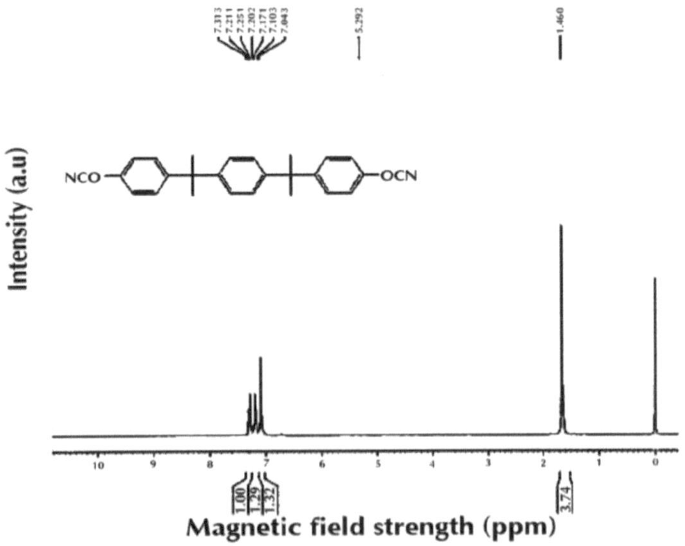

FIGURE 12.11 ^{1}H NMR spectrum of BCE.

Figure 12.11 Illustrates the ^{1}H NMR spectrum of BCE: (300 MHz, $CDCl_3$) δ (ppm): 7.1 (s, 4H, ArH), 7.2 (d, 4H, ArH), 7.3 (d, 4H, ArH), and 1.6 (s, 12H, aliphatic H).

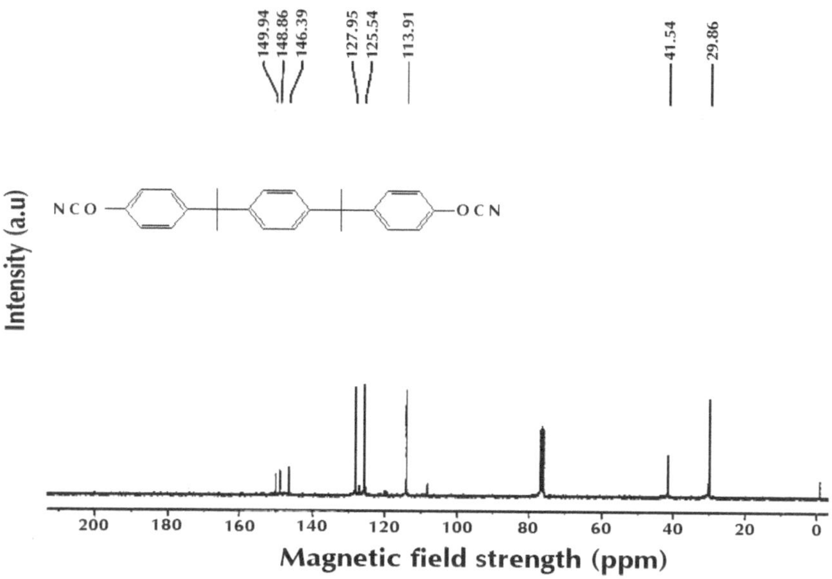

FIGURE 12.12 ^{13}C NMR spectrum of BCE.

Figure 12.12 Illustrates ^{13}C NMR spectrum of BCE: δ (ppm): 149, 148, 147, 127, 125, 113, 108 (aromatic carbon), 41, 25 (aliphatic carbon).

12.4 SYNTHESIS OF REINFORCEMENTS

12.4.1 SYNTHESIS OF OCTAANION

According to Laine et al., an octaanion solution $[(\text{—OSiO}_{1.5})_8]$ (Scheme 12.3) was prepared by mixing 32.2 mL (0.35 mol) of Me_4NOH (25 wt % in methanol), 15.6 mL (0.48 mol) of methanol, and 11.7 mL of deionized water followed by dropwise addition of 17 mL (0.0821 mol) of TEOS under N_2. The solution turned cloudy and was stirred at room temperature for overnight to obtain a clear solution of tetramethylammonium octaanion in quantitative yield [8].

12.4.2 SYNTHESIS OF OCTAKIS (DIMETHYLSILOXY) SILSESQUIOXANE (OCTAHYDRIDOCUBE, OHC)

The octaanion solution (58.4 mL) was added slowly to an ice-cooled solution of chlorodimethylsilane (33.5 mL) in hexane (200 mL) (Choi et al. 2003). After adding, the mixture was allowed to reach room temperature with a total constant stirring time of 60 min. The layers were separated, and the aqueous layer was washed with hexane (3 × 50 mL) and the total hexane portion was collected and dried with 5 g of $MgSO_4$. The total hexane volume was reduced to yield a white solid. The solid was washed with cold methanol (≈25 ml) and dried under vacuum to yield 3.52 g of octakis (dimethylsiloxy) silsesquioxane (OHC). The yield was 80% (Scheme 12.3).

12.4.3 SYNTHESIS OF OCTAKIS (DIMETHYLSILOXYPROPYL-GLYCIDYLETHER) SILSESQUIOXANE (OG-POSS)

An octahydrido cube (2 g, 1.95 mmol) was added to a 50 mL of round-bottom flask equipped with a magnetic bar and condenser [8]. The flask was evacuated and refilled with N_2. Toluene (25 mL) was added and the solution was stirred for about 5 min. Allylglycidylether (AGE) (2.3 mL, 19.45 mmol) was added to the solution followed by five drops of 2.0 mmol Pt(dvs). The reaction was continued for 12 h at 50°C and then cooled. 5 wt % activated charcoal was added to the reaction mixture. After 10 min of stirring, the mixture was filtered, and the removal of the solvents afforded an opaque viscous liquid (Scheme 12.3). The yield was 3.79 g (90.4%).

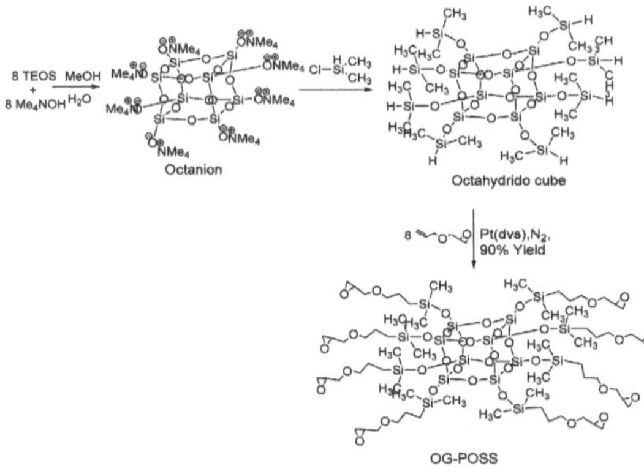

SCHEME 12.3 Synthesis of octakis (dimethylsiloxypropyl-glycidylether) silsesquioxane (OG-POSS).

12.4.4 SYNTHESIS OF SBA-15

Mesoporous silica SBA-15 was synthesized as per the procedure reported in [9]. The molar composition of the reaction mixture was 1 TEOS:0.017 P123:5.87 hydrochloric acid (HCl): 0.0025 1,3,5-trimethylbenzene (TMB): $183H_2O$. Briefly, 4 g pluronic polymer was dissolved with the addition of diluted 150 mL 1.6 M HCl at 40°C (pH < 2), with rapid stirring, the polymer was completely dissolved in 60 min and then 0.3 g of TMB (swelling agent) and 9.2 mL of TEOS (inorganic precursor) were added. The mixture was then transferred into a Teflon bottle. The bottle was tightly sealed with a Teflon tape insert. The bottle was aged at 80°C for 24 h. It is critical to make sure that the bottle was tightly sealed in order to prevent the solvent from evaporating. After 24 h, the mixture was cooled, and then filtered by vacuum filtration, and a white powder was obtained. The white powder was allowed to dry in air under vacuum for 24 h. The dried sample was calcinated in air at 550°C for 6 h with a heating rate 1°C/min. The final product is designated as SBA-15.

12.4.5 SYNTHESIS OF GLYCIDYL FUNCTIONALIZED SBA-15

Mesoporous silica was dispersed in ethanol under ultrasonic agitation for 30 min, and then the calculated amount of GPTMS was added to the mixture, and refluxed for 24 h under vigorous stirring. After the mixture was filtered with the Buchner funnel, the product was washed several times using ethanol, followed by hexane to remove the unreacted GPTMS, and dried in the vacuum oven at 50°C for 24 h (Scheme 12.4).

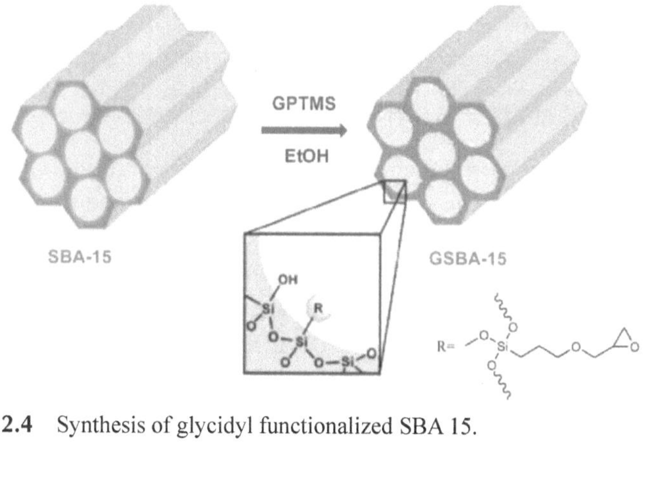

SCHEME 12.4 Synthesis of glycidyl functionalized SBA 15.

12.5 PREPARATION OF NANOCOMPOSITES

12.5.1.1 PREPARATION OF SBA-15/CE NANOCOMPOSITES

To 2 g of CE monomer (BCE) dissolved in 10 mL of 1,4-dioxane varying weight percentages of SBA-15 (5, 10, and 15 wt %) were added separately. The resulting product was stirred for 24 h at room temperature. These products were poured into silane coated glass plates and they were allowed to evaporate for 6 h at room temperature. All the samples were polymerized in a stepwise manner, at 100, 150, 200, and 250°C for 1 h to get SBA-15/CE nanocomposites.

12.5.1.2 PREPARATION OF POSS-AECE NANOCOMPOSITES

Total, 2 g of CE monomer was dissolved in 10 mL of 1,4-dioxane and the solution was cast in a glass plate. The solvent was allowed to evaporate at 60°C for 5 h followed by heat treatment of the precursor at 100, 150, and 200°C for 1 h and at 250°C for 2 h to obtain cured CEs.

Total, 2 g of CEs monomer was first dissolved in 10 mL of 1,4-dioxane. The desired amount of OG-POSS (5 and 10 wt %) was dissolved in a minimum amount of 1,4-dioxane. Instead of stirring, the homogeneous mixture was subjected to sonic waves in a sonicator for effective collision and distribution at 25°C for 60 min. The blends were cast on a glass plate and pretreated with dichlorodimethylsilane. After drying at 60°C for 5 h, the blended samples were cured at 100, 150, and 200°C for 1 h each and at 250°C for 2 h in an air oven. The red brown colored POSS reinforced CE (POSS-AECE) nanocomposites were obtained, with thicknesses ranging from 0.3 to 0.5 mm (Scheme 12.5).

12.5.1.3 PREPARATION OF CYANATE ESTER-SILICA (CE—SiO_2) HYBRID

The BCE and TEOS were taken in a 1:1 weight ratio, along with varying molar equivalents (0.5, 1.0, 1.5, and 2.0) of coupling agent (3-aminopropyl)triethoxysilane (APTES) or (3-glycidoxypropyl)methyldiethoxysilane (GPTMS) with respect to the BCE monomer. 0.2 g tetraethyl orthosilicate (TEOS) was dissolved in 4 mL of tetrahydrofuran (THF) and exact equivalents of dilute acid (0.1 N) was added. 0.2 g BCE and varying molar equivalents of APTES or GPTMS with respect to BCE were dissolved in 4 mL of THF separately in another flask. Both solutions were mixed and agitated for

60 min. Then the curing was performed at 100°C for 60 min, at 150°C for 120 min, and finally postcured at 200°C for 120 min (Schemes 12.6 and 12.7).

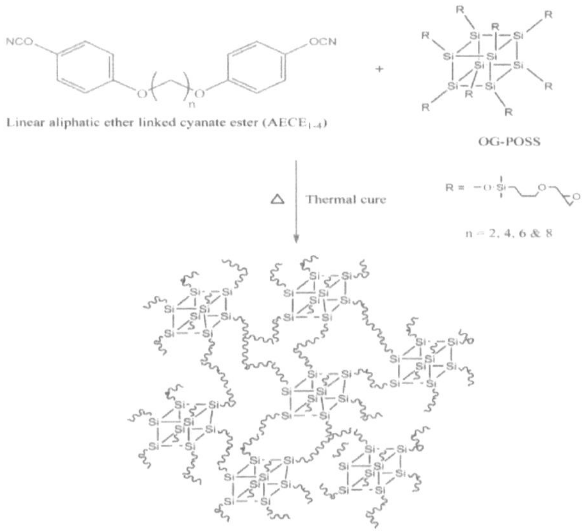

SCHEME 12.5 Synthesis of cyanate ester-based nanocomposites.

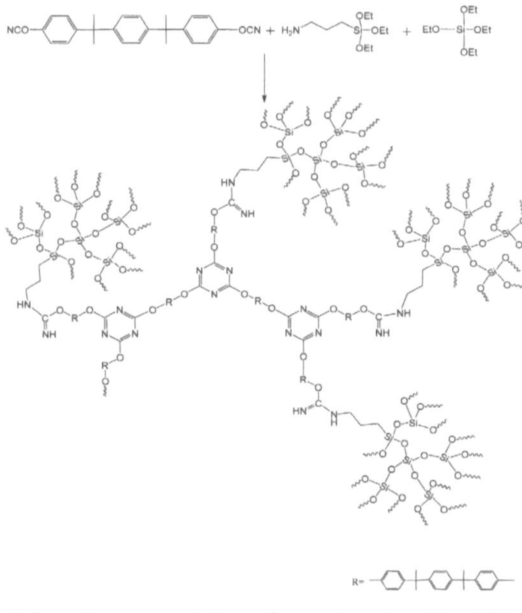

SCHEME 12.6 Schematic representation of cyanate ester silica (CE–SiO_2) hybrid using APTES.

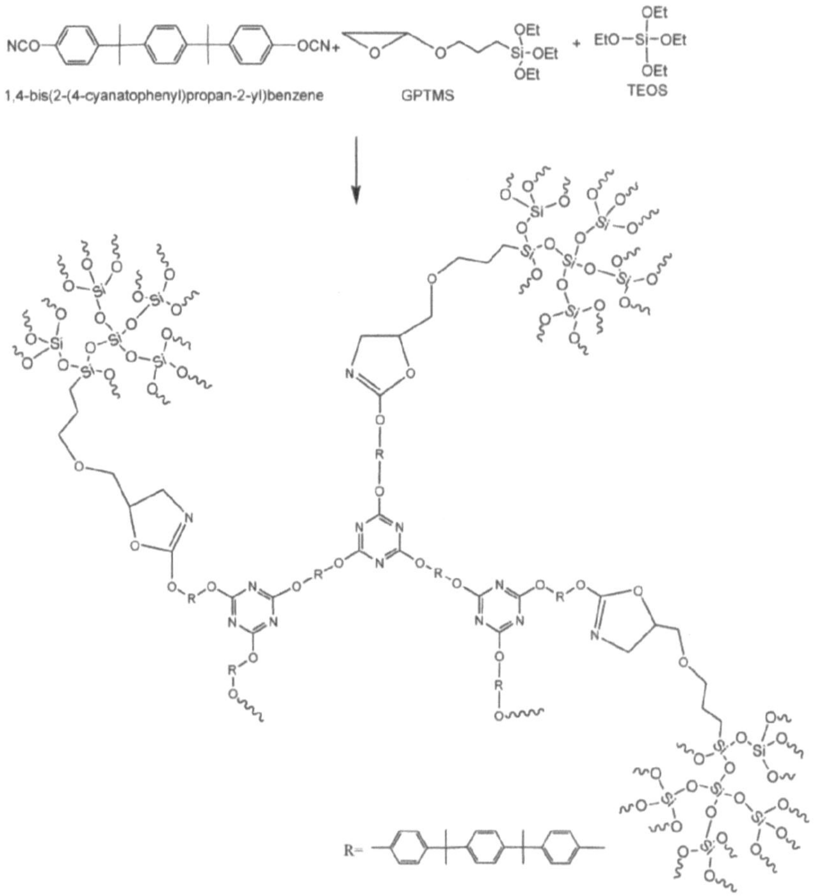

SCHEME 12.7 Schematic representation of cyanate ester silica (CE–SiO_2) hybrid using GPTMS.

12.6 CHARACTERIZATION OF POLYMER COMPOSITES

12.7 VIBRATION ANALYSIS

SBA-15 reinforced CE nanocomposites were prepared by incorporating various weight percentages of SBA-15 into BCE by thermal curing. The SBA-15 reinforced BCE nanocomposites were confirmed by the FTIR spectra. From the FTIR (Figure 12.13), the disappearance of bands at 2237 cm^{-1} (the—OCN group of the CE) and 954 cm^{-1} (the glycidyl group of SBA-15) and the appearance of a band at 1678 cm^{-1} corresponding to an oxazoline linkage, confirms the occurrence of a reaction between the OCN group of BCE and the glycidyl group of GPTMS. The presence of a band at 1078 cm^{-1} due to Si—O—Si confirms the presence of a silica network in the hybrid nanocomposites.

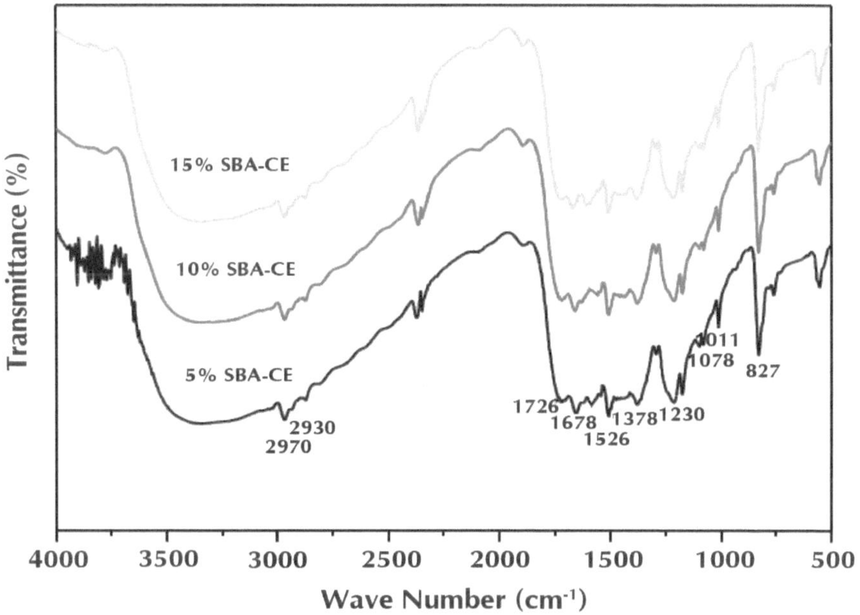

FIGURE 12.13 Fourier transform infrared spectra of various weight percentages SBA-15 reinforced BCE nanocomposites.

The linear aliphatic ether-linked CEs were synthesized via four steps approach. The POSS-AECE nanocomposites (Scheme 12.5) were prepared by incorporation of varying weight percentages of OG-POSS (5 and 10 wt %) into AECE resin. Figure 12.14 shows the FTIR spectra of varying composition of the POSS-AECE nanocomposites. The disappearance of bands at 2237 cm^{-1} (the —OCN group of the CE) and 914 cm^{-1} (the glycidyl group of OG-POSS) and the appearance of a band at 1678 cm^{-1} corresponding to an oxazoline link the glycidyl group of OG-POSS. The symmetric and asymmetric stretching frequencies of the —CH_2 group appeared at 2924 and 2868 cm^{-1}, respectively. The presence of a band at 1100 cm^{-1} due to Si—O—Si confirms the formation of a silica network in the hybrid.

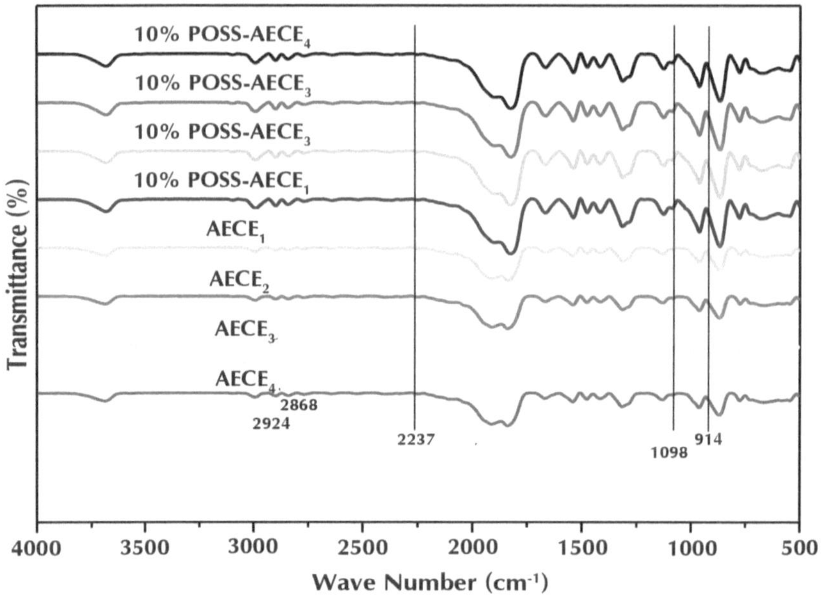

FIGURE 12.14 Fourier transform infrared spectra of varying composition of the POSS-AECE nanocomposites.

Schemes 12.6 and 12.7 illustrate the sequence of reactions involved in the preparation of CE–silica hybrid (CE—SiO_2) nanocomposites using APTES and GPTMS, respectively, as coupling agents. Table 12.1 summarizes the composition of reactants used for the preparation of CE—SiO_2 hybrids. The FTIR spectra of the CE—SiO_2 hybrid prepared using APTES as a coupling agent are shown in Figure 12.15. The disappearance of the band at 2237 cm^{-1} again corresponds to the —OCN group of the CE, and the appearance of a band at 1605 cm^{-1}, due to the —C═N linkage, confirms the occurrence of the addition-type reaction between OCN and the amino group of APTES. The presence of a band at 1100 cm^{-1} corresponding to Si—O—Si confirms the presence of silica in the hybrid nanomaterials. Furthermore, the appearance of bands at 1365 and 1565 cm^{-1} indicate that the curing reaction of BCE proceeds via the formation of a cyanurate ring. Figure 12.16 shows the FTIR spectra of the CE—SiO_2 hybrids prepared using GPTMS as the coupling agent. The disappearance of bands at 2237 cm^{-1} (the —OCN group of the CE) and 914 cm^{-1} (the glycidyl group of GPTMS) and the appearance of a band at 1678 cm^{-1} corresponding to an oxazoline linkage, confirm the occurrence of a reaction between the OCN group of BCE and the glycidyl group of GPTMS. The presence of a band at 1100 cm^{-1} due to Si—O—Si confirms the presence of a silica network in the hybrid.

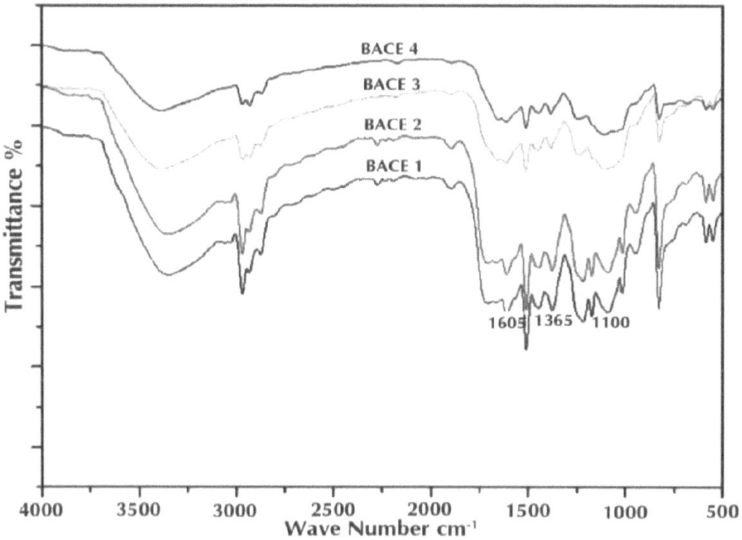

FIGURE 12.15 Fourier transform infrared spectra of cyanate ester silica (CE—SiO_2) hybrid using APTES (a) 1:05:1, (b) 1:1:1, (c) 1:1.5:1, and (d) 1:2:1 (BCE: APTES: TEOS).

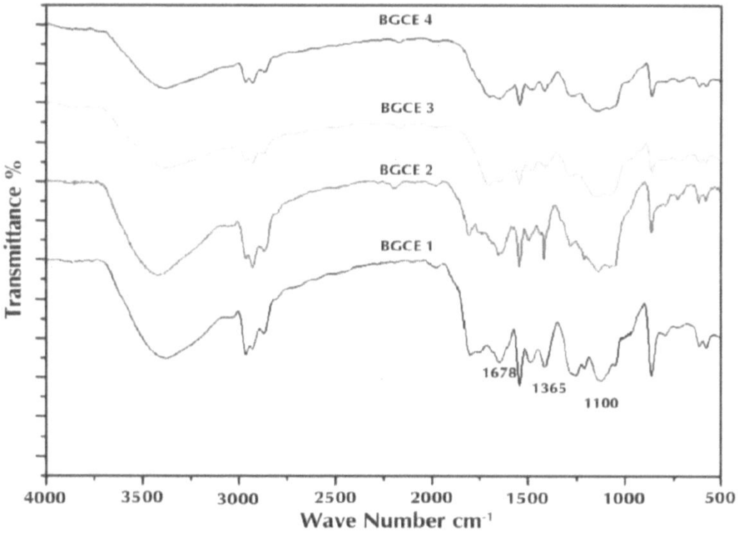

FIGURE 12.16 Fourier transform infrared (FT-IR) spectra of cyanate ester silica (CE—SiO_2) hybrid using GPTMS (a) 1:0.5:1, (b) 1:1:1, (c) 1:1.5:1, and (d) 1:2:1 (BCE: GPTMS: TEOS).

TABLE 12.1 Curing Time Cycle of Neat BCE and CE—SiO_2 Hybrid Nanocomposites

Description	BCE (mg)	APTES (mg)	GPTMS (mg)	TEOS (mg)	Ratio (BCE: *CA:TEOS)	1,4-Dioxane (mL)	0.1 M HCl_{aq} (μL)	Curing cycle Temp (°C)/(min)	Appearance
Neat BCE	200	0	0	0	–	4	0	120 + 180°C/120 + 120 min	Transparent
$BACE_1$	200	45	0	200	1:0.5:1	4	83	120 + 180°C/120 + 120 min	Transparent
$BACE_2$	200	90	0	200	1:1:1	4	96	120 + 180°C/120 + 120 min	Transparent
$BACE_3$	200	135	0	200	1:1.5:1	4	110	120 + 180°C/120 + 120 min	Translucent
$BACE_4$	200	180	0	200	1:2:1	4	123	120 + 180°C/120 + 120 min	Translucent
$BGCE_1$	200	0	60	200	1:0.5:1	4	83	120 + 180°C/120 + 120 min	Transparent
$BGCE_2$	200	0	119	200	1:1:1	4	96	120 + 180°C/120 + 120 min	Transparent
$BGCE_3$	200	0	179	200	1:1.5:1	4	110	120 + 180°C/120 + 120 min	Translucent
$BGCE_4$	200	0	238	200	1:2:1	4	123	120 + 180°C/120 + 120 min	Translucent

*CA: Coupling agent

12.8 THERMAL ANALYSIS

The values of T_g of neat BCE and GSBA-15 reinforced BCE nanocomposites were obtained from the DSC analysis, and are presented in Table 12.2. The T_g values of the SBA-15/CE nanocomposites are increased with an increase in the amount of GSBA-15. The neat BCE has a T_g value of 182°C, whereas 15 wt % of GSBA-15 incorporated BCE nanocomposites exhibited a T_g value of 203°C. The organic domains in SBA-15 also participate in the thermal polymerization, and strongly interact with the BCE matrix. In addition, the rigid mesoporous silica restricted the mobility of the polymer chains and rotational barrier of the PBZ molecules, thereby improving the T_g values of the resultant nanocomposites.

TABLE 12.2 Thermal and Dielectric Properties of SBA-15/CE Nanocomposites

Experiments	T_g (°C)	10% Weight loss (°C)	50% Weight loss (°C)	Char yield (at 700°C) (%)	Dielectric constant (ε)
SBA-15	–	–	–	94.1	–
GSBA-15	–	–	–	83.5	–
Neat BCE	182	264	353	8.8	3.33
5 wt % SBA-15/CE	191	275	365	13.2	2.90
10 wt % SBA-15/CE	196	278	371	14.7	2.53
15 wt % SBA-15/CE	203	281	380	16.6	2.11

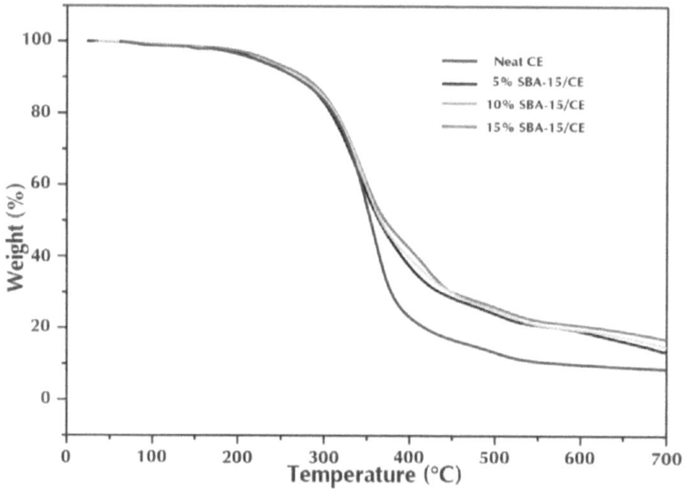

FIGURE 12.17 Thermogravimetric analysis (TGA) thermogram of neat BCE and SBA-15/CE nanocomposites.

The effect of SBA-15 on the thermal degradation of the BCE was studied using the thermogravimetric analysis (TGA) in the temperature range of ambient to 700°C in a nitrogen atmosphere. From the TGA curves shown in Figure 12.17, it was observed that the thermal degradation of the neat BCE and its composite films occur through a single degradation step, which indicates that a good phase interconnection exists between the mesoporous silica and BCE matrix, and also confirms the successful chemical graft of the GPTMS onto the mesoporous silica materials. It is clearly seen from Figure 12.17 that the thermal stability of the BCE nanocomposites is increased by the incorporation of mesoporous silica. The increase in weight residues ultimately increases the thermal stability. The improvement of thermal stability is mainly due to the chemical interaction between the BCE matrix and the mesoporous silica. The enhanced thermal stability may be explained as due to the partial ionic nature and the high bond energy of Si—O—Si, which may cause delay in degradation [10]. The char yield of the hybrids increase with increasing silica content. The increasing char residue can reduce the formation of combustible gases and decrease the exothermic nature of the pyrolysis reaction, which in turn, lowers the thermal conductivity and inhibits the combustion of the burning moieties [11]. Furthermore, it may also be ascertained that the thermal stability of organic materials can be improved by introducing inorganic components such as silica, on the basis of the fact that these materials have inherently good thermal stability and flame retardant behavior [12, 13].

The values of T_g for the neat AECEs and POSS-AECE nanocomposites are presented in Table 12.3. The POSS-AECE nanocomposites exhibit enhanced values of T_g compared to that of neat AECE. The T_g value of neat AECE and 5, 10 wt % POSS-CE nanocomposites are presented in Table 12.3. The increase in the values of T_g of the POSS-AECE nanocomposites may be explained by the following reasons. First, the incorporation of POSS increases the cross-linking density of the resulting nanocomposites, and increases the rigidity of the nanocomposite system. Second, the incorporation of POSS creates the porosities (free volume) in the nanocomposites; this effect has, however, been counteracted by an increase in the cross-linking density.

Among the POSS-AECE systems, the $AECE_1$-based system possesses the highest value of T_g and the $AECE_4$-based system exhibits the lowest value of T_g. The values of other systems lie between these two systems. The T_g value of POSS-AECE decreased linearly with an increasing chain length. This may be due to the flexible aliphatic backbone, which influences the T_g

of the hybrid systems, by increasing the internal rotations and thermal motion of the polymer chains. The decrease in the values of T_g may be due to an increase in the alkoxy chain length, which imparts flexibility.

The TGA was used to ascertain the thermal stability of neat AECE and OG-POSS reinforced AECE nanocomposites, and the data obtained from the TGA are presented in Figure 12.18 and Table 12.3. The POSS-AECE nanocomposites exhibited higher thermal stability and higher char yield than those of neat AECE resin. Generally, an improvement in the thermal stability of hybrid composites is related to the degree of interaction between the polymer matrix and the inorganic phase, and to the fact that the inorganic component has an inherently good thermal stability. In addition, the presence of the rigid cubic siloxane and partial ionic nature of inorganic Si—O—Si skeleton also contributed to the thermal stability of the hybrid nanocomposites. The POSS core gave an additional heat capacity, which stabilizes the materials against thermal decomposition. The loss of organic materials from the

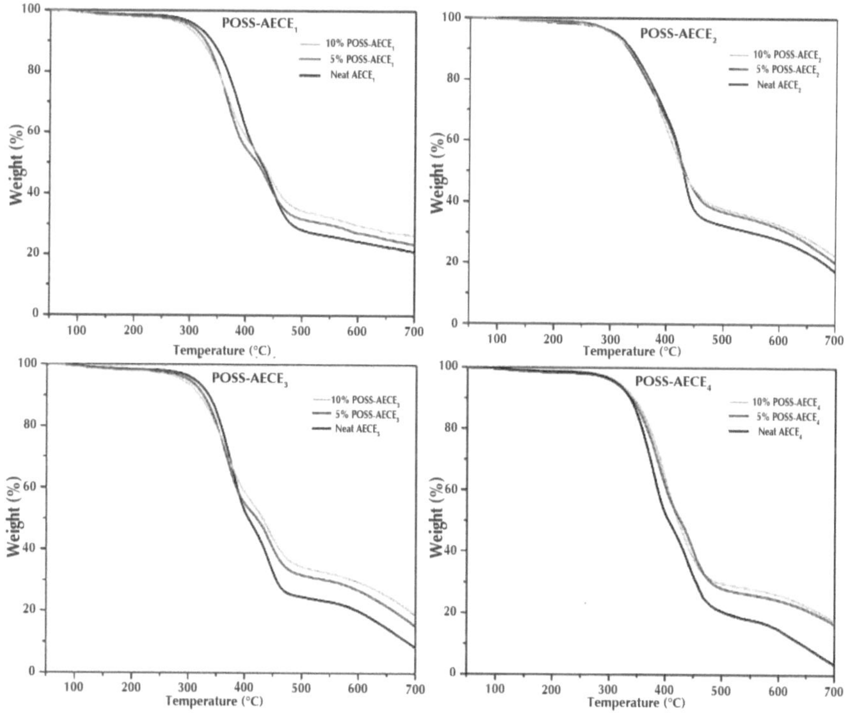

FIGURE 12.18 Thermogravimetric analysis thermogram of neat AECE and POSS-AECE nanocomposites.

segmental decomposition through gaseous fragments could be reduced by a well dispersed POSS core in the AECE matrix. Further, it was also observed that with the incorporation of OG-POSS into the AECE resin, reduced both the volatile decomposition and the polymer flammability. Further, when POSS degrades, it leaves an inert silica layer, which can form a protective layer on the surface of the material preventing further oxidation of the inner part of the matrix [14]. Among the four systems ($AECE_{1-4}$), the $AECE_1$ based POSS incorporated system possesses highest thermal stability, and the POSS incorporated $AECE_4$ system shows the lowest thermal stability; this is due to the higher aliphatic chain present in the system, which induces free rotational movement and reduces the thermal stability of the resulting nanocomposites.

TABLE 12.3 Thermal and Dielectric Properties of Neat AECE and POSS-AECE Nanocomposites

Experiments	T_g (°C)	10% weight loss (°C)	50% weight loss (°C)	Char yield (at 700°C) (%)	Dielectric constant(έ)
$AECE_1$	220	336	417	20.1	4.2
5% POSS-$AECE_1$	228	351	425	23.5	3.6
10% POSS-$AECE_1$	235	375	430	27.2	3.1
$AECE_2$	217	330	423	16.2	3.8
5% POSS-$AECE_2$	222	329	427	19.0	3.5
10% POSS-$AECE_2$	230	331	429	21.4	2.9
$AECE_3$	210	328	410	10.2	3.6
5% POSS-$AECE_3$	223	325	420	15.1	3.0
10% POSS-$AECE_3$	228	321	427	19.4	2.6
$AECE_4$	203	336	420	5.1	3.2
5% POSS-$AECE_4$	211	338	425	14.8	2.9
10% POSS-$AECE_4$	221	341	440	16.0	2.4

The thermal properties of the neat BCE and CE—SiO_2 hybrid were characterized using the DSC and TGA. The values of T_g of the neat BCE and hybrid nanomaterials are presented in Table 12.4. From the DSC thermogram it is observed that the hybrids $BACE_2$ (1:1:1) and $BGCE_2$ (1:1:1) show higher T_g values than those of the other hybrid systems (1:0.5:1, 1:1.5:1 and 1:2:1), which might be due to the equivalent possibilities of cyclotrimerization and covalent bond formation between the organic and inorganic phases compared with the other systems. This may be attributed to the fact that the coupling agent influences the thermal properties of the hybrid nanomaterials. In the case of the $BACE_1$ system, the composition of the coupling agent (1:0.5:1) is less, and hence, the covalent bond formation between the organic and inorganic phases will be meager, compared with that of the $BACE_2$ system (1:1:1), which leads to the lower value of T_g. In the case of the $BACE_4$ system, the covalent bond formation between the organic and inorganic phases will be high, through adduct formation between the OCN and NH_2 group; hence, only a lesser percentage of free OCN groups are available for the formation of the triazine ring; this may be the reason for the lower T_g of the $BACE_4$ system. Similar results were observed in the case of CE-SiO_2, obtained using GPTMS as a coupling agent. Among the two CE—SiO_2 hybrids, the hybrid obtained using GPTMS as a coupling agent possesses a lower T_g, which is due to the presence of the flexible oxazoline linkage.

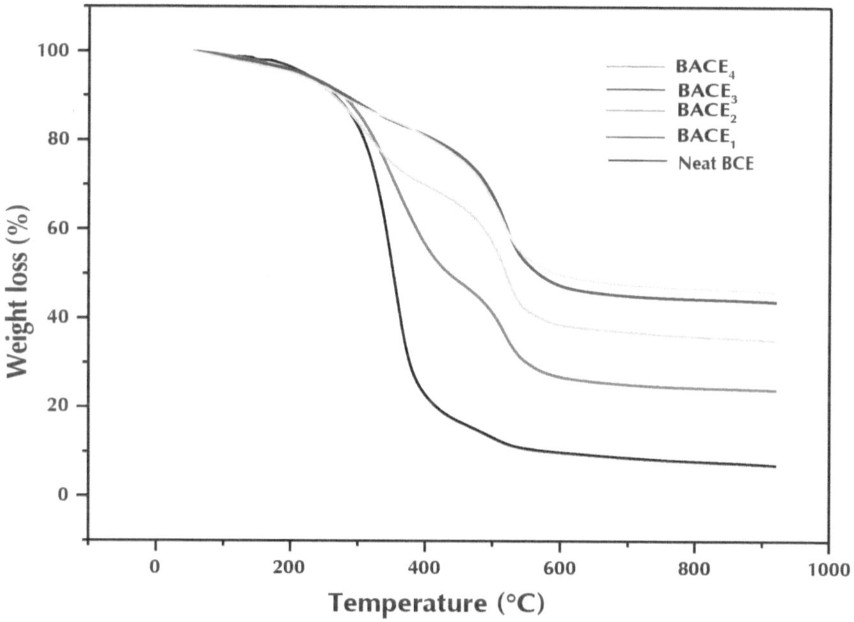

FIGURE 12.19 Thermogravimetric analysis thermogram of CE—SiO_2 hybrid nanomaterials using APTES.

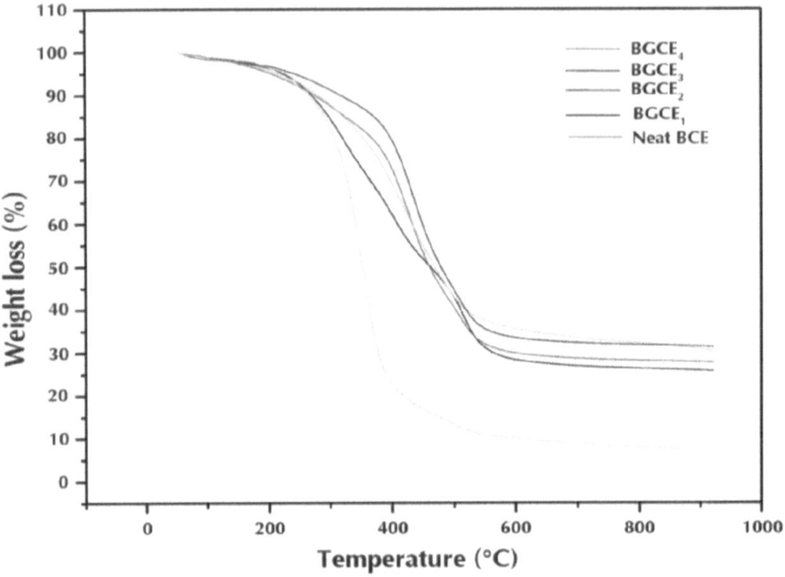

FIGURE 12.20 Thermogravimetric analysis thermogram of CE—SiO_2 hybrid nanomaterials using GPTMS.

The TGA of all the systems is shown in Table 12.3 and Figures 12.19 and 12.20. The CE—SiO_2 hybrid showed higher thermal stability (as evidence by the thermal degradation temperature) when compared to that of the cured

TABLE 12.4 Thermal Properties of Neat BCE and CE—SiO_2 Nanocomposites

Experiments	Glass transition temperature (T_g °C)	10% weight loss at temp (°C)	50% weight loss at temp (°C)	Char yield (at 900°C) (%)	Dielectric constant (ε)
Neat BCE	182	264	353	7.1	3.33
$BACE_1$	206	273	436	24.0	3.10
$BACE_2$	250	276	520	35.1	2.96
$BACE_3$	208	281	568	43.8	2.85
$BACE_4$	206	290	597	46.0	2.66
$BGCE_1$	185	271	458	25.7	3.05
$BGCE_2$	205	275	462	27.7	2.83
$BGCE_3$	194	283	471	30.4	2.61
$BGCE_4$	192	317	485	31.3	2.40

neat BCE, due to the highly thermally stable silica network in the hybrid. The CE—SiO_2 hybrid prepared using APTES as a coupling agent, shows a double degradation behavior (i.e., two steps), whereas the cured hybrid obtained from the GPTMS system, shows a single degradation behavior. This may be due to the cleavage of a thermally weak adducting linkage that occurs first, followed by the cleavage of the triazine ring. The char yield of CE—SiO_2 hybrids is higher than that of neat BCE due to the presence of the silica network in the systems.

12.9 DIELECTRIC CONSTANT

Figure 12.21 displays the values of the dielectric constants measured at a frequency of 1.0 MHz as a function of the mesoporous silica content. It is very interesting to observe that the values of the dielectric constant were reduced from 3.34 to around 2.90 by incorporating 5 wt % SBA-15 into the composite film, and the dielectric constant gradually decreases with an increase in the amount of the mesoporous silica materials continuously. For the composite film containing 10 wt % SBA-15, the dielectric constant exhibits a

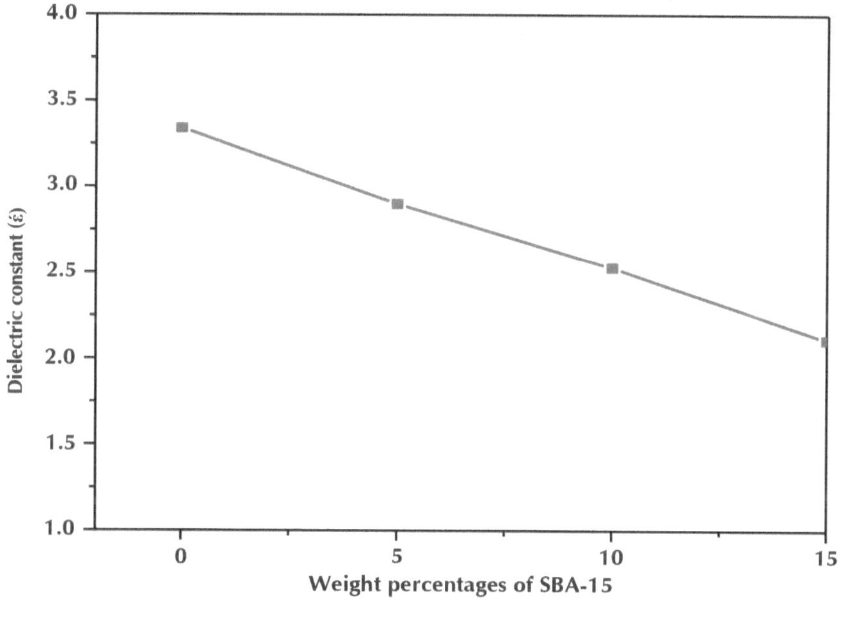

FIGURE 12.21 Dielectric constant of SBA-15/CE nanocomposites.

minimum value of 2.53. However, the dielectric constant of the SBA-15/CE composite films continuously decreases until the content of SBA-15 reaches 15 wt %, at which the composite film achieves a minimum dielectric constant ($k = 2.11$). All the measurements have been repeated five times, and the deviations of the testing data were controlled with ±0.05. Therefore, it is concluded that the reduction in the value of the dielectric constant resulted from the incorporation of air voids ($k = 1$) when the mesoporous silica materials are introduced into the composite films. The dielectric constant of a composite film decreases with an increase in the interaction force.

A lower dielectric constant is one of the most desirable properties for next-generation electronic devices. Table 12.3 gives the values of the dielectric constant of different POSS-AECE nanocomposites with varying percentages of POSS concentration. The dielectric constant of the hybrid decreases with an increase in the amount of POSS. The reduction in the dielectric constant of the POSS-AECE hybrids may be explained in terms of the creation of pores and a free volume increase by the presence of the rigid and large POSS structure. The signal propagation delay time of integrated circuits is proportional to the square root of the dielectric constant of the matrix, and the signal propagation loss is proportional to the square root of the dielectric constant and dissipation factor of the matrix. Thus, a material with a low dielectric constant and low dissipation factor will reduce the signal propagation delay time and loss [15], and improve the effective and efficient functioning of electronic instruments.

Among the POSS-AECE systems studied, the $AECE_4$-based system possesses the lowest dielectric constant, and the $AECE_1$-based system exhibits the highest dielectric constant. The values of the other two systems lie between these two systems. This may be due to the higher aliphatic chain that reduces the polarity and enhances the free rotation of the molecules.

The interlayer dielectric (ILD) surrounds and insulates the interconnect wiring. As the line spacing of the interconnects decreases, the associated higher resistance and capacitive coupling cause an increasing problem of signal delay, known as resistor-capacitor (RC) delay, of the circuit itself. Lowering the k value of the ILD decreases the RC delay, lowers power consumption, and reduces "cross-talk" between nearby interconnects [16]. Hence, the development of new low-k materials is warranted for advanced microelectronic applications. In this direction, a new $CE—SiO_2$ hybrid material is developed in the present study, and its insulating behavior is assessed. Table 12.4 and Figure 12.22 show the values of the dielectric constant of neat BCE and $CE—SiO_2$ hybrid nanocomposites. From the data it is inferred that the incorporation of silica into the BCE lowered the values of the dielectric constant, when compared to that of neat BCE. This is attributed to the presence of low polar Si—O—Si present in the hybrid nanocomposites.

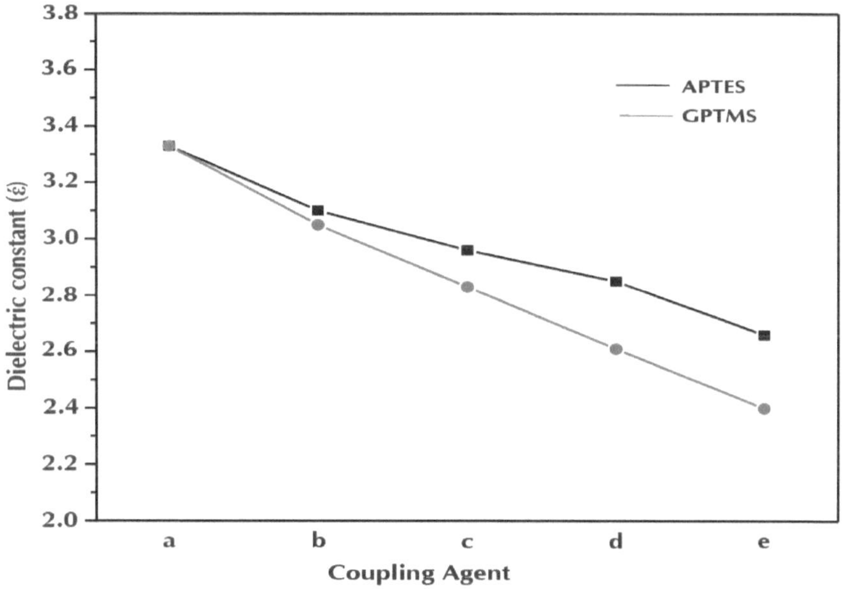

FIGURE 12.22 Dielectric properties of neat BCE and CE—SiO_2 hybrid nanocomposites (a) neat BCE (b) 1:0.5:1, (c) 1:1:1, (d) 1:1.5:1, and (e) 1:2:1.

12.10 MORPHOLOGICAL PROPERTIES

12.10.1 SBA-15/CE NANOCOMPOSITES

The mesoporous structure of the SBA-15/CE nanocomposites is retained even after the stepwise thermal polymerization of the BCE monomer, which is evidenced by the TEM images (Fig. 12.23a and b). The TEM images of the SBA-15/CE nanocomposites are presented in Figure 12.23(a) and (b). They show a homogeneous morphology, and there was no significant aggregation noticed in the hybrid nanocomposites even at the maximum loading of GSBA-15 (15 wt %).

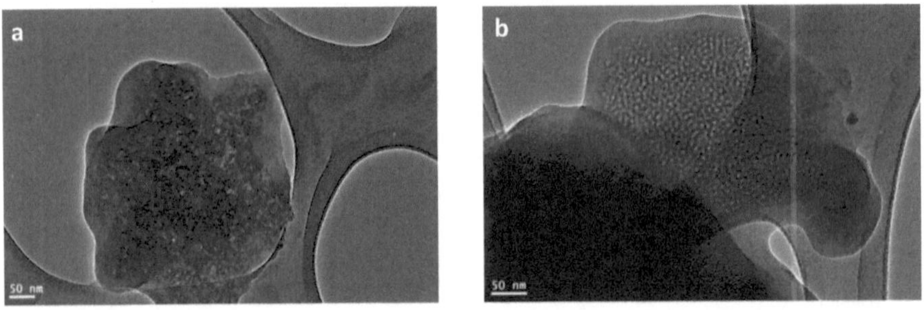

FIGURE 12.23 TEM images of the (a) 5 wt % SBA-15/CE, (b) 15 wt % SBA-15/CE.

12.10.2 POSS-AECE NANOCOMPOSITES

The scanning electron microscope is used to investigate the morphology of neat AECE and POSS reinforced AECE nanocomposites. Figure 12.24 shows the SEM micrograph of neat AECE and POSS-AECE nanocomposites. The

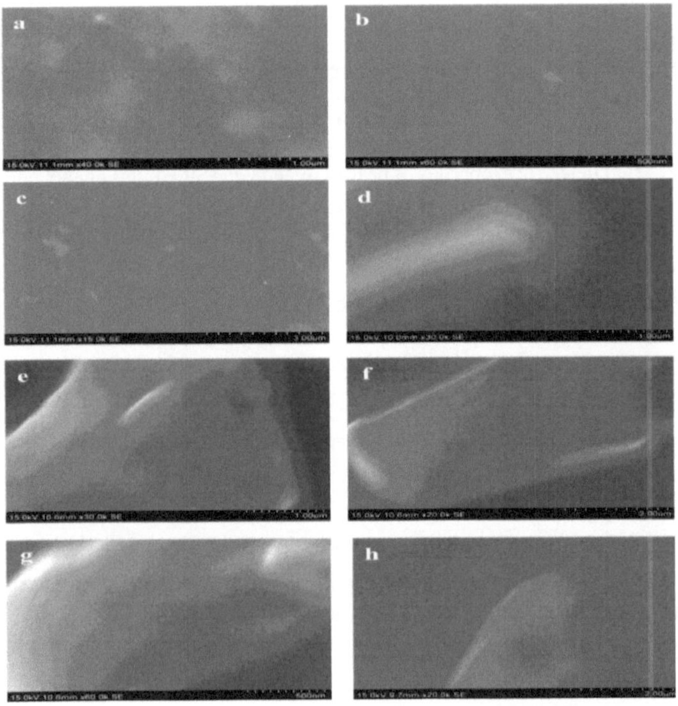

FIGURE 12.24 SEM micrograph of (a) neat $AECE_1$, (b) 10% POSS-$AECE_1$, (c) $AECE_2$, (d) 10% POSS-$AECE_2$, (e) neat $AECE_3$, (f) 10% POSS-$AECE_3$, (g) neat $AECE_4$, and (h) 10% POSS-$AECE_4$.

SEM images indicated the smooth and homogeneous morphology of the neat AECE and POSS-AECE systems. The POSS incorporated system exhibited a featureless morphology and no discernable phase separation was observed. This indicates the good miscibility (compatibility) of POSS with AECE, and also confirms the molecular-level dispersion of POSS in the AECE nanocomposites. The homogeneous morphology could be ascribed to the formation of covalent bonding between POSS and CE. This was further confirmed by the TEM analysis.

Figure 12.25(a–d) represents the TEM micrographs of 10 wt % of OG-POSS reinforced AECE nanocomposites. It shows the homogeneous morphology, and no localized domains were observed. Only a few darker points were observed at approximately about 50 nm in the polymer matrix, which represents the dispersion of OG-POSS in the AECE matrix. An observation indicated that the nanocomposite is a material with a particle size of the dispersed phase having at least one dimension of less than 100 nm [17]. Thus, the POSS moieties are well dispersed at a nanometer scale in the AECE matrix to form a POSS-AECE network.

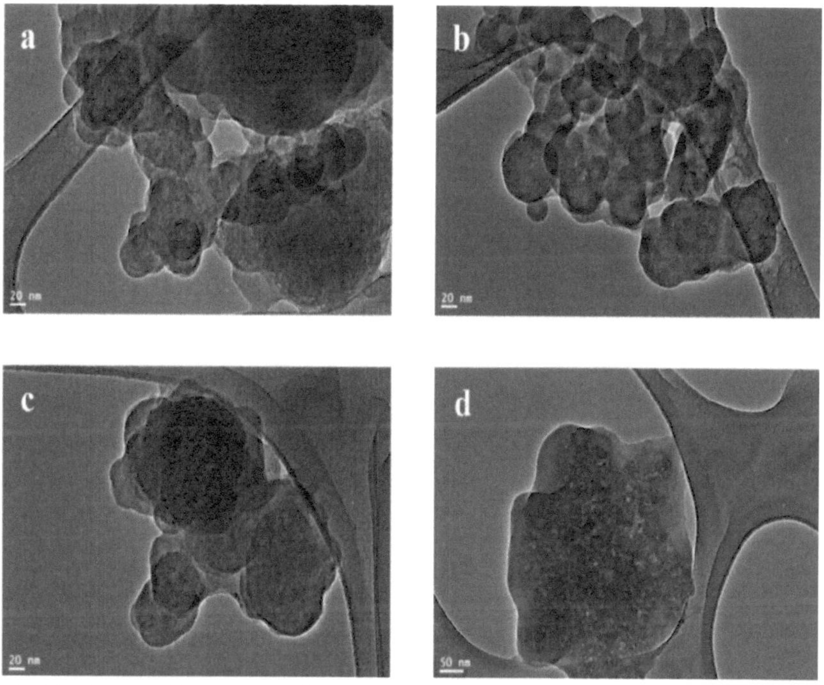

FIGURE 12.25 TEM image of (a) 10% POSS-$AECE_1$, (b) 10% POSS-$AECE_2$, (c) 10% POSS-$AECE_3$, and (d) 10% POSS-$AECE_4$.

12.10.3 NEAT BCE AND CE—SiO_2 HYBRID NANOMATERIALS

The surface topology of the neat BCE and CE—SiO_2 hybrid nanomaterial was characterized using AFM techniques; the AFM images of neat CE (Fig. 12.26a) shows a uniform and smooth surface morphology. The AFM images of the hybrid $BACE_2$ (1:1:1) and $BGCE_2$ (1:1:1) are presented in Figure 12.26(b) and (c), respectively. The surface of the hybrid thin film is quite smooth and has little visible defects, indicating that the synthesized hybrids have a molecular-level dispersion of organic and inorganic networks, and the silica particles (5–10 nm) are distributed uniformly at the nanoscale in the organic phase. This is further supported by the SEM analysis (Fig. 12.27 shows the SEM images of neat BCE, $BACE_2$, and $BGCE_2$), from which it can be seen that the hybrid exhibits excellent structural uniformity, with no cracks or flaws.

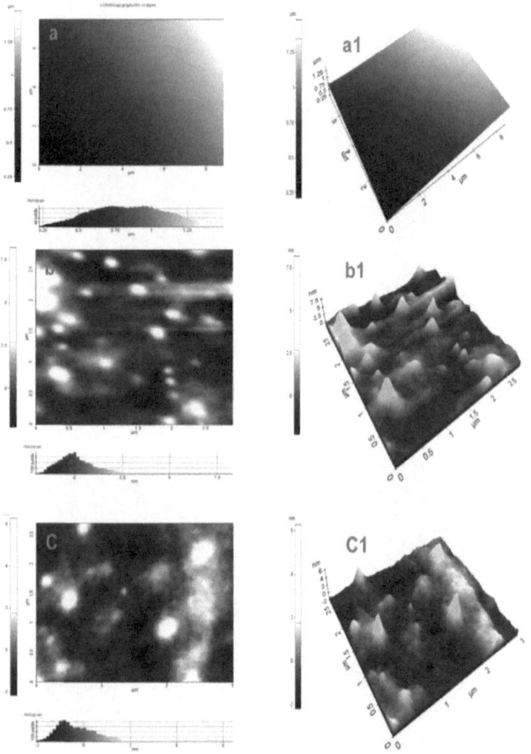

FIGURE 12.26 AFM image of CE—SiO_2 hybrid (a) 2D image of neat BCE and (a1) 3D image of neat BCE, (b) 2D image of $BACE_2$ and (b1) 3D image of $BACE_2$, and (c) 2D image of $BGCE_2$ and (c1) 3D image of $BGCE_2$.

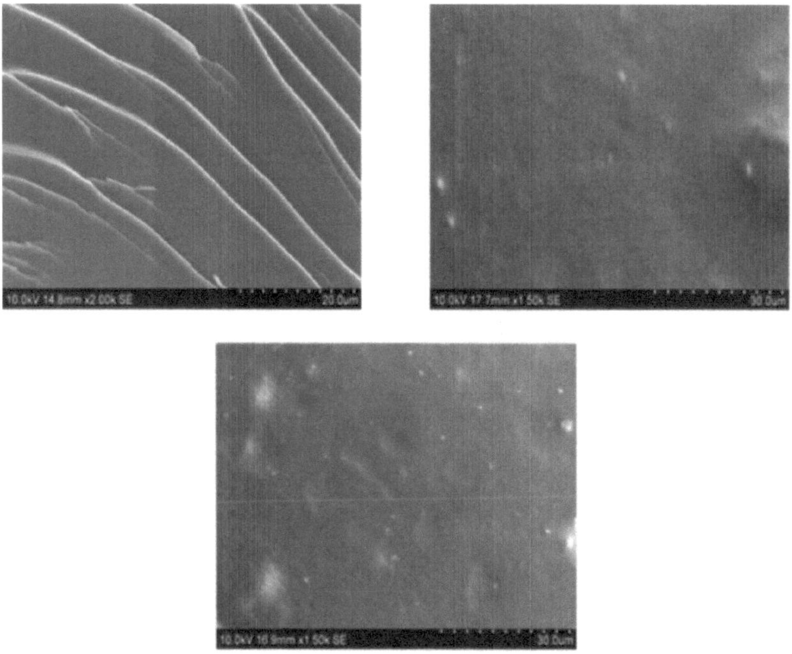

FIGURE.12.27 SEM image of CE–SiO_2 hybrid (a) neat BCE, (b) $BACE_2$, and (c) $BGCE_2$.

12.11 CONCLUSION

A new series of linear aliphatic ether linked CE-based POSS-AECE and aromatic cyanate ester (BCE)-based SBA-15 and BCE based SiO_2 organic inorganic hybrid nanocomposites have been developed. The formation of nanocomposites was confirmed by different analytical techniques. The data obtained from the thermal analysis indicate that $BACE_4$ nanocomposites exhibit excellent thermal properties such as higher T_g, thermal stability, and higher char yield than those of other nanocomposites system. The value of dielectric constant ($\acute{\varepsilon}$) of SBA-15 incorporated BCE is lower when compared to that of other nanocomposites developed in the present work.

ACKNOWLEDGMENT

The authors wish to thank P. Prabunathan, Department of Chemical Engineering, Anna University, Chennai, for his support extended to prepare this manuscript.

KEYWORDS

- **cyanate esters**
- **nanocomposites**
- **low-k material**
- **polymer**
- **epoxy resins**

BIBLIOGRAPHY

1. Huang, Y.; Economy, J. New High Strength Low-K Spin-On Thin Films for IC Application. *Macromolecules*. **2006**, *39*, 1850–1853.
2. Watanabe, Y.; Shibasaki, Y.; Ando, S.; Ueda, M. Synthesis and Characterization of Novel Low-k Polyimides from Aromatic Dianhydrides and Aromatic Diamine Containing Phenylene Ether and Perfluorobiphenyl Units. *Polym. J.* **2006**, *38*, 79–84.
3. Maier, G. Low Dielectric Constant Polymers for Microelectronics. *Prog. Polym. Sci.* **2001**, *26*, 3–65.
4. Zhang, C.; Guang, S.; Zhu, X.; Xu, H.; Liu, X.; Jiang, M.Mechanism of Dielectric Constant Variation of POSS-Based Organic−Inorganic Molecular Hybrids. *J. Phys. Chem. C*. **2010**, *114*, 22455–22461.
5. Su, K.; Bujalski, D.R.; Eguchi, K.; Gordon, G.V.; Li Ou, D.; Chevalier, P.; Hu, S.; Boisvert, R.P. Low-K Interlayer Dielectric Materials: Synthesis and Properties of Alkoxy-Functional Silsesquioxanes. *Chem. Mater*. **2005**, *17*, 2520–2529.
6. Fu, G.D.; Zhang, Y.; Kang, E.T.; Neoh, K.G. Nanoporous Ultra-Low-K Fluoropolymer Composite Films via Plasma Polymerization of Allylpentafluorobenzene and Magnetron Sputtering of Poly(Tetrafluoroethylene). *Adv. Mater*. **2004,** *16*, 839–842.
7. Nunomura, S.; Kog, K.; Shiratani, M.; Watanabe, Y.; Morisada, Y.; Matsukil, N.; Ikedal, S. Fabrication of Nanoparticle Composite Porous Films Having Ultralow Dielectric Constant. Jpn. *J. Appl. Phys*. **2005**, 44, 1509–1511.
8. Miller, R.D.; In Search of Low K Dielectrics. *Science*. **1999**, 286, 421–423.
9. Vinoba, M.; Kim, D.H.; Lim, K.S.; Jeong, S.K.; Lee, S.W.; Alagar, M. Biomimetic Sequestration of CO2 and Reformation to CaCO3 Using Bovine Carbonic Anhydrase Immobilized on SBA-15. *Energ. Fuels*. 2011, 25, 438–445.
10. Devaraju, S.; Vengatesan, M.R.; Ashok Kumar, A.; Alagar, M. Polybenzoxazine-Silica (PBZ-SiO_2) Hybrid Nanocomposites through In Situ Sol-Gel Method. *J. Sol-Gel Sci. Techn*. **2011**, *60*, 33–40.
11. Devaraju, S.; Vengatesan, M.R.; Selvi, M.; Ashok Kumar, A.; Alagar, M. Synthesis and Characterization of Bisphenol-A Ether Diamine Based Polyimide POSS Nanocomposites for Low K Dielectric and Flame-Retardant Applications. *High Perform. Polym*. Vol. **2012**, *24*, 84–96.

12. Choi, M.H.; Chung, I.J. Mechanical and Thermal Properties of Phenolic Resin-Layered Silicate Nanocomposites Synthesized by Melt Intercalation. J. *Appl. Polym. Sci.* **2003**, *90*, pp. 2316–2321.
13. Dean, K.; Krstina, J.; Tian, W.; Varley, R.J. Effect of Ultrasonic Dispersion Methods on Thermal and Mechanical Properties of Organoclay Epoxy Nanocomposites. *Macromol. Mater. Eng.* **2007**, 292, 415–427.
14. Devaraju, S.; Prabunathan, P.; Selvi, M.; Alagar, M. Low Dielectric and Low Surface Free Energy flexible Linear Aliphatic Alkoxy Core Bridged Bisphenol Cyanate Ester Based POSS Nanocomposites. *Front. Chem.* **2013**, *1*, 1–10.
15. Nagendiran, S.; Alagar, M.; Hamerton, I.; Octasilsesquioxane-Reinforced DGEBA and TGDDM Epoxy Nanocomposites: Characterization of Thermal, Dielectric and Morphological Properties. *Acta Mater.* **2010**, 58, 3345–3356.
16. Laine, R.M. Nanobuilding Blocks Based on the $[OSiO_{1.5}]_x$ (x = 6, 8, 10) Octasilsesquioxanes. *J. Mater. Chem.* **2005**, *15*, 3725–3744.
17. Komarneni, S. Nanocomposites. *J. Mater. Chem.* **1992**, 2, 1219–1230.

CHAPTER 13

HISTORY OF ANTIFOULING COATING AND FUTURE PROSPECTS FOR NANOMETAL/POLYMER COATINGS IN ANTIFOULING TECHNOLOGY

K. PRIYA DASAN

*Material Chemistry Division, SAS, VIT University, Vellore, Tamil Nadu 632014, India. *E-mail: priyajeenetd@gmail.com*

CONTENTS

ABSTRACT

Biofouling has been a matter of huge concern in terms of economy and environment. Though many techniques are known for countering this, the coatings technology involving biocides dominates the market. Tributyltin (TBT)-based coatings were the most widely used and popular among the antifouling coating industry till its toxic nature to marine systems led to a ban by International Maritime Organisation (IMO) on its application in antifouling coatings. This promoted research for new technologies in antifoul coatings, and still the search is at nascent stage. Previous studies have shown that antimicrobial formulations in the form of nanoparticles could be used as effective bactericidal materials. The bactericidal effect of these metal nanoparticles has been attributed to their small sizes. However, to understand the potential of nanometals as biocides, a deep understanding of fouling technology and antifouling coatings are very much required. The present chapter looks into the fundamental aspects of biofouling, the antifouling techniques involved, the IMO ban on TBT, and other biocides. The chapter also discusses the potential of nanometals as biocides in place of conventional metal-based coatings in market.

13.1 INTRODUCTION

Biofouling can be defined as the undesirable accumulation of microorganisms, plants, algae, and/or animals on wetted structures. Biofouling is found in almost all circumstances where water-based liquids are in contact with other materials. Industrially important examples include membrane systems, such as membrane bioreactors and reverse osmosis, spiral wound membranes, cooling water cycles and power stations, and oil pipelines carrying oils with entrained water, especially those carrying used oils, cutting oils, soluble oil, or hydraulic oil (Fig. 13.1). The process of biological fouling occurs mainly in two stages. Figure 13.2 shows the schematic representation of fouling stages.

Stage one involves the formation of microfouling, where biofilm formation and bacterial adhesion takes place. This includes an initial accumulation of adsorbed organics, the settlement and growth of pioneering bacteria creating a biofilm matrix. The second stage is macrofouling, where attachment of larger organisms, which include species, such as barnacles, mussels, polycheate worms, bryozoans, and seaweed. Together, these organisms form a fouling community. The microfouling starts with the formation of a slime

FIGURE 13.1 Industrially critical biofouling prone sites.

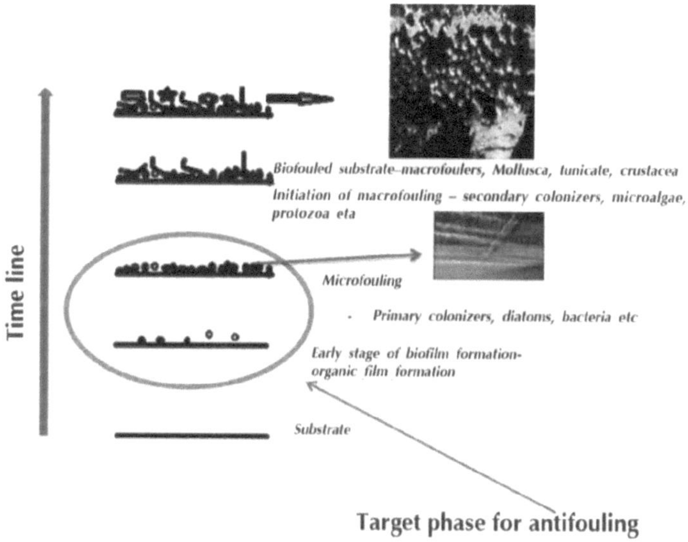

FIGURE 13.2 Schematic representation of fouling stages.

layer consisting of bacteria that accumulate on the surface of the structure under concern. Organic molecules, such as polysaccharides, proteins, and proteoglycans, and possibly inorganic compounds rapidly accumulate on every surface giving rise to the so-called conditioning film. This process is essentially governed by physical forces, such as Brownian motion, electrostatic interaction, and Van Der Waals forces. Rapidly developing bacteria and single-cell diatoms settle on this modified surface. These species are first adsorbed reversibly, again mainly by a physical process, and afterwards adhere and form, together with protozoa and rotifers, a microbial biofilm. Though the film formation and its adherence depends on a number of factors, temperature and pH plays a major role in slime formation. For example, diatoms attachment depends on the pH of the hull coating, meanwhile the biofilm-causing bacteria *Vibrio alginolyticus*, is sensitive to temperature changes and pH.

Once the biofilm is fully established, macrofouling will be initiated. The existence of adhesive exudates, such as polysaccharides, proteins, lipids and nucleic acids, and the roughness of irregular microbial colonies help to trap more particles and organisms. These include algal spores, barnacle cyprids, marine fungi, and protozoa, some of which may be attracted by sensory stimuli. This sometimes extends to the formation of more complex community that includes multicellular primary producers, grazers, and decomposers. This stage also includes the settlement and the growth of larger marine invertebrates together with the growth of macroalgae (seaweeds). Contact and colonization between the microorganism (biofilm actors) and the surface is promoted by the movement of water through Brownian motion, sedimentation, and convective transport, although organisms can also actively seek out substrates due to propulsion using flagella. Bacteria and other colonizing microorganisms secrete extracellular polymeric substances to envelope and anchor them to the substrate, thereby altering the local surface chemistry, which further stimulate growth, such as the recruitment and settlement of macroorganisms.

Typical characteristics of macrofoulers are: fast metamorphosis, rapid growth rates, low degree of substrate preference, and high adaptability to different environments. Macro biofouling are of two types. Calcareous or hard fouling are formed by barnacles, bryozoans, molluscs, tube worms, and zebra mussels. The noncalcareous fouling includes: algae, slimes, hydroids, sponges, and seaweed. The formation of macrofouling community depends on a large number of factors, such as the duration of exposure to water, temperature, surface of structure under concern, speed in case of ship hulls, etc. The type of fouling community will develop depending on the surface property of the structure and the ecosystem.

The presence of different molecules and organisms in the film influences the settlement of subsequent organisms. For example, the population which grows on moored structures is frequently different from that observed on ships.[1] The moored structures contains a larger portion of mussels and soft bodied liable to be removed at high speed. In temperate waters, mussels constitute the major fouling on bouys, although shellfishes are not usually observed on ships unless they lie idle in harbor. The fouling, which occurs on sound equipment is similar to that of ships bottoms. Barnacles, tube worms, tunicates, hydroids, and bryozoa are the chief offenders. Algae are relativley unimmportant since the installation is commonly located too far below the surface to favor plant growth. In a survey of the condition on naval vessels recently made by the naval research laboratory, a few sound domes were found free of fouling. Barnacles were rarely absent even from ships which were docked at short intervals or were cruising on norhtern waters. Tube worms were characterestic on ships from South Pacific.[1]

13.2 ECONOMIC LOSS AND ENVIRONMENTAL ISSUES

Biofouling is one of the most serious problems currently facing worldwide maritime domains. Besides the huge economic loss due to fouling, scientists world over are more concerned about the environmental issues related to the fouling. Fouling causes huge material and economic costs in maintenance of mariculture, shipping industries, naval vessels, and seawater pipelines. Biofouling is especially economically significant on ships' hulls, where high levels of fouling can reduce the performance of the vessel and increase its fuel requirements. Individually small, accumulated biofoulers can form enormous masses that severely diminish ships' maneuverability and carrying capacity. Even minor biofouling has a significant impact on the overall profitability of the vessel's operations when considered across a fleet and a vessel's 25–30 year lifetime. Slime formation resulting in biofouling can results in an increase in fuel consumption between 1 and 2%. The impact of macro-biofouling on fuel consumption greatly varies depending on the nature of the unwanted guest. While seaweed cause sea fuel consumption increase of up to 10%, shells barnacles, oysters, and mussels can cause a massive increase of 40%. In addition to fuel penalties in the short and long term, extensive biofouling will eventually lead to hull corrosion. Biofouling has huge impact on fuel efficiency and bunker fuel cost implications.[2] Loss of speed from moderate fouling can range between 10 and 18%. A month and a thick layer of slime in two months, by the end of those two months of sailing, it would be requiring 110 tonnes of fuel per day to maintain the same cruising speed. International Paint estimates that an extra 72 million tonnes of fuel would be burned each year. If this scenario was flipped, and the savings were not realized, the increased fuel consumption would lead to the production and release into the environment an estimated extra 210 million tonnes of carbon dioxide and 5.6 million tonnes of sulfur dioxide. An increase of the frequency of dry-docking operations, that is, time is lost and resources are wasted when remedial measures are applied. A large amount of toxic wastes is also generated during this process.[3,4]

Another major issue related to fouling is the migration of species resulting in loss of native species and invasion by alien species.[5,6] For example, the zebra mussel, the species listed in the top 10 of BIMCO's (Baltic and International Maritime Council) "most unwanted" species, is one among the many invasive species that biofouling has caused in introducing to native ecosystems the world over. Zebra mussels are aggressive biofoulers and the species proliferates in a wide variety of environments, thereby exposing shipping communities to the threat of zebra mussel biofouling as that were previously safe—and therefore unprepared.

13.3 ANTIFOULING TECHNIQUES

The antifouling technique is as long as human civilization. There is some evidence that metal sheets on wooden vessels were probably used in the 1500–300 BC period.[7] Greeks and Romans used similar approaches sometimes, including arsenic and sulfur mixed with oils to prevent against the attack of shipworms.[8] From the 13th to 15th century pitch, blended with several other components, such as oils, resin, and tallow, were widely used. One of the first attested reference about underwater use of copper was in 1618, during the reign of the Danish King Christian IV, mentioning the use of copper for sheltering keel and rudder. In the same period, copper (copper sulfide or a copper/arsenic compound) was used as an antifoulant in a British patent.[9] In the second part of 1700s copper was widely used, especially in British Navy.[10] After the introduction of iron ships at the end of the 18th century, the use of copper sheathing was drastically reduced due to its corrosive effects on iron. Several alternatives were tried thereafter, including sheathings of zinc, lead, nickel, arsenic, galvanized iron and alloys of antimony, zinc, and tin, followed by wooden sheathing, which was then coppered.[12,13]

Consequently, in this period, a variety of paints based on the mixing of one or more toxicants in a "polymeric" matrix started to be developed. So, by the late 18th and into the 19th centuries, coatings containing copper, arsenic, and mercury were increasingly applied to vessel hulls.

13.4 POLYMER BASED ANTIFOULING COATINGS

The commercial antifouling coatings are mainly of two types: those containing biocides and the ones without biocide. Among these two, the biocide containing coatings has gained much popularity in terms of commercial success. The commercially available biocide fouling coatings are of three types: soluble matrix antifouling paints and insoluble matrix antifouling paints. Figure 13.3 shows the mechanism of working of soluble matrix antifouling coatings. Insoluble matrix antifouling paints are based on insoluble resins, such as chlorinated rubber, vinyl, or acrylic groups. In these types of paints, only the biocides are leached out leaving behind a porous paint film skeleton and as the thickness of the porous layer increases, the leaching rate of the biocide is reduced. The paint skeleton remaining after the leaching process is over is relatively weak, making recoating with fresh paint difficult. The effective life of these paints is about 12 months.[8]

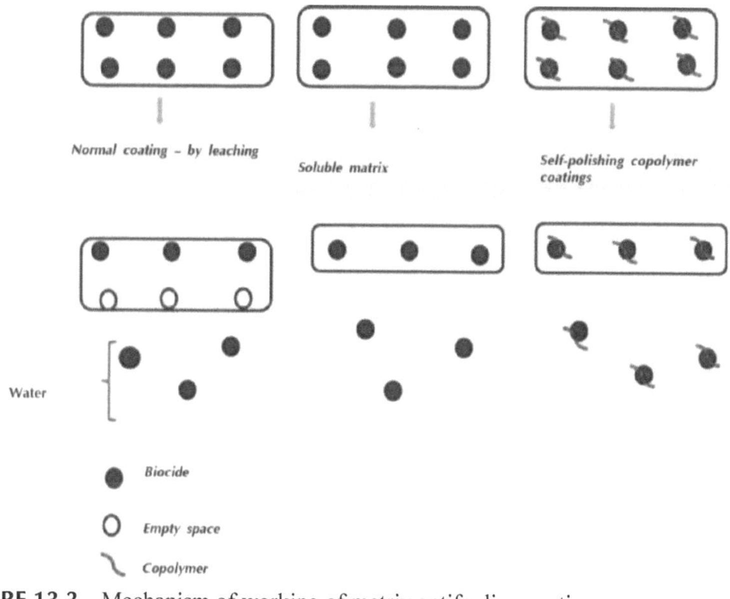

FIGURE 13.3 Mechanism of working of matrix antifouling coatings.

Soluble matrix antifouling paints are the conventional free association paints. The biocide is physically dispersed and subsequently released from the paint matrix. Sea water penetrates into the paint film and leaches out the biocide. The water-soluble biocides are typically dispersed in a slightly soluble matrix usually made of resin, plasticizers, or synthetic polymers.[11] The life span of these paints is usually between 12 and 18 months. In the eroding/ablative or controlled depletion polymer (CDP) coatings, in addition to the leaching of the biocide, the paint matrix is continually worn of by a dissolution/erosion process, which increases the leaching rate of the biocide. The most common biocide used in these paints is copper either as a metal or as a compound. To improve the efficiency, booster biocides are frequently incorporated into the paint matrix. The commercially available antifouling coatings can be broadly classified as self-polishing copolymer (SPC) antifoulings, self-polishing antifoulings, and CDP antifoulings. All have differing effects on vessel hull roughness and differing abilities to control fouling. The different technologies are determined by the biocide delivery mechanism employed.

Self-polishing copolymers undergo hydrolysis in sea water, which results in thinner leached layers with excellent control of biocide release. Water migrates into the paint film and the surface reacts with sea water ions and becomes soluble. The surface then dissolves, releasing biocides, and film thickness is lost. Reaction/solution continues with the film getting thinner through polishing. The coating stops working when it has all polished away. There are three main SPC technologies currently available: copper acrylate, zinc acrylate, and silyl acrylate. In CDP technology, water migrates into the paint film while dissolved rosin and biocides leach into the sea. Insolubles are left in the leached layer. The dissolution rate continuously declines and the leached layer grows until it is too thick to allow further water penetration. CDP antifoulings are not as effective as SPC systems and their generally thick leached layers limit performance and negatively affect recoatability. Such products, however, have a place as the lowest cost per square meter "value for money" antifoulings and are suitable for use in low fouling areas or for ships with short dry docking intervals.

13.5 MECHANISM BEHIND ORGANIC INORGANIC COATINGS

Chemical reactions and diffusion phenomena are key mechanisms in the performance of biocide-based antifouling paints. The performance of these coatings are markedly affected by sea water conditions.[12] These paints are based on the release of several biocides, which are linked or, more often, embedded in a film-forming organic matrix. Sea water has to penetrate into the paint, dissolve such biocides and diffuse out into the bulk phase again. To avoid the buildup of long diffusion paths and consequently decreasing release rates, the organic matrix is designed for slow reaction with sea water (and sea water ions) within the paint pores. Once this reaction has reached a certain conversion at the sea water–paint interface, the binder phase is released, thus controlling the thickness of the biocide-depleted layer (leached layer). Many references to the influence of sea water parameters on the performance of antifouling paints can be found in the open literature.[10–12] The salinity value influences the dissolution of the most typical biocidal pigment (Cu_2O) particles,[13] the reaction of important binder components, such as rosin[20] and the cleavage of the TBT groups in TBT–SPC paints.[14] The influence of temperature is also significant as it affects the rate of all chemical reactions, dissolution rates, and transport processes associated with the activity of chemically active A/F paints. The effect of sea water pH on the release rate of TBT groups from TBT-SPC paints was measured by Hong-Xi et al.[15] and subsequently used by Kiil et al.[16] in the modelling and analysis of such paints. In these studies, the effect of pH on the dissolution rate of Cu_2O pigment particles was also reported. The influence of pH is even more important in the case of rosin-based paints, as reported by Woods Hole Oceanographic Institution[17] and Rascio et al.[18] The solubility of rosin is increased dramatically with increasing pH values. It is most likely that sea water ions, pH, and temperature will also play a significant role in the reactions associated with the current tin-free biocide-based coatings because these are based on mechanisms similar to those of TBT-SPC paints. In addition, the severity of the biofouling and, consequently, the antifouling requirements, and the environmental fate of the released toxicants are affected by most of these parameters. Despite these facts, most studies dealing with the development of new chemically active antifouling binders or coatings lack studies on the behavior of such systems in waters under conditions different from the "standard" or "average" ones. This could eventually lead to biocide-based paints performing excellently under certain conditions, but failing in waters with different characteristics. Consequently, it is useful to characterize the environment faced by antifouling coatings by determining the range of values of the most significant sea water variables.

13.6 BIOCIDES IN COATINGS—ENVIRONMENTAL ISSUES

Copper and tin-based antifouling paints ruled the coating industry till a few years back. The possibility of the bioaccumulative aspect of the compound in some ducks, fish, and seals, and the resultant threat of it eventually entering the food chain and appearing on people's plates has been a huge concern. It has been shown that extremely low concentrations of TBT cause defective shell growth in the oyster *Crassostrea gigas* (20 ng/l) and imposex, development of male characteristics in female genitalia, in the dog-whelk *Nucella* sp. (1 ng/l).[19,20] Malformation have been observed in many other species and the IMO also reports accumulation in mammals and debilitation of the immunological defenses in fishes. However, TBT-based paint was also extremely toxic to nontarget organisms. It is an endocrine-disrupting chemical, which results in disruptions in reproduction. In 1988, the problem was brought to the attention of the Marine Environment Protection Committee (MEPC) of the IMO and the United Nations Agency concerned with the safety of shipping and the prevention of marine pollution. As a result, IMO in 1990 adopted a resolution recommending governments to adopt measures to eliminate antifouling paints containing TBT. In the 1990s, the MEPC continued to review the environmental issues surrounding antifouling systems, and in November 1999, IMO adopted an assembly resolution that called on the MEPC to develop an instrument, legally binding throughout the world, to address the harmful effects of antifouling systems used on ships. Their solution called for a global prohibition on the application of organo-tin compounds, which act as biocides in antifouling systems on ships by 1 January 2003, and a complete prohibition by 1 January 2008. In October 2001, IMO adopted a new International Convention on the Control of Harmful Antifouling Systems on Ships, which prohibit the use of harmful organotins in antifouling paints used on ships and establish a mechanism to prevent the potential future use of other harmful substances in antifouling systems. The European Union (EU) also passed legislation that bans the application of tin-containing coatings and prohibits vessels with tin-based antifouling from entering EU ports.

The impact of the less known biocides has also been studied. Several algal toxic compounds have been tested worldwide including chlorothalonil, dichlofl uanid, Irgarol 1051, 2,3,5,6-tetrachloro-4-sulfuronyl pyridine (TCMS) pyridine, thiocyanatomethyl thiobenzothiazole (TCMTB), diuron, dichloro-octylisothiazolin (DCOIT, Sea Nine 211), zinc and copper pyrithione (zinc and copper omadine), and zineb.[21,22] These are often herbicides (e.g., Irgarol 1051 and diuron, but also fungicides) that have negative effects on the growth rate of photosynthetic organisms. Legislation now exists in some countries to regulate the use of some "booster" biocides in antifouling coatings mainly Diuron, Irgarol 1051, etc. In the UK, a review of booster biocides in 2000 resulted in only four biocides gaining approval (dichlofl uanid, DCOIT (Trade name: Sea Nine 211), zinc pyrithione, and zineb). Approvals of chlorothalonil, diuron, and Irgarol 1051 were revoked due to their high toxicity at low concentrations and their persistence in the environment.[23] Irgarol 1051 and diuron are also banned in Denmark (DEPA, 2008), and diuron is banned in the Netherlands. The use of Irgarol 1051 in antifouling paints is not permitted in Australia as it was not granted approval for use as an antifouling biocide by the Australian Pesticides and Veterinary Medicines Authority, when its presence was detected and the risks it posed assessed in the 1990s. Applications for approval were submitted to the European Union for eleven AF biocides, including copper (II) oxide, copper thiocyanate, and Irgarol 1051, but not diuron.[24]

13.7 ALTERNATIVE TO TBT ANTIFOULING TECHNOLOGY

Some techniques that gained lots of popularity in this field were nonstick coatings, which contain no biocide but have extremely slippery surface—preventing fouling occurring and making it easier to clean when it does. They are suitable for vessels with minimum speed of 30 knots. However, damage to coating is difficult to repair. Light fouling occurs but is easily removed with high-pressure hose in annual dry dock visits. Periodic cleaning of hull is most appropriate for ships operating in both sea and fresh water and in areas where few organisms attach to hull. Research on use of natural compounds is in early stages, but active metabolites (e.g., ceratinamine and mauritiamine) have been identified and new biocides have been synthesized. Enzymes can break the sticking of bacteria (the first phase of fouling's growth) to the hull; while the concept of hydrophilic coating has been inspired by the preference of fouling to stick to hydrophobic surfaces, such as rocks and vessels. The organisms have no grip on hydrophilic "wettish" surfaces. Creating a difference in electrical charge between the hull and sea water unleashes chemical process which prevents fouling. This technology is shown to be more effective than tin-free paint in preventing fouling, but system is easily damaged and expensive. Prickly coatings include coatings with microscopic prickles. Effectiveness depends on length and distribution of prickles, but have been shown to prevent attachment of barnacles and algae with no harm to environment.

13.8 NANOTECHNOLOGY IN ANTIFOULING COATINGS

Nanomaterials can play an important role in antibacterial applications primarily due to their large surface area and size/shape-dependent physicochemical properties.[23] Aluminum oxide (Al_2O_3), silicon dioxide (SiO_2), titanium dioxide (TiO_2), and zinc oxide (ZnO) are some of the antimicrobial additives. Polymeric materials with great structure tailorability and flexibility have pronounced potential to inhibit aggregation of nanometals and form uniform surface coatings on various substrates. As discussed earlier, the controlled release rate possible with polymeric-based coatings reduce cytotoxicity.[36] More importantly, they can be designed to resist bacteria adhesion and enhance bactericidal properties.[37–39] Thus, it is profitable to combine metal and polymer matrix to form multifunctional composite coatings for antibacterial applications. It is noted that polymer/nanometal composites with different shapes serve various chemical and physical functions. Firstly, they act as stabilizers for the metal nanoparticles and prevent nanometal from aggregation in solutions or on surfaces. Secondly, polymers function as linkers for nanometals, which is directly loaded or in situ synthesized in antibacterial composite coatings. Thirdly, polymers can be used as matrix to control silver ion release by changing the interaction between polymers and nanometal, as well as its concentration. Despite these factors the role of nanometal in antifouling techniques has been very less. However, the results from the applications of these nanometals in other antimicrobial applications, such as food packaging, medicine, textiles, etc., provide possible potential of these methods in antifouling techniques.

A large number of metal nanoparticles are used in antimicrobial coating industry. Huge investments are being made by different companies for the development of these nano-based coatings for efficient and superior antifouling properties. Nanomaterials have unique properties compared to their bulk counterparts. For this reason, nanotechnology has attracted a great deal of attention from the scientific community. Metal oxide nanomaterials like ZnO and CuO have been used industrially for several purposes, including cosmetics, paints, plastics, and textiles. A common feature that these nanoparticles exhibit is their antimicrobial behavior against various microorganisms. Nanoparticles are expected to exhibit enhanced antimicrobial properties when compared to the bulk metal due to a much larger surface area to volume ratio. Nanomaterials, the building blocks of the nanotechnology revolution, are defined by having at least one dimension less than 100 nanometers (nm) in size. Due to this small size, the nanomaterials exhibit new and improved properties over their "bulk material" counterparts.

These coatings inhibit the growth of odor-causing bacteria, fungi, and algae. Antimicrobial coatings demand for indoor air quality is expected to grow at a compound annual growth rate (CAGR) of 12.8% from 2012 to 2018 Brajesh Kumar et al.[25] In case of mold formation, various coatings and products containing antimicrobial substances are being put to use as an aftercare method. Most of the buildings, where mold aftercare is necessary, are older than five years and hence the market is expected to grow at a high CAGR of 13.1% from 2012 to 2018 Kushwahw et al.[26] The developments of coatings with antimicrobial properties have huge commercial demand. Inorganic nanoparticles with antimicrobial activity are emerging as a new class of coating materials to fulfil the increasing general demands for hygiene in daily life. Silver has been used extensively in many bactericidal fields (Silver and Phung).[27] The synthesis of metal nanoparticles stabilized polymers is interesting, since these materials offer tremendous options for combining properties stemming from both the inorganic components and the polymers (Slawson et al.[28]; Yamanaka et al.[29]; Becker).[30]

One of the most extensively studied and commercially important nanometal for antimicrobial applications is nanosilver. The biocidial effect of Ag^+, with its broad spectrum of activity, including bacterial, fungal, and viral agents can be achieved at submicromolar concentration. Micromolar levels of Ag+ have been reported to affect the respiratory systems or interfere with the membrane permeability to protons and phosphates. As per literature, nanosilver was shown to destabilize the outer membrane, collapse the plasma membrane potential, and deplete the levels of intracellular ATP. Though the mode of action of nanosilver and Ag^+ were similar, the effective concentration of nanosilver and Ag^+ ions were at nanomolar and micromolar levels, respectively. This may be the reason for the better antibacterial property of nanosilver compared to silver ions, that is, silver nitrate/HBPE. Similar trend has been reported by many researchers (Atlanta and Becker[30]; Percivel et al.)[31] Some studies reported that silver nanoparticle (AgNP) size, shape, surface charge, surface coating, solution chemistry, and solubility affect its bactericidal property by Carlson et al.[32] However, the extent to which these factors affect toxicity directly by influencing particle-specific biological effects or indirectly by affecting silver ion release remains an open question. The applications of silver polymeric nanocomposites as advanced antimicrobial agents have also been reviewed recently. Silver ions are known with their potent toxicity against broad range of bacteria, fungi, and viruses. Moreover, silver-based antimicrobials demonstrated positive properties, such as thermal and chemical stability, environmental safety, and low toxicity to human cells. These characters make silver-based materials suitable for wide varieties of applications. A recent study shows the influence of nanosilver particles for the protection of archaeological stones against fungal and bacterial proliferation. This study highlighted the synthesis of nanosilver particles (AgNPs) using the biogenic volatiles of the bacterial strain *Nesterenkonia halobia*. The antimicrobial activities of AgNPs were evaluated against the Gram-positive bacterial strain *Streptomyces parvullus* and fungal strain *Apergillus niger*. Furthermore, the silver particles were mixed with two types of consolidation polymers and were used to coat the external surfaces of sandstone and limestone blocks. The stones treated with silicon polymer loaded with AgNPs showed an elevated antimicrobial potentiality against *Aspergillus niger* and *S. parvullus*. Scan electron microscope (SEM) and electron dispersive X-ray spectroscopy (EDX) analysis of treated stones demonstrated the existence of nanocomposite structures containing the elemental silver. Polymers functionalized with AgNPs can be used not only as potent biocides but also for the consolidation of the historic monuments and artifacts. The most common nanocomposites used as antimicrobial films for food packaging are based on metallic silver, which is well known for its strong toxicity to a wide range of microorganisms, as ionic silver and as metallic silver with high temperature stability and low volatility. Silver nanoparticles have a high antimicrobial effect because they can insert within the cell membrane due to their small size and release silver ions locally from the silver particle.

ZnO has generated a lot of interest among researchers as potent antimicrobial agent. The antifungal activity of ZnO nanorods prepared by the chemical solution method against Candid a albicans was studied extensively by Halliwell and Gutteridge.[33] In the study, ZnO nanorods have been deposited on glass substrates using the chemical solution method. The as-grown samples were characterized by scanning electronmicroscopy (SEM) and X-ray diffraction (XRD). XRD showed ZnO nanorods grown in (002) orientation. The antifungal results indicated that ZnO nanorod arrays exhibit stable properties after two months and play an important role in the grow the inhibitory of *Candida albicans*. The antimicrobial activity of ZnO is considered to be due to the generation of hydrogen peroxide (H_2O_2) from its surface. The antimicrobial properties of copper are well established, especially when metals are applied in the form of nanoparticles (NPs). Copper NPs or colloids agglomeration can result in the decrease of their antimicrobial and antifungi properties. The problem of NPs stability was solved by the development of silica nanospheres containing immobilized silver or copper NPs. The authors claimed that the nanospheres can be applied as the effective antibacterial or antifungi additives for architectural paints and impregnates.

The most important advantage of these additives lies in the stability of copper NPs as well as in the harmlessness to human beings. The properties of silica nanospheres containing immobilized silver or copper NPs were tested by using SEM, energy dispersive X-ray spectroscopy (EDS), atomic absorption spectroscopy (AAS), and photon correlation spectroscopy. The antimicrobial properties of these nanospheres were investigated using *E. coli*, Staphylococcus aureus as well as *A. niger, Paecilomycesvarioti, Penicillium funiculosum*, and *Chaetomium globosum* in comparison with pure silica nanospheres. The result of antimicrobial tests of silicone acrylic architectural paint and impregnates containing these additives were also discussed in the paper. The interior paints investigated have been formulated on the basis of aqueous acrylic dispersion, rutile titanium dioxide, extenders, and special additives as photocatalytic nano zinc oxide and different types of photocatalytic anatase titanium dioxide. Organic dye Orange II was used as an indicator for the reactivity of photocatalytic surfaces. The absorbance change of Orange II solution was measured by a photometer. An agar plate method was used for the evaluation of antimicrobial effect of the coatings. The effectiveness of coatings was demonstrated using the following bacteria relevant to hygiene: *E coli*, Staphylococcus aureus, Pseudomonas aeruginosa, fungi *A. niger*, and *Penicillium chrysogenum*. The nanoform of ZnO was found to be the best photocatalytic agent and also the best and broadest spectrum antimicrobial agent for these interior paints. Statistically significant differences between the control coatings and the coatings with nanoform of ZnO were found using rank-sum test. Highly ionic nanoparticulate metal oxides can be prepared with extremely high surface areas and unusual crystal morphologies having numerous edge/corner and reactive surface sites.[34–36]

The AP–MgO/X2 nanoparticles were found to show biocidal activity against certain vegetative Gram-positive bacteria, Gram-negative bacteria and the spores.[37] AP–MgO nanoparticles are found to possess many properties that are desirable for a potent disinfectant.[38,39] Because of their high surface area and enhanced surface reactivity, the nanocrystals adsorb and carry a high load of active halogens. Their extremely small size allows many particles to cover the bacteria cells to a high extent and bring halogen in an active form in high concentration in proximity to the cell. Standard bacteriological tests have shown excellent activity against *E. coli* and *Bacillus megaterium* and a good activity against spores of *Bacillus subtilis*.

There is limited information available about the antimicrobial activity of nano CuO. As CuO is cheaper than silver, easily mixes with polymers and relatively stable in terms of both chemical and physical properties, it finds a wide application.[40, 41] It is suggested that highly ionic nanoparticulate metal oxides, such as CuO, may find potential application as antimicrobial agents as they can be prepared with extremely high surface areas and unusual crystal morphologies[42].CuO nanoparticles were effective in killing a range of bacterial pathogens involved in hospital-acquired infections. But a high concentration of nano-CuO is required to achieve a bactericidal effect[43]. Aluminum oxide NPs have wide-range applications in industrial and personal care products. The growth-inhibitory effect of alumina NPs over a wide concentration range (10–1000 μg/mL) on *E. coli* have been studied[44]. Fourier transform infrared studies have shown differences in structure between nanoparticles treated and untreated cells.

TiO_2 is a well-known antimicrobial agent manly due to its photocatalytic generation of strong oxidizing power when illuminated with UV light at wavelength of less than 385 nm. TiO_2 particles catalyze the killing of bacteria on illumination by near-UV light.[44] The generation of active free hydroxyl radicals (OH) by photoexcited TiO_2 particles is probably responsible for the antibacterial activity. The antimicrobial effect of TiO_2 photocatalyst on *E. coli* in water and its photocatalytic activity against fungi and bacteria has been demonstrated.[45] There are limited studies on the antifungal activity of metal nanoparticles. The fungistatic and fungicidal effects of the silver NPs against selected pathogenic yeasts causing invasive life-threatening fungal infections in intensive care patients has been studied. However, the effect of these nanoparticles on the biofilm formation needs to be critically analyzed.

13.9 CONCLUSION

The biofilm formation and the fouling of maritime structures has been a huge concern from the time man started civilization. Though many methods have been adopted from time to time with relative success, the application of TBT was the most popular for a very long time. Though TBT-based antifouling coatings played a major role in the shipping industry the toxicity caused by these metal particles raised lots of concern. This was followed by ban by IMO on metal-based coatings. Frantic efforts have been going on in coatings industry to find an environment-friendly solution. It has been established that the inorganic–organic polymeric coatings can be the future for the antifouling techniques. The expansion of nanotechnology has however been not much studied the antifouling techniques. However the successful application of these metal nanoparticles in combination with polymeric base materials in fields such as medicine, textiles, interior paints, etc., shows a possible huge potential for these combination in antifouling techniques. Meanwhile, a close watch on the reports concerning the toxicity level of these nanometals is a requisite to predict the possible revolution in the field of antifouling coatings with nanometals.

KEYWORDS

- **biofouling**
- **nanoparticles**
- **microfouling**
- **macrofouling**
- **polymer coatings**

BIBLIOGRAPHY

1. Yebra, D.M.; Kiil, S.; Dam-Johansen, K. Antifouling Technology—Past, Present and Future Steps Towards Efficient and Environmentally Friendly Antifouling Coatings. *Prog. Org. Coat.* **2004**, *50*, 75–104.
2. Rascio, V.J.D. Antifouling Coatings: Where Do We Go from Here. *Corros. Rev.* 2000, *18*, 133–154.
3. Abbott, A.; Abel, P.D.; Arnold, D.W.; Milne, A. Cost–Benefit Analysis of the Use of TBT: the Case for a Treatment Approach. *Sci. Total Environ.* **2000**, *258*, 5–19.
4. Rouhi, A.M. The Squeeze of Tributyltin. *Chem. Eng. News* **1998**, *27*, 41–42.
5. Brancato, M.S.; OCEANS'99 MTS/IEEE, *Riding the Crest into the 21st Century*, 1999, vol. 2, pp. 676.
6. Reise, K.; Gollasch, S.; Wolff, W.J. Introduced Marine Species of the North Sea Coasts. *Helgol. Mar. Res.* 1999, *52*, 219–234.
7. Lunn, I. *Antifouling: A Brief Introduction to the Origins and Developments of the Marine Antifouling Industry.* BCA Publications, Thame, UK, 1974.
8. Callow M.E. Ship Fouling: Problems and Solutions. *Chem. Ind.* **1990**, *5*, 123–127.
9. WHOI, *Marine Fouling and its Prevention.* Woods Hole Oceanographic Institute, Annapolis, Iselin, COD, USA, 1952.
10. Lewis, J.A.; Marine Biofouling and its Prevention on Underwater Surfaces. *Mater. Forum.* **1998**, *22*, 41–61.
11. *CEPE Antifouling Working Group*, Final Report, EC Project No. 96/559/3040/DEB/E2, 1999.
12. Kiil, S.; Weinell, C.E.; Pedersen, M.S.; Dam-Johansen, K.; Arias Codolar, S. Erratum: Dynamic Simulations of a Self-Polishing Antifouling Paint Exposed to Seawater. *J. Coat. Technol.* **2002**, *74*, 45–54.
13. Ferry, J.D.; Carritt, D.E. Action of Antifouling Paints. Solubility and Rate of Solution of Cuprous Oxide in Seawater. *Ind. Eng. Chem.* **1946**, *38*, 612–617.

14. Kiil, S.; Weinell, C.E.; Pedersen, M.S.; Dam-Johansen, K. Analysis of Self-Polishing Antifouling Paints Using Rotary Experiments and Mathematical Modeling. *Ind. Eng. Chem. Res.* **2001**, *40*, 3906–3920.
15. Yuanghui, W.; Hongxi, C.; Meiying, Y.; Huai-Ming, G.; Jing-Hao, G. Studies on the Hydrolysis of Organotin Polymers. I. Hydrolytic Rates of Poly(Tributyltin Methacrylate) and Poly(Tributyltin Methacrylate-Co-Methyl Methacrylate). Fujian Shifan Daxue Xuebao, (in Chinese) **1988**, *4*, 61–68.
16. Kiil, S.; Weinell, C.E.; Pedersen, M.S.; Dam-Johansen, K. Mathematical Modelling of a Self-Polishing Antifouling Paint Exposed to Seawater: A Parameter Study. *Chem. Eng. Res. Des.* **2002**, *80*, 45–52.
17. Cullen, V. Down to the Sea for Science 75 Years Of Ocean Research Education and Exploration at the Wood Hole Oceanographic Institution, 2005. Woods Hole Oceanographic Institution (WHOI), US Naval Institute, Annapolis, Iselin, COD, 1952.
18. Giúdice, C.A.; Del Amo, B.; Rascio, V.J.D. The Use of Calcium Resinate in the Formulation of Soluble Matrix Antifouling Paints Based on Cuprous Oxide. *Prog. Org. Coat.* **1988**, *16*, 165–176.
19. G. Swain, *Proceedings of the International Symposium on Sea Water Drag Reduction*, The Naval Undersea Warfare Center, Newport, 1998, pp. 155–161.
20. Evans, S.M.; Leksono, T.; McKinnel, P.D. Tributyltin Pollution: A Diminishing Problem Following Legislation Limiting the Use of TBT-Based Anti-Fouling Paints. *Mar. Pollut. Bull.* **1995**, *30*, 14–21.
21. Reed, R.H.; Moffat, L.; Copper Toxicity and Copper Tolerance in *Enteromorpha compressa* (L.) Grev. *J. Exp. Mar. Biol. Ecol.* **1983**, *69*, 85–103.
22. N. Voulvoulis. Antifouling Paint Booster Biocides: Occurrence and Partitioning in Water and Sediments. Konstantinou I Eds.; The Handbook of Environmental Chemistry—5.0 Antifouling Paint Biocides. Springer, New York, 2006, pp 155–170.
23. Cresswell, T.; Richards, J.P.; Glegg, G.A.; Readman J.W. The Impact of Legislation on the Usage and Environmental Concentrations of Irgarol 1051 in UK Coastal Waters. *Mar. Pollut. Bull.* **2006**, *52*, 1169–1175.
24. Pereira, M. Ankjaergaard, C. Advances in Marine Antifouling Coatings and Technologies. Woods Head Publishing, Cambridge, UK, 2009, pp. 240.
25. Kumar, B.; Smita, K., Cumbal, L.; Debut, A.; Pathak, R.N. Sonochemical Synthesis of Silver Nanoparticles Using Starch: A Comparison. *Bioinorg. Chem. Apply.* **2014**, *2014*, 1–14.
26. Kushwaha, K.; Saini, A.; Saraswat, P.; Agarwal, M.K.; Saxena J. Colorful World of Microbes: Carotenoids and their Applications. *Advan. Biology*. **2014**, 837891, 1–13.
27. Silver, S.; Phung L.T.; Bacterial Heavy Metal Resistance: New Surprises. *Annu. Rev. Microbiol.* **1996**, *50*, 753–789.
28. Slawson, R.M.; Van Dyke, M.I.; Lee, H.; Trevors, J.T. Germanium and Silver Resistance, Accumulation, and Toxicity in Microorganisms. *Plasmid.* **1992**, *27*, 72–79.
29. Yamanaka, M.; Hara, K,; Kudo, J. Bactericidal Actions of a Silver Ion Solution on *Escherichia coli*, Studied by Energy-Filtering Transmission Electron Microscopy and Proteomic Analysis. *Appl. Environ. Microbiol.* **2005**, *71*, 7589–7593.
30. Becker, R.O. Silver Ions in the Treatment of Local Infections. *Metal. Based. Drugs.* **1999**, *6*, 297–300.
31. Percival, S L.; Bowler, P.G.; Russell, D. Bacterial Resistance to Silver in Wound Care. *J. Hosp. Infect.* **2005**, *60*, 1–7.

32. Carlson, C.; Hussain, S.M.; Schrand, A.M.; Braydich-Stolle, L.K.; Hess, K.; Jones, R. L.; Schlager, J.J. Unique Cellular Interaction of Silver Nanoparticles: Size-Dependent Generation of Reactive Oxygen Species. *J. Phys. Chem. B.* **2008**, *112*, 13608–13619.
33. Halliwell, B.B.; Gutteridge, J.M.C. *Free Radicals in Biology and Medicine*. Oxford University Press, Oxford, UK, 1999.
34. Hochmannovaa, L.; Vytrasovab, J. Photocatalytic and Antimicrobial Effects of Interior Paints. *Prog. Org. Coat.* **2010**, *67*, 1–5
35. Klabunde, K.J.; Stark, J.; Koper, O.; Mohs, C.; Park, D.; Decker, S.; Jiang, Y.; Lagadic, I.; Zhang, D. Nanocrystals as Stoichiometric Reagents with Unique Surface Chemistry. *J. Phys. Chem.* **1996**, *100*, 12142–12153.
36. Huang, L. Controllable Preparation of Nano-MgO and Investigation of its Bactericidal Properties. *J. Inorg. Biochem.* **2005**, *99*, 986–993.
37. Richards, R.; Li, W.; Decker, S.; Davidson, C.; Koper, O.; Zaikovski, V.; Volodin, A.; Rieker, T.; Klabunde, K. Consolidation of Metal Oxide Nanocrystals. Reactive Pellets with Controllable Pore Structure that Represent a New Family of Porous, Inorganic Materials. *J. Am. Chem. Soc.* **2000**, *122*, 4921–4925.
38. Koper, O.; Klabunde, J.; Marchin, G.; Klabunde, K.J.; Stoimenov, P.; Bohra, L. Nanoscale Powders and Formulations with Biocidal Activity Toward Spores and Vegetative Cells of *Bacillus* Species, Viruses, and Toxins. *Curr. Microbiol.* **2002**, *44*, 49–55.
39. Xu, J.F.; Ji, W.; Shen, Z.X.; Tang, S.H.; Ye, X.R.; Jia, D.Z.; Xin, X.Q. Preparation and Characterization of CuO Nanocrystals. *J. Solid State Chem.* **1999**, *147*, 516–519.
40. Kwak, K.; Kim, C. Viscosity and Thermal Conductivity of Copper Oxide Nanofluid Dispersed in Ethylene Glycol. *Korea–Australia Rheol. J.* **2005**, *17*, 35–40.
41. Cava, R.J. Structural Chemistry and the Local Charge Picture of Copper Oxide Superconductors. *Science*. **1990**, *247*, 656–662.
42. Tranquada, J.M.; Sternlieb, B.J.; Axe, J.D.; Nakamura, Y.; Uchida, S. Evidence for Stripe Correlations of Spins and Holes in Copper Oxide Superconductors. *Nature*. **1995**, *375*, 561–565.
43. Stoimenov, P.K. Metal Oxide Nanoparticles as Bactericidal Agents. *Langmuir*. **2002**, *18*, 6679–6686.
44. Ireland, J.C.; Klostermann, P.; Rice, E.W.; Clark, R.M. Inactivation of *Escherichia coli* by Titanium Dioxide Photocatalytic Oxidation. *Appl. Environ. Microbiol.* **1993**, *59*, 1668–1670.
45. Matsunaga, T.; Tomada, R.; Nakajima, T.; Wake, H. Photochemical Sterilization of Microbial Cells by Semiconductor Powders. *FEMS Microbiol. Lett.* **1998**, *29*, 211–214.

INDEX

C

D

E

G

H

I

J

K

L

M